RISING SEAS

RISING SEAS

PAST, PRESENT, FUTURE

Vivien Gornitz

Columbia University Press *New York*

Columbia University Press
Publishers Since 1893
New York Chichester, West Sussex
cup.columbia.edu

Library of Congress Cataloging-in-Publication Data
Gornitz, Vivien.
Rising seas : past, present, future / Vivien Gornitz.
p. cm.
Includes bibliographical references and index.
ISBN 978-0-231-14738-5 (cloth : alk. paper) — ISBN 978-0-231-14739-2 (pbk. : alk. paper) —
ISBN 978-0-231-51920-5 (ebook)
1. Sea level—Climatic factors. 2. Floods. 3. Climatic changes. I. Title.
GC89.G67 2012
551.45′8—dc23
2012036495

♾

Columbia University Press books are printed on permanent and durable acid-free paper.
This book is printed on paper with recycled content.
Printed in the United States of America
c 10 9 8 7 6 5 4 3 2 1
p 10 9 8 7 6 5 4 3 2

COVER PHOTO: Marc Yankus
COVER DESIGN: Shaina Andrews

Contents

Preface

Growing amounts of greenhouse gases in the atmosphere generated by human activities are already affecting the Earth's climate, more noticeably near the poles and on high mountains. The specter of rising sea level is of growing concern to the tens of millions of people living along the world's shorelines and on small, low-elevation islands. However, ample evidence from the geologic past indicates that the average height of the world's oceans has varied considerably. Within the last million years alone, the Earth's climate has alternated dramatically at least eight times between glaciations, when vast ice sheets blanketed northern Europe and much of North America, and interglacials—warm periods much like today. During the last glaciation, which climaxed around 21,000 years ago, the global sea level lowered by approximately 120 meters as compared with today, and during the last interglacial ~125,000 years ago, it stood 4–6 meters (or more) above present levels. The vast ice masses blanketing North America and Scandinavia during the last Ice Age locked up the equivalent of 120 meters (394 feet) of ocean water. After the ice melted, the sea reached nearly its present height by 7,000 to 6,000 years ago, fluctuating at most by a few feet since then.

However, the pace of sea level rise has quickened since the late nineteenth century and may be climbing even faster during the last 20 years, closely paralleling the rise in global temperature. Could future greenhouse gas–induced global warming push the Earth's climate system into an unstable mode, triggering a catastrophic meltdown of the polar ice sheets? Would major coastal cities with millions of inhabitants, like New York City, Los Angeles, London, Tokyo, Shanghai, Hong Kong, Mumbai (Bombay), and Bangkok, to name just a few, be inundated?

While a major destabilization of the polar ice sheets appears unlikely within the next 100–200 years, it cannot be ruled out entirely. Therefore, a certain sense of urgency motivates our need to gain a better understanding of the causes and consequences of sea level rise. *Rising Seas: Past, Present, Future* addresses these issues and provides the basic background for appreciating why sea level rise truly matters. The book begins with an introduction to the oceans—their physical properties and short-term surface fluctuations due to meteorological processes, the tides, storms, and currents. The introductory chapter sets the stage for chapter 2, which explores the multitude of longer-term oceanic, atmospheric, and geological processes that govern sea level over periods ranging from years to "deep time"—i.e., tens of millions of years or more.

A much longer-term perspective, extending well beyond the 100–150-year-long period of instrumental records, provides a deeper understanding of the observed twentieth-century changes and those anticipated for the near future. Bounds on past sea level variability can be set by estimating ancient sea levels from natural "archives" such as ocean sediments, corals, and ice cores, and running computer models. Thus, by peering into the distant past, we improve our ability to anticipate potential scenarios of sea level change and thereby plan ahead more effectively. Through an examination of the paleoclimate record, chapters 3, 4, and 5 document sea level changes at several critical periods in the geologic past (for example, the mid-Pliocene warm period, the Eemian—the last interglacial, and after the last Ice Age) that can serve as possible analogs of a warmer future with higher sea levels. The major sea level incursion that followed the end of the last Ice Age occupies chapter 5. A closer look at periods of accelerated sea level rise, such as the "meltwater pulses" of the last deglaciation, provides glimpses of potential worst-case scenarios of ice sheet collapse.

Chapter 6 continues with observations of twentieth-century trends from tide gauges, and more recently, satellite data, as well as atmospheric-ocean general circulation model (AOGCMs) projections for the coming century (chapter 7). Close connections exist between the greenhouse effect, evidence for a warming world, and rising sea level. Current sea level trends exceed those of past centuries. Comparison of the longer tide-gauge records with the latest satellite altimetry data suggests further recent sea level acceleration. This apparent speedup is worrisome and may portend even greater changes ahead. Some recent studies suggest that the latest Intergovernmental Panel on Climate Change (IPCC) sea level rise projections may already be too low. How much and how soon will the Greenland and Antarctic ice sheets melt? The climate system is quite sensitive to minor perturbations, and computer models may miss important signals. Given the limited ability of current global climate models to describe dynamic ice sheet behavior, which could lead to large jumps in future sea level, paleoclimate analogs could offer some important insights. (However, analogs also have drawbacks. Because past and present climatic and environmental conditions cannot be exactly replicated, these differences may preclude completely accurate future projections.)

Even today, coastal residents in low-lying areas face flood hazards from major storms. As sea level rises, the surges generated by large cyclones will reach higher and inundate more land area. Another risk is that the intensity and frequency of such extreme meteorological events could also increase in a warmer world. Chapter 8 investigates the impacts of changing storm patterns and sea level rise on coastal regions—areas and populations at risk and examples of vulnerable cities and regions.

Chapter 9 tackles the larger and more difficult issue of adapting to sea level rise. Various ways of living with a rising sea are outlined, ranging from purely defensive "hard" engineering solutions to sensible land use planning, and as a last resort, inland migration. Ultimately, mitigating the effects of sea level rise will depend on curbing greenhouse gas emissions. Chapter 10 briefly summarizes several means of limiting carbon emissions to round out the picture. It also ties together some of the main themes of the book— for example, the changes in sea level already under way, what the past teaches us about future outcomes, and what we can do to avert a watery future.

As this book goes to press, New York City, Long Island, and New Jersey are recovering from the devastating effects of Hurricane Sandy, which struck our shores late on October 29, 2012. This superstorm, with maximum wind gusts near 150 kilometers per hour (90 miles per hour) at landfall, 8 kilometers (5 miles) southwest of Atlantic City, NJ, caused extensive flooding in low-lying neighborhoods of lower Manhattan, Brooklyn, Queens, and Staten Island (New York City), neighboring Hoboken and Jersey City, as well as numerous seaside communities, including Atlantic City, Seaside Heights, and Point Pleasant, NJ. At the Battery tide gauge, lower Manhattan flood waters topped 13.88 feet, setting a historic record![1] Sandy's destructiveness resulted

1. The 13.88-foot water level (MLLW) at the NYC Battery tide gauge on Oct. 29, 2012 tops the recorded data since 1856 and probably exceeds that of the 1821 hurricane, which was reported to be around 13 feet.

from a rare chain of events: the collision between a tropical cyclone and a winter storm, and a blocking pattern in the jet stream that pushed the storm westward toward the New Jersey coast. Extreme flood levels in New York Harbor were caused by strong winds on the right-hand side of the hurricane that pushed water landward at high tide and full moon, and the funneling effects of the sharp bend between the New Jersey and Long Island shorelines. Since sea level in New York City has climbed 43.2 centimeters (17 inches, or 1.4 feet) since 1856, a storm of the same intensity as Sandy, had it occurred over a century ago, would have caused much less flooding and associated damage.

The global sea level trend between 1993 and 2012 now stands at 3.1 ± 0.4 millimeters/year, based on satellite altimetry (http://sealevel.colorado.edu), in spite of a drop of 5 millimeters per year between 2010 and 2011, attributed to a very strong La Niña event that dumped heavy rains over Australia, northern South America, and southeast Asia. This temporarily lowered the ocean mass (Boening, C. et al., 2012. The 2011 La Niña: So strong, the oceans fell. *Geophysical Research Letters* 39(19), doi:10.1029/2012GL053055).

Meanwhile, atmospheric CO_2 levels are 391 parts per million and temperatures are 0.67°C (1.21°F) above the twentieth-century average, tying 2012 with 2005 for the warmest September since 1880 (http://www.ncdc.noaa.gov/sotc/global/2012/9), accessed Oct. 22, 2012). Nine of the 10 warmest years in the modern meteorological record have occurred since 2000 (http://giss.nasa.gov/research/news/20120119/). In September 2012, Arctic sea ice shrank to its lowest summer extent since 1979, when satellite observations of the region began. (In the Antarctic, by contrast, strong circumpolar winds pushed sea ice outward to a record high September extent). The Greenland ice sheet melted over 97 percent of its surface for several days in July, 2012 (http://www.nasa.gov/topics/earth/features/greenland-melt.html; last updated July 24, 2012). Although the last such large-scale melting occurred in 1889, recent years have seen a growing number of summer surface melt days.

These 2012 climate and sea level trends reinforce the conclusions presented in *Rising Seas* that we are moving toward an increasingly warmer and, quite likely, a more aquatic planet. We would do well to take heed and begin preparations to stem the oncoming tide.

Vivien Gornitz
November 12, 2012

Acknowledgments

Rising Seas: Past, Present, Future represents the culmination of a long journey into research on global sea level change that initially began years ago at the suggestion of James Hansen, Head, NASA Goddard Institute for Space Studies (GISS), New York City. Numerous colleagues have subsequently contributed many perceptive insights and much advice. In particular, my sincere thanks are extended to Nick Christie-Blick of the Lamont Doherty Earth Observatory (LDEO) at Columbia University, Steven Pekar of LDEO, and Ken Miller and Jim Wright of the Department of Earth and Planetary Sciences at Rutgers University for their help in elucidating sequence stratigraphy and backstripping. I also appreciate the contributions of Gavin Schmidt, GISS, for beneficial discussion on oxygen isotopes; Mark Chandler, GISS, for insightful conversations on mid-Pliocene climates and sea levels; Allegra LeGrande, GISS, and Anders Carlson, University of Wisconsin, for information on sea level change and the 8,200-year cold event; Robin Bell, LDEO, for useful discussion on ice sheet behavior; and David Rind, GISS, for clarifying phasing of Quaternary climate variability.

I wish to thank Patrick Fitzgerald, Columbia University Press, for inviting me to write this book and for his many helpful comments along the way. Thanks are also extended to Bridget Flannery-McCoy, Assistant Editor, Columbia University Press, for her expert editorial support, as well as the helpful contributions of the CUP Art Department. Appreciation is also expressed to Mr. José Mendoza for drafting assistance and to Cynthia Rosenzweig, GISS and Center for Climate Systems Research, Columbia University, for her encouragement and understanding during this lengthy undertaking. I also thank the reviewers for their constructive suggestions that helped to improve this book.

RISING SEAS

Not only do the tides advance and retreat in their eternal rhythms, but the level of the sea itself is never at rest. It rises or falls as the glaciers melt or grow, as the floor of the deep ocean basin shifts under its increasing load of sediments, or as the earth's crust along the continental margins warps up or down in adjustment to strain and tension. Today a little more land belongs to the sea, tomorrow a little less. Always the edge of the sea remains an elusive and indefinable boundary.
—RACHEL CARSON, *THE EDGE OF THE SEA*

1 The Ever-Changing Ocean

In the seemingly limitless extent of the ocean, the ancient sages envisioned a state of primeval formlessness out of which all life emerged. While the ancients had no concrete evidence for their intuitive insights, we now know from the geological record of sedimentary rocks that oceans have existed on Earth for at least 3.8 billion years, possibly more. Life has been present on Earth for almost as long. The first living cells most probably did originate in the ocean, although it has been debated whether this occurred in a shallow tidal pool or in the deep ocean at hydrothermal vents. Yet, in spite of the ocean's antiquity on this planet, tiny changes in its shape, depth, and volume from one moment to another have led to major reconfigurations over geological eons. As the mean water level of the ocean varies over time, the boundary between land and sea shifts back and forth in an endless battle between rock, sand, and waves.

Standing on the beach on a balmy, sunny day, as the tides gently roll in and out on their daily round with the Moon, today's beachgoer is blissfully unaware of the subtle, almost imperceptible changes in the oceans' average elevation that are currently taking place. But over longer periods of time, perhaps even within a person's lifetime and even more evident over the course of several generations, the rising sea leaves a clearer mark. Small low-lying slands and some tidal marshes may become submerged. Barrier beache may gradually migrate landward as sand is eroded from the seaward margin and redeposited on the bay side. Grassy tidal wetlands may also migrate farther inland, occupying higher ground. Inhabitants of coastal cities may notice that port facilities and shorefront neighborhoods are flooded more frequently during storms. Coastal **aquifers** may become more saline. Yet the sea displays a host of contrasting moods ranging from mirror-calm stillness to stormy towering waves that can sink large oceangoing vessels and pound fiercely against the shore, undercutting cliffs and washing away expensive ocean-view homes.

The oceans are in constant motion on timescales ranging from minutes to millennia and eons, their heights rising and falling daily with the tides, with the waves and surge of a passing storm, more slowly with the changing seasons, and almost imperceptibly with gradual alterations of the Earth's climate and configuration of the ocean basins. Currents, waves, and wind constantly reshape the shoreline, at the dynamic interface between land, sea, and air. The ephemeral changes of the ocean surface ride upon the longer-term trends in sea level, which are the main focus of this book. Although barely noticeable from day to day, even from one year to the next, these tiny shifts grow into major differences over extended periods of time. What drives these changes and how will they affect those who live near the sea, now and in the future? We need to place the 20th-to-21st-century transition that is already under way into a longer-term context extending well beyond the period of instrumental records. By reconstructing ancient sea levels using natural "archives" of environmental change recorded in ocean sediments, corals, and ice cores, and running computer models, scientists can place bounds on past sea level variability. Thus, a glance backward may teach us vital lessons that will help us anticipate coming trends and prepare better for the future. Placing today's changes in sea level and climate into a much longer, geologic-scale time frame enables us to draw parallels between sea level during previous warm periods and possible future behavior of the ice sheets and oceans.

As greenhouse gases accumulate in the Earth's atmosphere and signs of global warming become increasingly apparent, hundreds of millions of people living near the coasts of the world and on small, low-lying islands face the prospect of rising sea level. Dramatic changes in sea level are nothing new in this planet's history, however. Sea level fluctuated roughly 120 meters between **glacial and interglacial periods**, with even greater differences further back in geologic time. Of great concern is the possibility that human-induced global warming could trigger a major meltdown of the polar ice sheets, submerging major coastal cities and low-lying islands.

Our task will be to gain a closer understanding of the natural processes that govern variations in the seas' average elevation and of why anthropogenic climate change may cause the elevation to increase in the future. Before embarking on this exploration of past sea levels and future prospects, we briefly review processes underlining shorter-term ocean variations, because they can locally magnify the effects of global sea level rise, affecting the safety and well-being of coastal residents.

WATERWORLD

The oceans cover 71 percent of the Earth's surface, or 362 million square kilometers (140 million square miles). Thus, our planet rightfully deserves to be called "waterworld." Earth is the only planet in the solar system on which water can exist at the surface simultaneously as a liquid (water), a solid (ice), and a gas (atmospheric water vapor). The oceans contain an enormous quantity of water—1.34 billion cubic kilometers (0.32 billion cubic miles), or 96.5 percent of the total at or near the Earth's surface (table 1.1).[1] The polar ice caps and glaciers constitute the second-largest reservoir (24 million cubic kilometers or 5.8 million cubic miles)—a mere 1.74 percent of the total. The transfer of large quantities of water from oceans to ice sheets and vice versa between past glacial and interglacial periods has led to major changes in global sea level. Chapter 7 explores the possibility that future warming could melt a significant volume of the Greenland and/or Antarctic Ice Sheets. Groundwater stores an amount of water roughly comparable to

Table 1.1 Distribution of the Earth's Water

Water Source	Water Volume (Millions of cubic kilometers)	Percent
Oceans	1,338	96.5
Ice caps, Glaciers, Permanent Ice	24.06	1.74
Groundwater	23.40	1.70
Soils	0.0165	0.001
Permafrost	0.3	0.022
Lakes	0.176	0.013
Atmosphere	0.013	0.001
Other (incl. Rivers, Wetlands)	0.015	0.001
Total	1,386	100.0

Sources: Gleick, 1996; Shiklomanov, 1997.

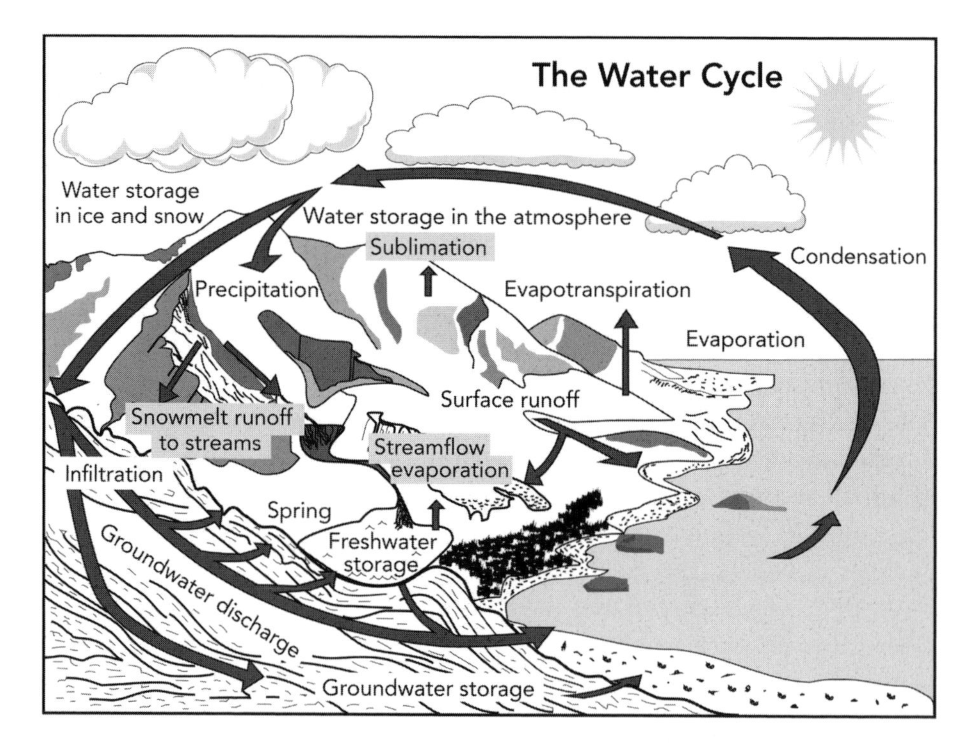

Figure 1.1 The water cycle, illustrating major water reservoirs and movement of water between reservoirs. (After USGS.)

that in ice—23.4 million cubic kilometers or 5.6 million cubic miles. Remaining freshwater sources hold a vital (in terms of our needs) yet tiny residual fraction of the total (table 1.1). Among the freshwater reservoirs, 68.7 percent are locked up in ice sheets and glaciers, 30.1 percent occur in groundwater, and the remaining 1.2 percent are distributed in lakes, rivers, soils, and swamps.[2]

Figure 1.1 schematically illustrates the major reservoirs of water and their movements around the Earth. These movements constitute the **hydrological cycle**. Water evaporates from the oceans and land and rises into the atmosphere. The moisture condenses into tiny water droplets or ice crystals as it cools. These coalesce into clouds. The atmosphere becomes saturated, holding all the water vapor it can at a given temperature. Eventually, when water droplets or ice crystals grow large and heavy enough, they will fall back to the ground as precipitation. Of the total moisture evaporated, close to 80 percent is precipitated over the oceans while the balance falls over land. While most water evaporated at sea precipitates over the ocean, around 10 percent is transported landward by wind. Over land, water is also evaporated from the soils, lakes, and reservoirs or **evapotranspired** by plants. This water falls back on land as rain, snow, and hail, and also runs off in rivers, returning to the ocean. A substantial fraction percolates downward beneath the surface, forming groundwater that is often confined for long periods in underground aquifers but eventually flows back to the ocean, thereby completing the cycle.

The world's oceans are separated by continents and physiographic basins and are conventionally divided into the Atlantic, Pacific, Indian, Arctic, and Southern. The Pacific Ocean—the largest—covers an area of 181 million square kilometers, followed by the Atlantic Ocean (94 million square kilometers), the Indian Ocean (74 million square kilometers), and the Arctic Ocean (12 million square kilometers). The average depth of the oceans is 3,700 meters (12,100 feet). The **continental shelves** surrounding the continents are the shallowest portions of the oceans, with a depth of less than 130–135 meters (430–443 feet)[3] (fig. 1.2). These actually represent submerged extensions of the continents. Typically, continental shelves vary in width from hundreds of meters to more than 1,000 kilometers. The continental shelf is bounded by a much steeper **continental slope** that grades seaward into a more gently sloping **continental rise**, and ultimately the smooth, relatively featureless **abyssal plains** (fig. 1.2).

The deep ocean is intersected by a nearly continuous submerged mountain chain that is 65,000 kilometers (40,000 miles) long, known as the **mid-ocean ridge**, rising 2–3 kilometers (1.2–1.9 miles) above the adjacent ocean floor and encircling the seafloor like the seams on a baseball. A narrow rift valley at its crest is the locus of active volcanism and seismicity. Fresh **lavas**

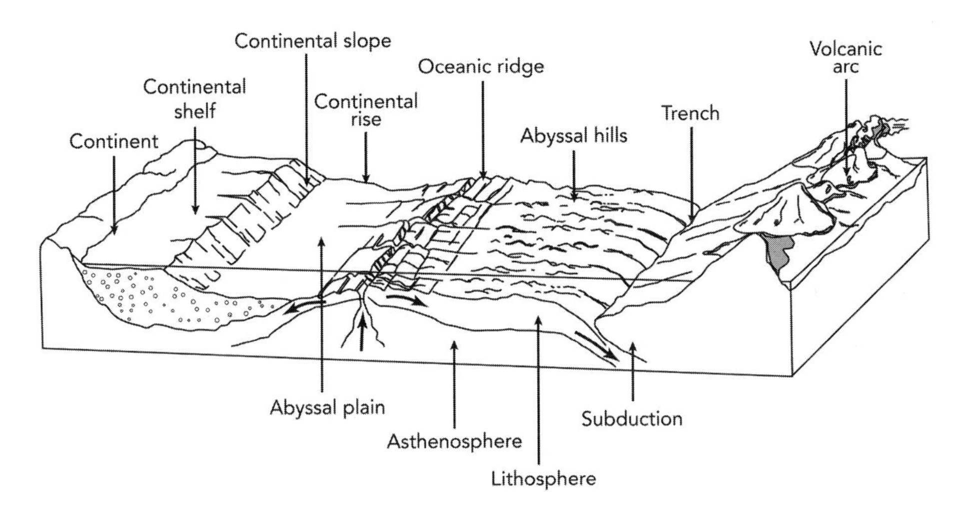

Figure 1.2 Major physical features of the ocean. Also illustrated are the processes of seafloor spreading and subduction of oceanic lithosphere. (Adapted from Thurman, 1997.)

emerge from the mid-ocean ridges along fissures, as the ocean gradually separates over millions of years, in a process called **seafloor spreading**. North America pulls away from Europe at an average rate of around 2.5 centimeters (1 inch) per year, roughly the rate at which fingernails grow.[4] Other mid-ocean ridges spread faster, such as the East Pacific Rise, which is moving at 8–13 centimeters (3–5 inches) per year.

The deepest parts of the ocean lie in the deep-sea trenches that encircle the Pacific Ocean, parts of the Caribbean Sea, and the eastern Indian Ocean. The Mariana Trench in the Pacific is the deepest, reaching 11,000 meters (36,000 feet)[5] beneath the ocean surface. Oceanic crust is dragged down into the Earth's mantle at the deep-sea trenches, in a process known as subduction. **Subduction zones** are sites of active volcanism and earthquakes (see chapter 2).

Seafloor spreading and subduction lead to fragmentation and movement of large segments of the **lithosphere** (or plates), according to the unifying concept of **plate tectonics**.[6] Gradually over tens of millions of years or more, these geologic forces ultimately reconfigure the ocean basins and continents, change the mean ocean depth, uplift mountains, and even alter the Earth's climate. The relationship between these internal geological processes and changes in climate and sea level will be examined further in subsequent chapters.

CURRENTS AND COUNTERCURRENTS

As the dominant reservoir of water on the Earth's surface, the oceans play an important role in influencing the climate. They accomplish this both by physically transporting warm water toward the poles and by transferring vast quantities of heat energy via large-scale ocean currents. Winds blowing across the water surface create basin-wide ocean circulation systems. The Earth's rotation deflects air in motion to the right in the Northern Hemisphere and to the left in the Southern Hemisphere. This process is called the **Coriolis effect**, after its discoverer, Gaspard Gustave de Coriolis (1792–1843).[7] Wind-driven ocean currents are deflected in an analogous manner, generating quasi-circular patterns called gyres (fig. 1.3). Set into motion by the Sun's heat and the prevailing winds, the currents flow around a pile of elevated water toward the center of the gyre. The Coriolis effect causes the surface currents to veer toward the right in the Northern Hemisphere. The inward-directed Coriolis force is balanced by an outward-directed force created by the elevated water and gravity. The balance of these two opposing forces creates a **geostrophic current** that flows around the gyre.

In the Atlantic Ocean, the North Equatorial Current, driven by the easterly trade winds, pushes warm tropical water toward the Caribbean and then turns north. Warm water from the Gulf of Mexico flows through the Florida Straits between Cuba and the Florida Keys, merges with warm Caribbean water and flows north into the Gulf. The Gulf Stream is a mighty ocean river that transports some 55 million cubic meters of water (1.9 billion cubic feet) per second,[8] equivalent to more than 500 times the flow of the Amazon River. It extends farther northeast as the North Atlantic and Norwegian Currents, which help maintain a relatively balmy climate in the British Isles and northwestern Europe. The Gulf Stream and North Atlantic Currents, however, may provide only a part of the warmth.[9] Since the British Isles and Scandinavia are surrounded by the sea, they already experience a fairly mild maritime climate. Furthermore, the southwesterly winds blowing across a warmer part of the Atlantic Ocean also bring considerable heat toward Europe.

A portion of the North Atlantic Current splits and flows south into the Canary Current off the coasts of Spain, Portugal, and northwest Africa. This finally joins the North Equatorial Current, thus closing the loop. A similar gyre circulates in the opposite direction in the South Atlantic Ocean. In the Pacific Ocean, the Pacific North Equatorial Current flows west, then turns north, joining the Kuroshio Current. The Kuroshio flows northeast into the North Pacific Current, which subsequently turns southward and becomes the California Current, ultimately rejoining the North Equatorial Current. A gyre analogous to the one in the South Atlantic rotates counterclockwise in the southern Pacific Ocean. A smaller gyre is also present in the Indian Ocean.

The overall north-south distribution of the continents influences ocean circulation by diverting the eastward-flowing warm equatorial currents toward the north and south along the western boundaries of the ocean basins (fig. 1.3). Two such boundary currents have already been mentioned, namely the Gulf Stream off the Atlantic coast of North America and the Kuroshio Current off eastern Asia. In general, western boundary currents tend to move faster than those along eastern ocean basin margins for several reasons, including the Coriolis effect, which increases toward higher latitudes, the general circulation of the atmosphere, and friction between continental edges and the ocean currents. However, no continents obstruct the Antarctic Circumpolar Current, which flows from west to east around Antarctica, driven by the prevailing westerly winds.

In addition to the wind-driven surface currents, the contrasts in ocean temperature and salinity between tropics and poles set into motion a deep ocean circulation that is triggered by density variations. As water warms,

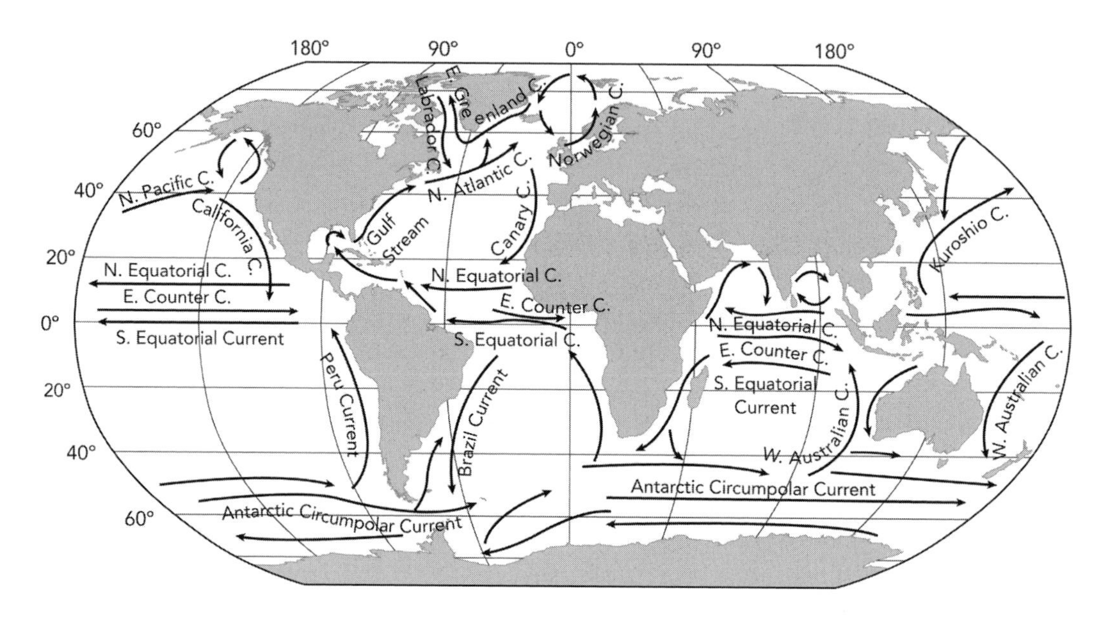

Figure 1.3 Major wind-driven surface currents of the world's oceans. (Modified from Thurman, 1997.)

its density decreases. On the other hand, as it becomes colder and saltier, the density increases. Therefore, at higher temperatures and lower salinity, water becomes less dense, more buoyant, and rises to the surface; at lower temperatures and high salinity, it sinks.

The salinity of the ocean can vary for a number of different reasons. Near hot, arid desert regions, such as off the coast of northern Chile, southern Peru, or northwest Africa, high rates of evaporation leave saltier water behind. Not surprisingly, the areas of highest marine salinity closely correspond to oceans adjacent to the world's desert belts. By contrast, in areas of high precipitation near the equator, or near river deltas such as that of the Amazon that carry large volumes of freshwater, the ocean is diluted, becoming less salty. At high latitudes, water freezes during the long, frigid winters, forming a layer of sea ice. Salt is squeezed out as the ice crystallizes, making the remaining surface water much saltier. This cold and dense water tends to sink. In summer, the sea ice melts and freshwater runoff from melting snow and ice near the shore dilutes the ocean water, forming a "freshwater lid" or layer.

These ocean density variations initiate a worldwide pattern of deep ocean currents often referred to as the **thermohaline circulation** (or the global ocean conveyor system, or the oceanic conveyor belt) (fig. 1.4).[10] (The more comprehensive term **Meridional Overturning Circulation**, or **MOC**, includes other factors such as winds and tides, which are often difficult to separate from the density-driven currents). The coldest, densest water occurs at high latitudes. For example, as the North Atlantic and Norwegian Currents reach the Norwegian-Greenland Seas, the ocean water cools down from 10–12°C to only 2–4°C with a salinity of 3.48 to 3.51 percent[11] and becomes dense enough to sink as **North Atlantic Deep Water (NADW)**. This deep, dense submarine current is joined by additional cold sinking water off southeastern Greenland and the Labrador Sea. The NADW, with a volume equivalent to 100 Amazon Rivers,[12] flows south and can be traced all the way toward the Southern Ocean near Antarctica, where part of it mingles with an even colder, denser layer of water (the Antarctic Bottom Water) and joins the currents that flow around Antarctica. The deepwater flow then branches into the Indian Ocean and wraps around southern Australia into the Pacific Ocean. A return flow of warmer, less salty water travels at shallow to medium depths from the western Pacific and Indian Oceans around the Cape of Good Hope, South Africa, back into the Atlantic Ocean, eventually joins the Gulf Stream–North Atlantic Current and finally the NADW, completing the circuit. (The Gulf Stream is predominantly a wind-driven current, as described above, but around one-fifth can be attributed to the thermohaline circulation). Unlike the North Atlantic where the Gulf Stream carries saltier water north that sinks as it is chilled, surface salinity remains lower in the North Pacific, inhibiting deepwater formation there.[13]

Changes in the thermohaline circulation have been implicated in past (and perhaps future) abrupt climate changes (see also chapters 4 and 5). Many scientists believe that massive influxes of freshwater from glacial lakes and calving icebergs disrupted the NADW and initiated the Younger Dryas, a sudden cold spell between 12,800 and 11,600 years ago, toward the end of the **last ice age**. The reduced North Atlantic salinity prevented cold, dense water from sinking, thus weakening the great conveyor belt system and temporarily plunging the entire region back into near ice age conditions. The cold lasted decades to centuries before the conveyor system restarted.

Could such a process recur in the future, as the buildup of atmospheric greenhouse gases warms up the Earth? A warmer Earth would strengthen the hydrological cycle, producing greater amounts of rainfall and river runoff, thereby freshening the North Atlantic Ocean. Increased melting of glaciers and ice caps would enhance the freshening. Computer climate models reinforce this view, suggesting that ocean freshening and decreased surface water density would slow down the thermohaline circulation and cool the North Atlantic and northwestern Europe, even as the rest of the world continues to warm. However, recent observations of ocean density and sea surface heights yield conflicting results.[14]

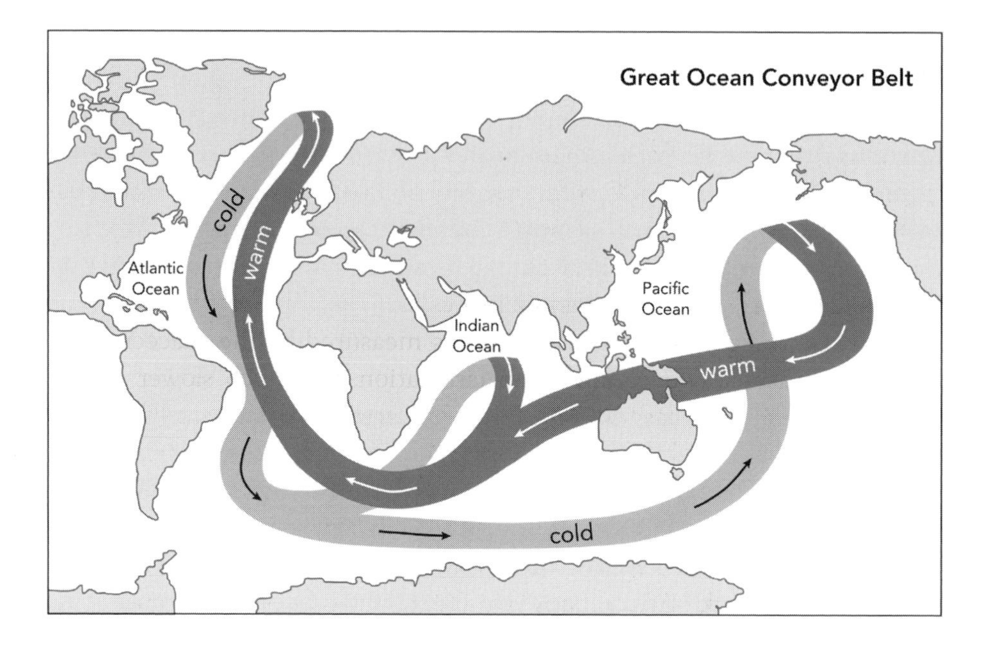

Figure 1.4 A simplified representation of the global ocean conveyor belt. (After W. Broecker, used with permission.)

As the thermohaline circulation changes, so do ocean surface height and sea level. At present, sea level in the North Atlantic is nearly one meter lower than in the North Pacific because of the formation of deep water in the former ocean, but not the latter.[15] While the ocean surface would quickly readjust to a shift in currents without permanently changing global mean sea level, any resulting longer-lasting climate trend could alter mean ocean height (see chapter 2).

Since the early 1990s, orbiting space satellites have been monitoring the seasonal and multi-year variations in ocean surface heights. TOPEX/Poseidon, a joint U.S.-French National Aeronautics Space Administration–Centre National d'Études Spatiales mission, which operated between 1992 and 2005, was followed by the U.S.-French Jason-1, launched on December 7, 2001 (fig. 1.5). Variations in ocean topography are mapped by precisely tracking the satellite's position along its orbital path and measuring its height relative to the center of the Earth with several independent instruments. In practice, the height of the satellite is determined with respect to a fixed reference frame, such as the **reference ellipsoid**. The reference ellipsoid is an idealized mathematical surface that closely approximates the shape of the Earth, taking into account the fact that the Earth bulges slightly at the equator and is flattened at the poles, due to the rotation on its axis. On-board radar altimeters measure the precise distance between the spacecraft and the top of the ocean by recording the time needed for a microwave pulse to travel from the orbiting craft to the sea surface and back. The elevation above the ocean can be readily calculated, by knowing the time and the speed of light, then correcting for atmospheric and instrumental effects. By subtracting the spacecraft's elevation above the ocean from its distance to the Earth's center, the sea surface height relative to the center of the Earth can be determined to within 4–5 centimeters (around 2 inches).

The ocean height at a given location as measured by the spacecraft reflects the continually evolving ocean circulation and much slower gravity variations. Repeated observations of the same area over many satellite cycles (the orbit of TOPEX/Poseidon, for example, repeated every 10 days) eliminates features linked to underwater bathymetry, leaving others that vary over time due to a dynamic ocean. In this manner, radar altimetry has become an important tool in providing a continuous view of an ever-changing sea surface topography, and in conjunction with other measurements has greatly improved our understanding of ocean circulation and climate phenomena

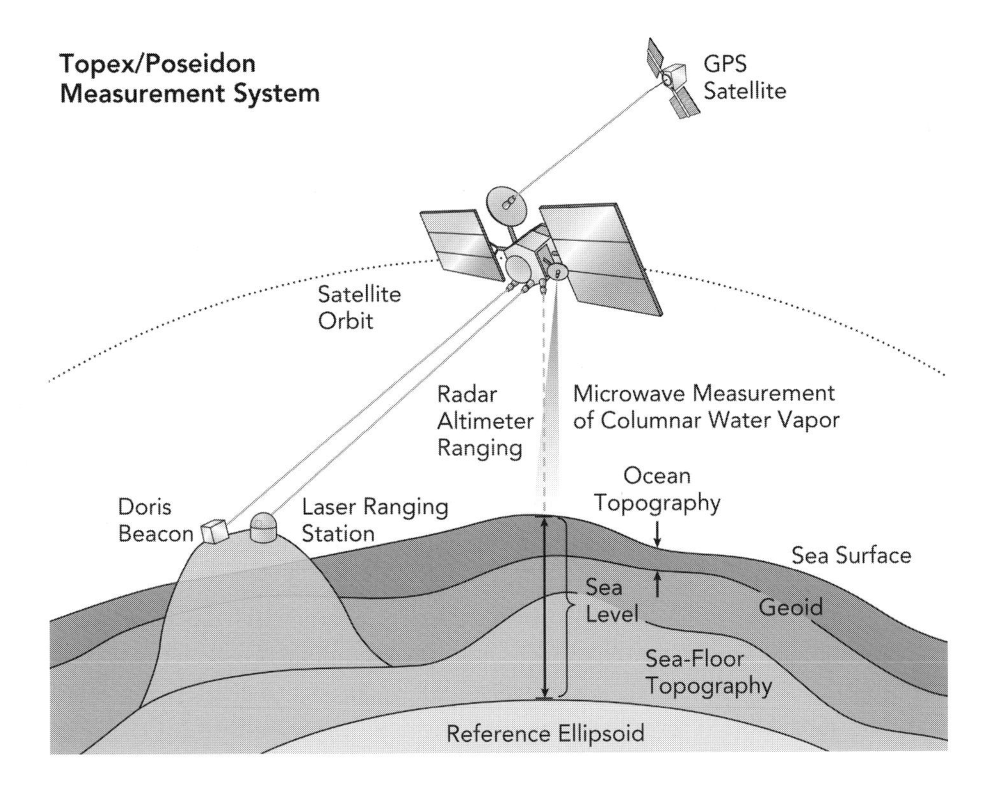

Figure 1.5 The TOPEX/Poseidon satellite altimeter. (NASA JPL.)

such as the **El Niño–Southern Oscillation (ENSO)**. By averaging the observed variations in sea surface height over all the oceans since 1992, year to year changes can be detected. These measurements show that global mean sea level is now increasing by 3.4 millimeters per year.[16] This recent satellite trend is greater than that calculated from the much longer record provided by tide gauges. Is this an early warning sign of global warming and greater amounts of ice melting? Possibly. However, the satellite record is still too short to rule out the possibility of a shorter-term climate oscillation (see chapter 6 for further discussion).

WAVES

Waves are the most conspicuous sign of a restless ocean that is always in motion. From gentle ripples, to swells, breakers, whitecaps, and towering rogue waves that can swamp large ships, these undulations are chiefly the product of wind action at the interface between air and water. Waves shape the configuration of the shoreline by transporting sand and sediment, by cutting into cliffs and marine terraces, and by pounding against rocky headlands, gradually wearing them down. Powerful storm waves can cause significant beach erosion and damage to man-made structures. They amplify the destructive effects of the storm surge.

Waves are set into motion by the wind blowing across the surface of the water. Energy generated by the moving air is transmitted to the water, resulting in the familiar periodic up and down motion we call waves. The top of a wave is the crest, separated by a low depression, or trough (fig. 1.6). The distance between successive crests or troughs is the **wavelength,** L. The vertical separation between crest and trough is the **wave height**, or amplitude, H. The time interval between successive crests or troughs is the **period,** T. The speed, S, or forward motion of the disturbance, equals the wavelength divided by the period, or:

$$(1)\quad \text{Speed} = \text{Wavelength/Period} \quad \text{or } S = L/T$$

Although the wave appears to be advancing, water particles essentially remain in place as successive wave crests and troughs pass. Imagine a cork bobbing on the water. It rises and moves forward on the crest, only to fall with the next trough, even moving backward until pushed up by the next crest. The cork's up-and-down motion describes a circle with a diameter equal to the wave height, as waves ripple forward beneath it. This idealized description assumes a smooth sine curve. Real waves have more angular

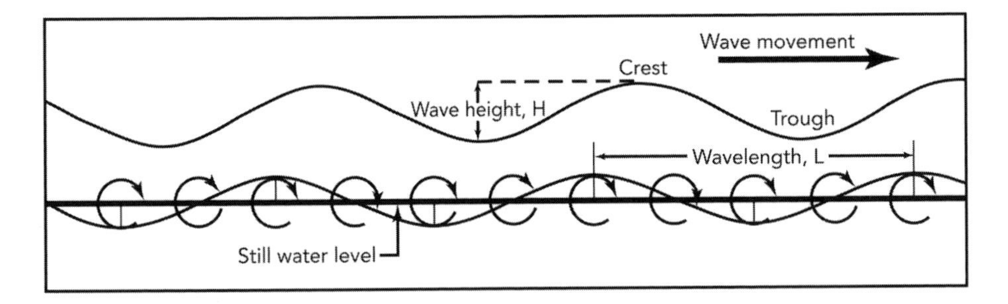

Figure 1.6 An ocean surface wave and its component parts.

crests than troughs, so that water particles (or bobbing corks) at the wave crests move faster than in the troughs, imparting a net forward motion.[17]

Wave height, *H*, is important because it is directly connected to the wave's energy and hence its ability to sculpt the shoreline or to wreak havoc during storms. In shallow water, the wave energy is directly proportional to the square of the wave height. Therefore, a doubling of wave height increases the corresponding energy by a factor of four. Surfers' delight can quickly transform into shore dwellers' nightmare!

Several factors help determine the height, hence energy, of waves. These include: (1) the wind speed, (2) wind duration (how long the wind blows), and (3) the **fetch** (the distance over which it blows in a given direction). Major storms can optimize all three factors, producing mighty waves.

Away from the storm, faster-moving waves eventually form a regular series of long- wavelength undulations that can travel with little loss of energy over extensive distances. Such waves are called **swell**. As waves approach shallower depths near the shore, they are slowed down by friction with the bottom but increase in height and steepness, curling forward until they break in the surf zone. Ideally this occurs when the water depth is 1.3 times the height of the breaking wave.[18] As their energy is ultimately dissipated along the shore, they act as an agent of erosion.

Water rises along the shore with each advancing wave and then pulls back to the sea. The forward rush is the **wave run-up** (fig. 1.7). The magnitude of the run-up depends on the slope of the surface, wave period, and height, as well as the astronomical tide and storm surge.

The continual pounding of waves and abrasion by sand and gravel against the base of a cliff, especially during storms, gradually undermines the stability of the cliff, causing it to retreat. Destructive wave action during storms may also cause beach erosion and damage to seaside homes (see also chapter 8).

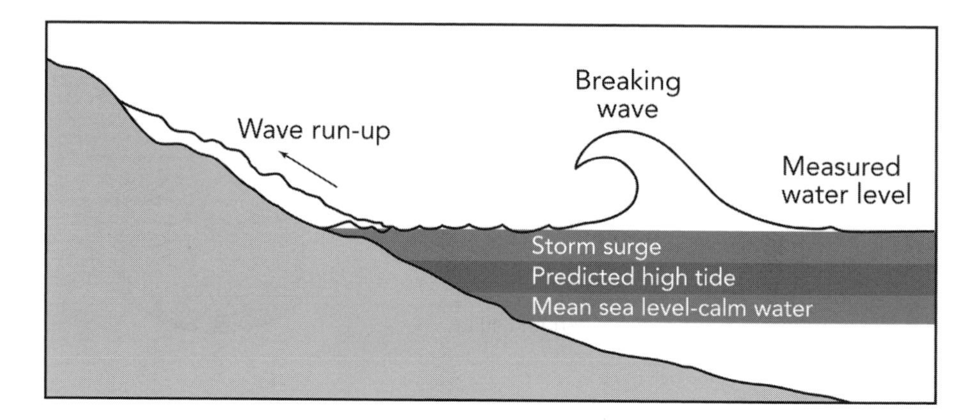

Figure 1.7 Schematic diagram of wave run-up.

TIDES

The sea rises and falls daily with predictable regularity, responding to the tidal pull of the Moon and Sun on the Earth. The basic physical forces that act in concert to produce the tides include: (1) the force of gravitation of the Moon (and Sun) on the Earth and (2) the centripetal force due to the revolution of the Earth and Moon around their common center of mass. These forces are modulated by the phase of the Moon, and to a lesser extent by the Earth's annual orbit around the Sun. Local conditions also alter the magnitude and timing of the tidal cycle—for example, the configuration of the shore, the underwater bathymetry, and the latitude.

According to Newton's Law of Gravitation, two particles are attracted toward each other by a force that is directly proportional to the product of their masses and inversely proportional to the square of the distance separating them:

(2) $F = G\, m_1 m_2 / r^2$

where G is the universal gravitational constant, 6.67×10^{-8} cm^3/g/sec^2, m_1 is the mass of the first particle (in grams), m_2 is the mass of the second particle, and r is the distance between them (in centimeters). In the Earth-Moon system, m_1 is the mass of the Moon, m_2 is the mass of the Earth, and r is the distance between their centers.

The Moon orbiting the Earth (or the Earth orbiting the Sun) is tied to its path by the force of gravity, just like a rock tied to a string that is swung in a circle by a child. If the string were to break, the rock would fly off in a straight

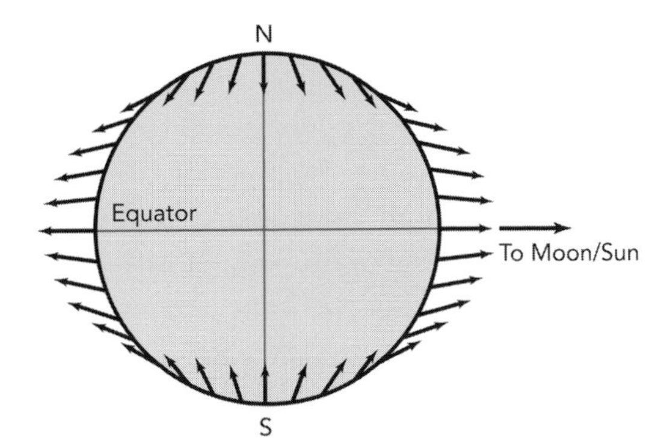

Figure 1.8 The distribution of the vertical tide-generating forces. The strongest tidal pull is in a direction toward and opposite the Moon (and Sun). The weakest tides are in a perpendicular direction.

line tangent to the circle. The string operates like a centripetal (center-seeking) force that keeps the rock on its circular path. Similarly, gravity "pulls" the Moon toward the Earth. If gravity were "turned off," the Moon would fly off into space along a straight-line trajectory. The centripetal force is the gravitational force required for the Earth and Moon to revolve around their common center of mass, located 4,700 kilometers (or 2,900 miles) from the Earth's center in the Earth-Moon direction, as they both orbit around the Sun. Every part of the Earth experiences the same centripetal force, since it acts as a rigid body. However, the gravitational attraction of the Moon on any mass on the Earth varies with its location, since the distance between that mass and the Moon is much less on the side of the Earth facing the Moon than on the opposite side.

The tide-raising force generated by the Moon (and to a lesser extent by the Sun) is the difference between the gravitational force exerted by the Moon on a mass located on the Earth's surface and the centripetal force. It is directly proportional to the mass of the Moon and inversely proportional to the cube of the Earth-Moon distance, or $G\ m_1/r^3$. The distance between the unit mass and Moon at a point on the Earth's surface directly facing the Moon is $r—R$, where R is the radius of the Earth, so that it experiences a stronger gravitational pull than a mass at the Earth's center (fig. 1.8). Therefore the gravitational force exceeds the centripetal force and the tide-raising force produces a bulge in the ocean water on the side of the Earth facing the Moon. On the opposite side of the Earth, the distance between the unit mass at the surface and Moon is $r + R$, so that the gravitational force is correspondingly less than the centripetal force. The net difference is equal and opposite, so that a tidal bulge occurs there as well (fig. 1.8).[19]

Although the gravitational force of the Sun on the Earth is approximately 177 times stronger than that of the Moon, owing to its overwhelmingly greater mass relative to the Moon, it is 390 times farther from the Earth than the Moon. However, since the tide-raising force varies as the cube of the distance, the Sun's effect on the tidal force is only 46 percent that of the Moon's.

As the Earth rotates upon its axis, the Moon orbits the Earth. After 24 hours, the point on the Earth that was directly under the Moon has lagged behind. In order to catch up, the Earth must turn for another 50 minutes to return to the same position relative to the Moon. Therefore, the tides return roughly one hour later each day.

Lunar and Solar Cycles and the Tides

The strengths of the tides fluctuate with the monthly and yearly cycles of the Moon and Sun. As the Moon orbits the Earth, its position relative to the Sun varies over a 29.5-day cycle, leading to the familiar phases: from new moon to first quarter, full, last quarter, and new moon again (fig. 1.9). Spring tides occur at new moon when the Sun and Moon are aligned on the same side relative to the Earth. Because the tidal forces of Sun and Moon reinforce each other, the range between mean high and low water is greatest then. Spring tides also occur at full moon, 14.8 days later, when Earth lies between the Moon and Sun. At first and last quarter, the tidal bulge created by the Moon lies at right angles to that produced by the Sun and their effects tend to cancel. Thus, neap tides have much smaller ranges.

The tides also vary in magnitude as the Moon moves toward or away from the Earth. Tidal ranges are magnified at **perigee** when the Moon comes closest to the Earth, once every 27.6 days. Conversely, they are much lower at lunar **apogee** when the Moon is most distant from the Earth. Tidal ranges are especially large when spring tides (at full or new moon) coincide with perigee; they are correspondingly much lower when neap tides (first and last quarter) coincide with apogee.[20]

The Earth's rotational axis is tilted at 23.5° with respect to the plane of its orbit around the Sun, called the **ecliptic**. This tilt is responsible for the seasons. The Earth's axis points directly toward the Sun during the summer in the Northern Hemisphere.[21] At the onset of Northern Hemisphere winter, the Earth's axis now tilts away from the Sun. (The seasons are exactly reversed in the Southern Hemisphere, such that Southern Hemisphere summer begins on December 22; there the axis points *toward* the Sun.)

The tides are also modulated by the Earth's orbit around the Sun. The Earth approaches the Sun most closely at **perihelion** in early January, and is farthest from the Sun (**aphelion**) in the northern summer months. Therefore, extra-strong tides can be expected in early January, especially if perihelion coincides with full or new moon at perigee.

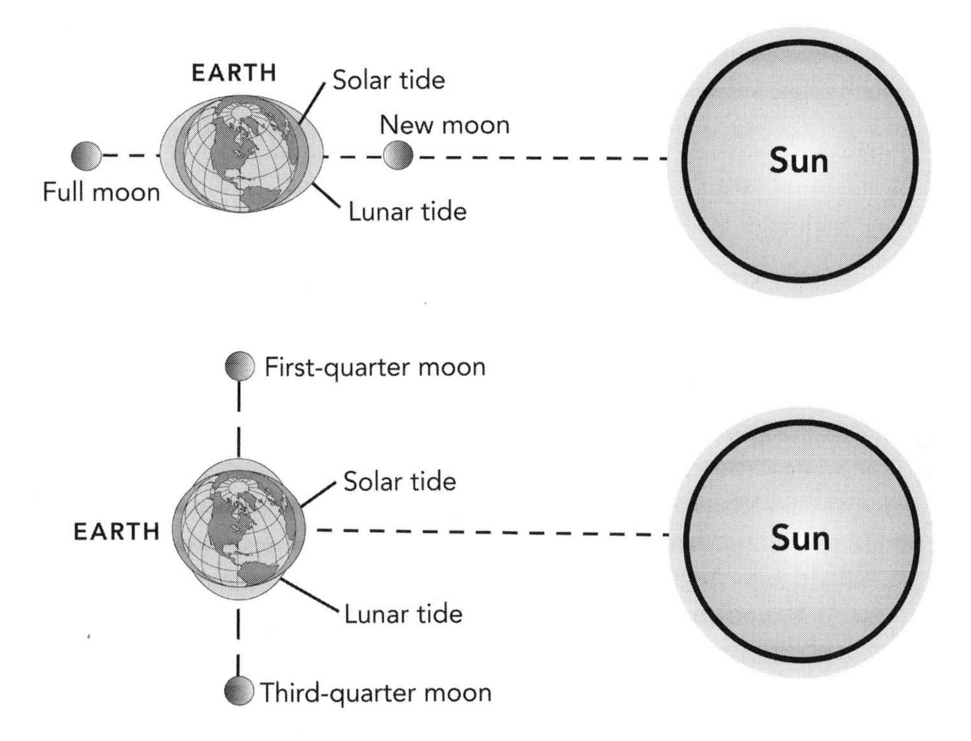

Figure 1.9 Phases of the Moon and the tides. Top: The tidal bulges are higher when the Earth, Moon, and Sun are aligned at full and new Moon (spring tides). Bottom: Tides are much lower at first- and last-quarter Moon, when the Moon is nearly at right angles to the Earth-Sun (neap tides). (Modified from Thurman, 1997; not drawn to scale.)

A longer tidal cycle—the **lunar nodal cycle**—arises from changes in the inclination of the Moon's orbit. The plane of the Moon's orbit makes an angle of 5 with respect to the ecliptic. However, over an 18.6-year cycle, this angle ranges from a maximum of 23.5° + 5°, or 28.5° to a minimum of 23.5° − 5°, or 18.5°.

Extreme Tidal Ranges

Certain coasts are notable for their extreme tidal range—the difference between the average daily high and low tide. Some well-known examples include the Bay of Fundy in Canada and Mont-St. Michel–Bay of St. Malo, France. The North Sea and Vancouver, Canada, also experience large tidal amplitudes. The Bay of Fundy, 258 kilometers (160 miles) long, holds the record for the world's largest tides. At the extreme northern end of Minas Basin, one of two narrow inlet basins, the maximum tidal range at perigee is around 17 meters (56 feet), diminishing to 2 meters (6.6 feet) at the entrance to the Bay of Fundy. By contrast, tidal ranges are very small in the Mediterranean, Baltic, and Caribbean Seas.

STORMY SEAS

Local sea level can vary dramatically within a few hours, above and beyond the rise and fall of the astronomical tides described above, because of approaching storms. Meteorological disturbances change the ocean height through wind action and variations in barometric pressure. Of these two forces, the wind exerts the more powerful effect.

A **storm surge** is the increase in ocean water level near the coast generated by a passing storm, above that resulting from the astronomical tides.[22] The sea responds to the storm in two different ways. As the air pressure drops, the load of the atmosphere on a column of water beneath the sea surface is reduced, causing the water to rise, and vice versa. For each drop in atmospheric pressure of 1 millibar, sea level rises by around 1 centimeter. This change is known as the **inverse barometer effect**. In nature, the inverse barometer effect is considerably modified by the complex interactions between the atmosphere and shallow water on continental shelves.

Another factor contributing to the surge is the drag or stress on the sea surface caused by the wind, measured as the horizontal force per unit area. The stress depends on the wind's speed and air density. The wind direction relative to the shore is important as well. Wind blowing toward the shore will pile up more water than wind blowing parallel to the shore. The wind's effect is amplified over wide expanses of shallow water, such as the North Sea or the northern end of the Bay of Bengal, off the coast of Bangladesh.

Two different types of storms generate strong surges that produce coastal flooding and beach erosion. They include the tropical cyclones that are most prevalent at low latitudes, and extratropical cyclones that are the familiar winter storm systems of the middle latitudes (called "**nor'easters**" in the eastern coastal United States). Tropical cyclones go by several different names, known as **hurricanes** in the Atlantic basin, **typhoons** in the western Pacific, and just plain **cyclones** in the Indian Subcontinent. Tropical cyclones are intense low-pressure systems that strengthen over the oceans, particularly where sea surface temperatures are at least 27°C (81°F). Although small in extent, they are very powerful. The storm surge produced by a hurricane is a dome of water raised by the combined effect of the low barometric pressure and the strong winds (maximum wind speeds of at least 119 kilometers per hour and surges of 1 to 2 meters). Because the highest-velocity winds flow counterclockwise around the eye of the hurricane in the Northern Hemisphere, as it moves forward, the strongest winds are concentrated in the upper-right-hand quadrant.

Although extratropical cyclones typically have much lower wind speeds than hurricanes, they can be quite destructive, because they cover a wider area (over 1,000 kilometers [621 miles] for extratropical cyclones) versus 100–150 kilometers [62–93 miles] for hurricanes). They also last longer, sometimes persisting over several tidal cycles at a given location.

Surges are influenced not only by meteorological conditions but also by the geometry of the coastline and underwater topography. The surge can be amplified by a wide continental shelf or where the coastline makes a right-angle bend, as for example, at the apex of the New York Bight[23] or the coast of Bangladesh. The effects of the surge are also intensified when it occurs nearly simultaneously with astronomical high tides, such as at new or full moon, and also at the solstices and equinoxes. On March 6 and 7, 1962, a powerful storm lashed the mid-Atlantic states from Virginia to New England with gale-force winds, high tides, and heavy rain or snow, causing extensive coastal flooding and beach erosion. The Ash Wednesday storm, as it later became known, lasted over five tidal cycles. Furthermore, the surge from this storm was especially strong because it coincided with new moon and perigee.[24] Hurricane Edna hit eastern Long Island and New England on September 11, 1954, causing significant damage there. The surge effects of this storm were also reinforced by its occurrence at full moon and perigee.[25]

Surges from tropical cyclones affect low-lying coasts and major river deltas of southeast Asia (e.g., Bangladesh, Thailand, Myanmar), the Gulf and Atlantic Coasts of the United States, and also the northeastern coast of Australia. Hundreds of thousands of people drowned during powerful cyclones in Bangladesh in 1970 and again in 1991. More than 8,000 people perished under a 4.6-meter (15-foot) hurricane surge that overwhelmed Galveston, Texas, on September 8–9, 1900. Another hurricane swept over Long Island and southern New England in September 1938, with a 6-meter-high surge, killing more than 700 people. More recently, Hurricane Katrina killed 1,500 people in coastal Mississippi and Louisiana on August 29, 2005, packing winds up to 205 kilometers per hour (125 miles per hour) and creating a surge up to 4.6 meters (15 feet) high.

QUASI-PERIODIC CLIMATE VARIATIONS: ENSO AND THE NAO

The ocean, as illustrated above, is constantly in motion. Waves wash up against the shore and retreat within minutes. The tides rise and ebb regularly during the course of a day. Other ocean phenomena play out over longer periods. **El Niño–Southern Oscillation (ENSO)** and the **North Atlantic Oscillation (NAO)** are two different examples of closely linked atmospheric and oceanographic processes that produce regional-scale changes in ocean height that last many months. They display similar quasi-periodic patterns that recur every few years, but with nowhere near the precise mathematical regularity of the astronomical tides. Although ENSO and the NAO generate variations in sea level that persist for months to a year or so, these changes are not permanent.

El Niño–Southern Oscillation (ENSO)

The cold waters along the coasts of Peru and Ecuador ordinarily teem with anchovies and other marine life. Around Christmas and for the first few months of the new year, a warm south-flowing current moderates the usually low water temperatures. Because of its timing, local fishermen have for centuries referred to this annual event as El Niño, after the Spanish for "the Christ Child." Every few years, the ocean waters become warmer than usual during the El Niño event. Whenever this happens, the fish harvests collapse and the marine ecosystem is severely disrupted. Oceanographers now realize that these local ecological effects are linked to a much wider chain of worldwide climate anomalies involving both oceans and atmosphere.

In the early 20th century, a British mathematician, Gilbert Walker, was investigating the relationship between periodic droughts and the failure of the Indian monsoon.[26] He soon discovered wide swings in atmospheric pressure and precipitation between opposite sides of the tropical Pacific and Indian Oceans that occurred every few years, which he termed the Southern Oscillation. When atmospheric pressure was high in Darwin, Australia, for example, it tended to be low in Tahiti, and vice versa. During times when the Darwin air pressure was high, the Australian–Indonesian–southeast Asian region was drier than normal and the Indian monsoon was often weak. Conversely, western South America experienced higher than average rainfall and poor fish harvests. This east-west oscillation in air pressure and rainfall forms part of an atmospheric circulation pattern, the Walker cell, in which warm, moist air near Indonesia rises, and as it cools, the water vapor condenses and produces heavy rainfall (fig. 1.10a, top). Reaching higher elevations, the air flows east, losing more moisture along the way and now, fairly dry and much cooler, sinks in the eastern Pacific. The descending arm of the cell joins the easterly trade winds, drawn toward the western Pacific by the gradient in surface air pressure between Tahiti and Australia. In the 1960s, the Norwegian-American meteorologist Jacob Bjerknes recognized that the Southern Oscil-

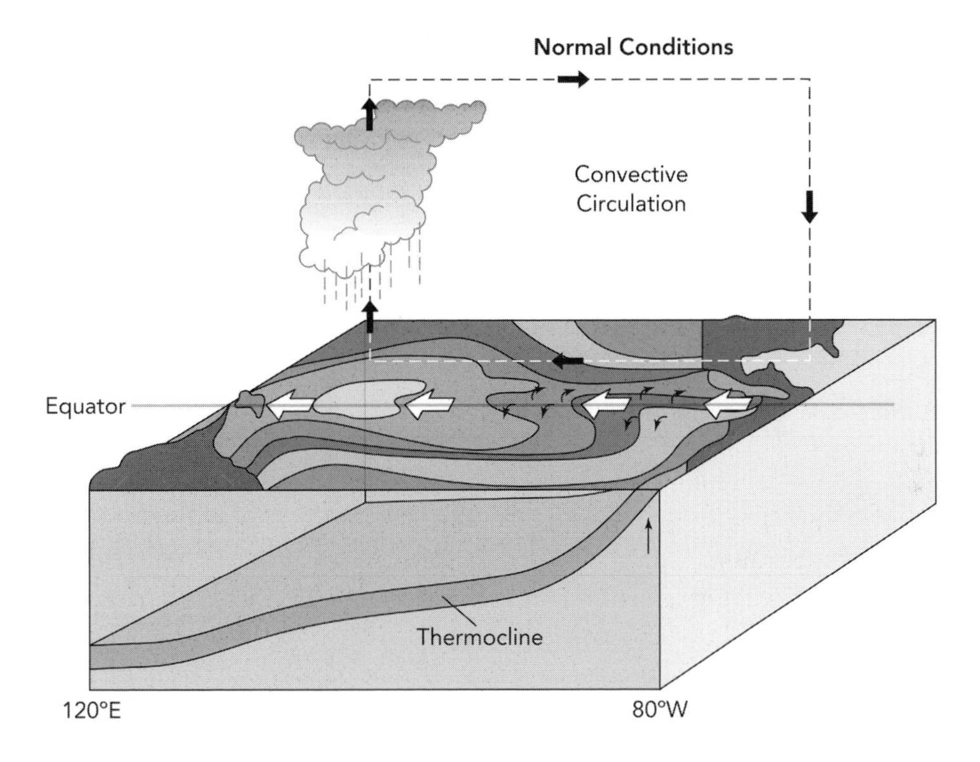

Figure 1.10a The "normal" state of the Pacific Ocean, showing the Walker circulation, the steep westward tilt of the thermocline, and the pooling of warm water toward the western Pacific. (Dr. M. J. McPhaden, Director, NOAA/PMEL/TAO Project Office, Seattle.)

lation and El Niño were different aspects of the same climate phenomenon, now called the El Niño–Southern Oscillation, or ENSO, that recurs roughly every 3 to 7 years. An El Niño event is recognized when monthly sea surface temperatures are at least 0.5°C above average for at least six months in the equatorial central Pacific.[27]

Usually, the easterly trade winds push warm surface water toward the western tropical Pacific, such that the sea surface near Indonesia rises half a meter relative to that of western South America. Temperatures at the sea surface are also up to 8°C higher than off the Pacific coast of South America, where colder water usually wells up from below.[28] During an El Niño, the trade winds slacken, the atmospheric pressure at Darwin rises, and the area of heavy rainfall centered over Indonesia shifts eastward. The weakened trades allow warmer water from the west Pacific to flow toward the coast of South America, creating a so-called "hot tongue."

At about the same time, other changes occur beneath the ocean surface. A thin warm layer of water near the surface is usually underlain by colder water

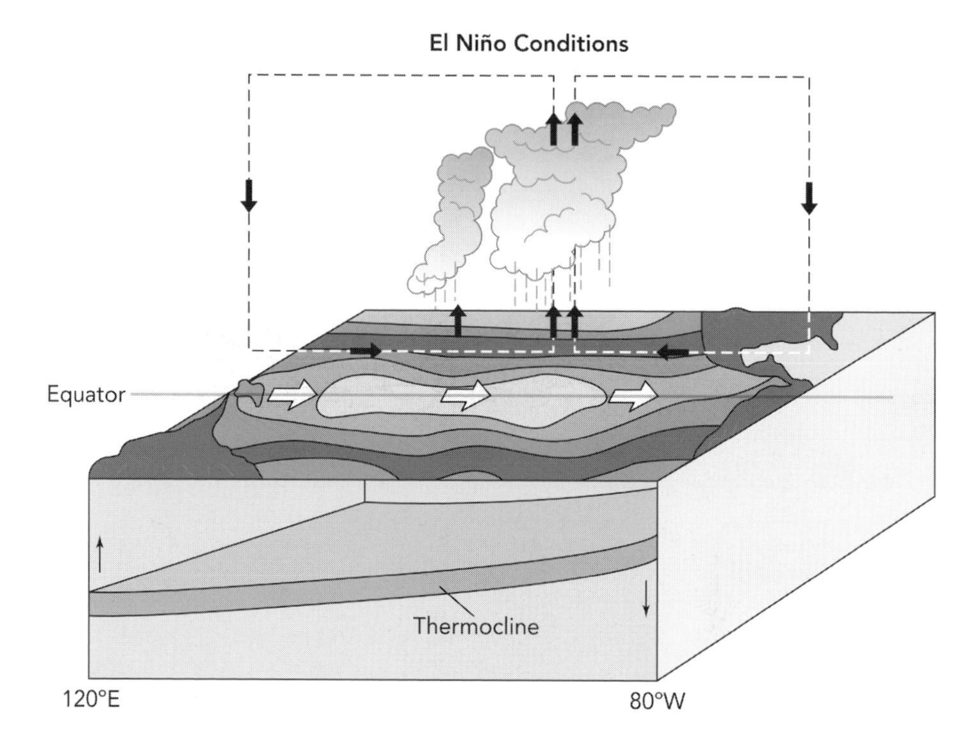

Figure 1.10b The El Niño phase of the ENSO cycle. The easterly trade winds have slackened, upward convection and the warm water pool have shifted toward the east, and the thermocline has become much shallower. (Dr. M. J. McPhaden, Director, NOAA/PMEL/TAO Project Office, Seattle.)

at depth. The boundary between these two layers is called the **thermocline**. Because of the upwelling of cold water, the thermocline in the eastern Pacific is generally fairly shallow, but it deepens with the influx of warmer water from the west as the El Niño condition develops (fig. 1.10b). Conversely, the thermocline in the western Pacific, which usually is much deeper because of the pool of warm water there, becomes shallower during an El Niño event (fig. 1.10b). These ocean changes also trigger eastward propagating pulses of warm water that diverge north and south along the west coasts of the Americas, causing sea level to rise. Major El Niño events (during the second half of the 20th century) appeared in 1957–1958, 1965–1966, 1972–1973,1982–1983, 1986–1988, 1991–1992, 1997–1998, and 2002–2003. Relatively weak El Niños occurred in 2006–2007 and again in 2009–2010.

During La Niña (the opposite phase of the ENSO cycle) the overall pattern reverses (fig. 1.10c). The trade winds strengthen, causing atmospheric pressure to decrease and enhance rainfall over Indonesia, southeast Asia, and Australia. As ocean water temperatures climb on the western side of

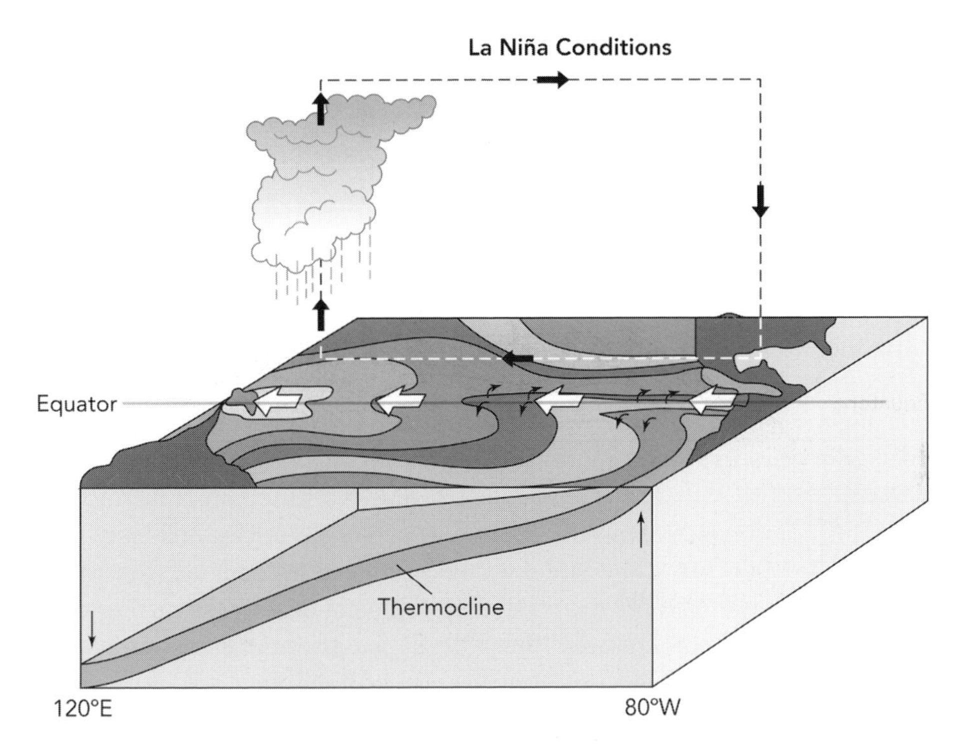

Figure 1.10c The La Niña phase of the ENSO cycle. This is a somewhat intensified form of the normal mode. (Dr. M. J. McPhaden, Director, NOAA/PMEL/TAO Project Office, Seattle.)

the Pacific basin, the thermocline deepens and sea level rises. Meanwhile in the eastern Pacific, sea surface temperatures drop, air pressure rises, and sea level falls. This behavior is a somewhat intensified form of the "normal" mode. Some recent La Niña years include 1954–1957, 1973–1974, 1974–1976, 1984–1985, 1988–1989, 1995–1996, 1998–2000, and 2007–2008.

Strong El Niño events produce many repercussions around the globe. The climate anomaly that disrupts the Peruvian fisheries also brings heavy rainfall across Peru, Ecuador, and northern Chile in December through February and to southern Brazil and Argentina a few months later. However, the Amazon Basin, northeast Brazil, and Central America tend to dry out. Wetter, cooler conditions prevail in California and the U.S. Southwest. High winter storm waves pound the coast of California, undermining sea cliffs and eroding beaches. Above-normal precipitation along the U.S. West Coast leads to widespread flooding and landslides. Northeast Africa is also wetter than usual, whereas southeast Africa can become parched. Drought conditions are common in Australia, southeast Asia, and even India, as the monsoon is usually (although not always) much weaker during strong El Niños.

Drought-stricken areas also experience more-frequent and more-intense forest fires. The above-average tropical ocean temperatures in many regions also affect marine ecosystems. For example, during the 1997–1998 event, massive widespread coral bleaching caused the mortality of 16 percent of the world's reef-building corals.[29] Not all El Niño climate impacts are negative, however. One benefit is the significantly reduced chance of major hurricanes striking the United States during El Niño years.[30] The climate impacts of La Niñas are roughly the opposite of those seen during El Niños, although the latter tend to produce more dramatic consequences.

An El Niño event typically generates sea levels that are tens of centimeters above average along the west coast of the Americas.[31] These anomalies persist for days to several months. The oceanic changes associated with El Niño give rise to eastward-traveling equatorial waves that bifurcate, moving both north and south along the shorelines of the Americas. Sea level rises because the waves bring in warmer water, creating a "hot tongue." The arrival of the warm water disrupts the normal upwelling of cold water, further driving up sea surface temperatures. The piling up and **thermal expansion** of the more buoyant warm water, as well as divergence of the equatorial waves both north and south produce higher than average sea levels along the western coasts of the Americas.

The propagation of these coastal waves north and south of the equator can be tracked by observing sea level variations on tide gauges at different locations, after correcting for the effects of local atmospheric pressure and long-term sea level trends due to earth motions. The 1997–1998 El Niño, the strongest of the 20th century, was also probably the most thoroughly studied such event. Tide gauges picked up a series of above-average sea level anomalies at San Francisco, California, starting in May 1997 and peaking in December 1997 into January 1998.[32] Somewhat earlier, tide gauges down south in La Libertad, Ecuador (2°12′ S latitude) had detected a similar pattern of sea level variations, which later appeared farther north at Tumaco, Colombia (1°50′ N), and as time progressed, showed up at Los Angeles, California (33°43′ N), and finally at San Francisco (37°48′ N). The waves traveled north along the coast of South America, from a point near the equator to California, at an estimated speed of around 2 meters per second, based upon the time it took the sea level signal to reach California from its apparent source.[33]

El Niño–related changes in sea surface height can also be followed from satellite radar altimeters like TOPEX/Poseidon. At least two distinct pulses of higher than normal ocean heights were tracked spreading from the equator toward both western North and South America during May–July and October–December 1997.[34] These closely coincided with some of the peaks seen in the tide gauge records.

North Atlantic Oscillation (NAO)

The Atlantic Ocean basin is considerably smaller than the Pacific basin. Nevertheless, the Atlantic experiences multi-year climate and ocean circulation variations that are somewhat reminiscent of the Pacific-based ENSO phenomenon, although the causative processes differ. The North Atlantic Oscillation (NAO) is another example of such climatic variability.[35] Although its clearest manifestations are meteorological in nature, i.e., changes in storm patterns, rainfall, sea and air temperatures, and sea ice, it also produces variations in regional sea level registered on tide gauges.

The NAO index has been defined as the difference in average winter (December–March) sea level air pressure between Lisbon, Portugal, and Stykkishólmur, Iceland.[36] (Other definitions are based on the pressure difference between the Azores, Bermuda, or Gibraltar and Iceland.) The atmospheric pressure surrounding Iceland is generally low, associated with damp, chilly weather, whereas the air pressure across a broad subtropical-to-temperate band of the Atlantic (stretching from Bermuda to the Azores and Portugal) is usually much higher. When the NAO index is positive, the Icelandic low is lower than normal, while the Azores high is above average, and the pressure gradient between them is stronger. As a result, the prevailing westerly winds across the North Atlantic are strengthened and bring with them more stormy weather that moves farther north-northeast than usual (fig. 1.11a). Thanks also to the moderating influence of the Gulf Stream and North Atlantic Current, milder, but wetter, than average winters occur in the British Isles and Scandinavia. On the other hand, stronger northerly winds over western Greenland and northeastern Canada produce colder than normal weather, whereas the southeastern United States becomes anomalously warm. Furthermore, because the storm systems over the eastern Atlantic move farther north than usual, the positive phase of the NAO tends to leave southern Europe, the Mediterranean region, North Africa, and the Middle East cooler and much drier than normal.

A year or several years later, as the pattern reverses during the negative phase of the NAO, the pressure difference between Lisbon and Iceland diminishes, northern Europe endures colder and drier winters, and southern Europe becomes wetter (fig. 1.11b). The NAO remained in a positive phase for most of the 1980s through the mid-1990s, giving rise to speculation that this has not only strengthened the recently observed global warming trend, but may even be a more permanent sign of regional climate change. During the 2000s, however, the NAO has reverted back to near-average conditions, so that it is more likely that the 1990s anomaly just expressed inter-annual climate variations.

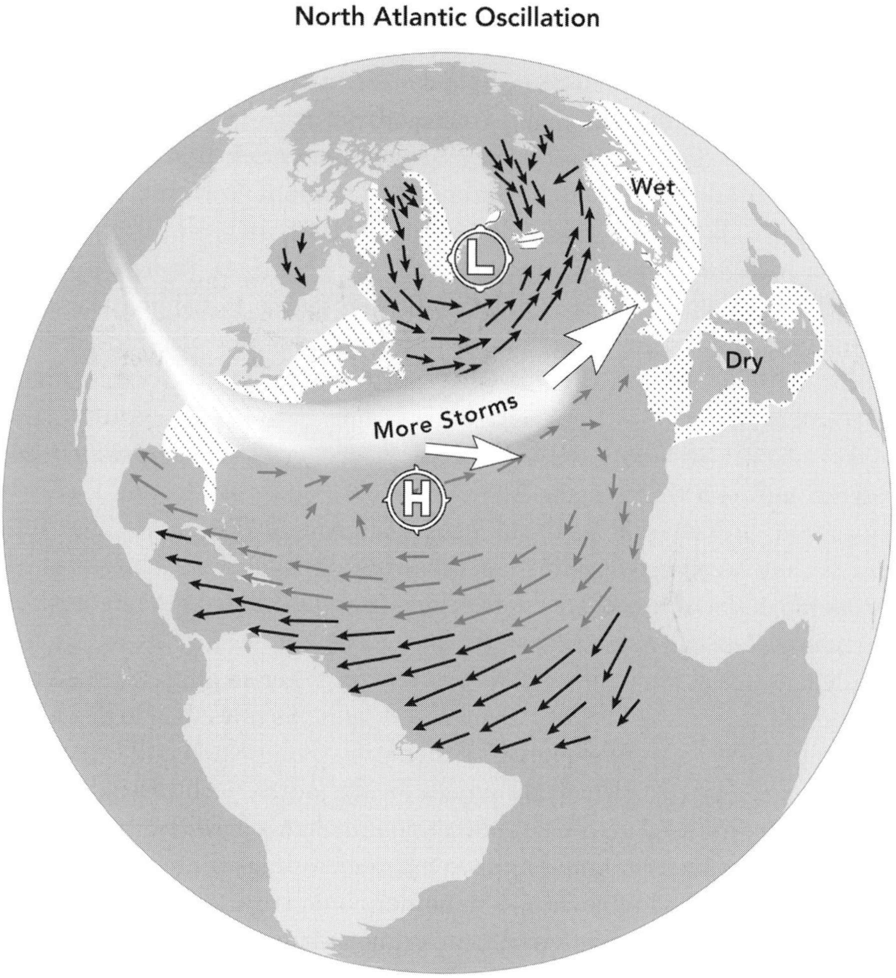

North Atlantic Oscillation

Wet

Dry

More Storms

L

H

Figure 1.11a Positive phase of the North Atlantic Oscillation. Intensified high pressure over the Azores, deeper low pressure over Iceland, northeasterly shift of storm tracks over the Atlantic Ocean, produce milder condition in northern Europe and drying of the Mediterranean and North Africa. (Martin Visbeck, IFM-GEOMAR, Kiel.)

The NAO, like ENSO, affects sea level in a number of regions, particularly in northwestern Europe, and to a lesser extent in the southeastern United States. In the positive phase of the NAO cycle, annual winter sea level in the North Sea and Baltic Sea is higher than average, and vice versa. During a positive NAO, northwestern Europe experiences increased winter storminess. The strong westerly winds generate large surges superimposed on the normal water level and tides.[37] Thus sea level rises, if only for the duration of

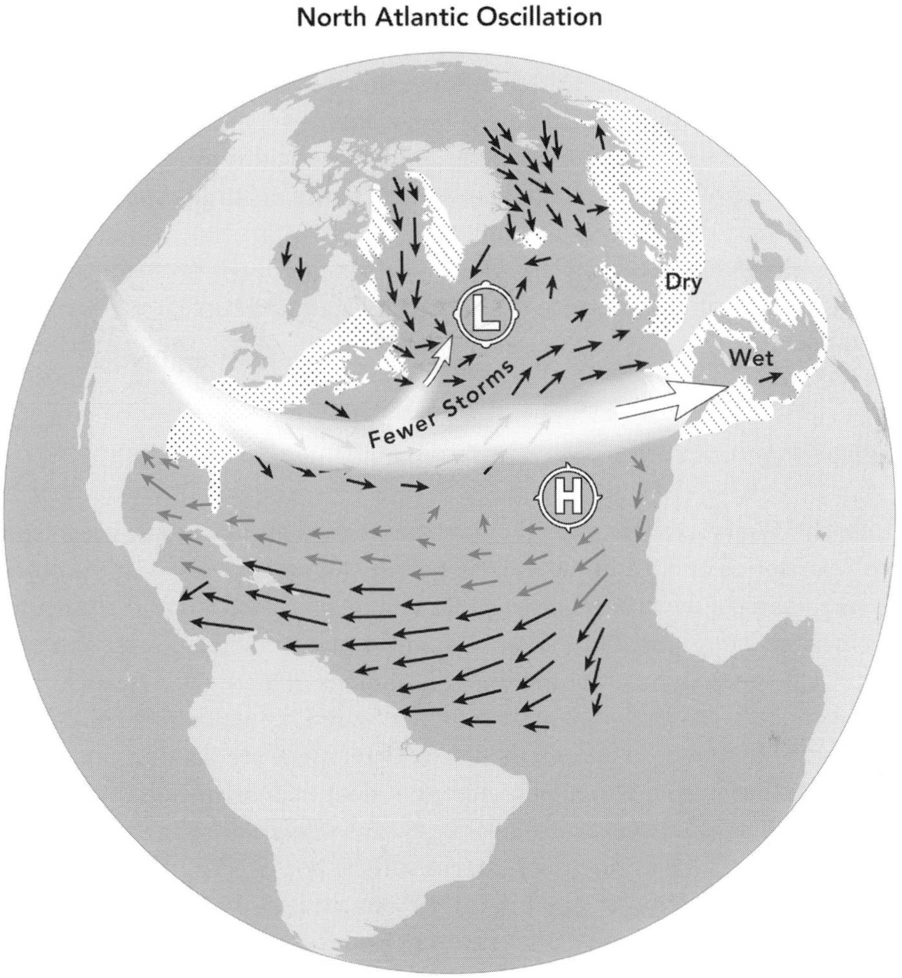

North Atlantic Oscillation

Figure 1.11b Negative phase of the North Atlantic Oscillation. Weaker high pressure over the Azores and a weaker low over Iceland. Storm tracks shift farther south, leaving northern Europe colder and drier and the Mediterranean and North Africa much wetter. (Martin Visbeck, IFM-GEOMAR, Kiel.)

the event. However, if more-frequent and stronger winter storms than usual occur, the winter-long sea level will remain above average. The changes in sea level resulting from the reduced atmospheric pressure and stronger winds are reinforced by warmer than average sea surface temperatures, which cause the upper layers of the ocean water to expand and rise.[38]

On the opposite side of the Atlantic Ocean, annual sea level fluctuations in the southeastern United States appear to correlate with the winter air pressure difference between Bermuda and Iceland (a proxy for the NAO).[39] The cause of this relationship is still unresolved, but it may partly stem from the NAO-associated atmospheric changes that also affect the wind-driven surface ocean currents. During the positive phase of the NAO, the enhanced high pressure over the Azores sets up a strong anticyclonic flow over much of the mid-Atlantic and drives storm tracks farther northeast than usual. At the same time, the anticyclonic flow may also cause the Gulf Stream to shift slightly closer to shore off Cape Hatteras, bringing warmer than usual sea surface temperatures closer to the U.S. East Coast and therefore raising sea level there.

THE DYNAMIC OCEAN

This brief overview illustrates the dynamic nature of the world's oceans on daily, seasonal, and multi-annual timescales. Large changes, now continuously monitored by orbiting spacecraft as well as ocean-based instruments, occur over varying distances as well. Atmospheric circulation affects the large-scale ocean gyres that move huge volumes of water and transfer vast quantities of heat from equatorial to higher latitudes, moderating the Earth's climate. Differences in ocean salinity and temperature also set into motion a planet-wide conveyor-belt system of deeper density-driven currents. The fairly short-lived, small-scale variations in water elevation produced by ocean currents, waves, the tides, storm surges, and even quasi-cyclical climate anomalies lasting several months to seasons (e.g., ENSO, NAO) do not generally lead to permanent changes in the average ocean height. However, their effects when superimposed on mean global sea level, affect the coastal zone and its inhabitants, as will be shown in later chapters. To understand the processes responsible for the longer trends in rising or falling sea levels that are the main focus of this book we now turn to chapter 2.

2

The Causes and Detection of Sea Level Change

An eroded cliff—the Orangeburg scarp—traverses much of the southeastern U.S. Atlantic Coastal Plain from North Carolina into Florida. The escarpment, a relict of ancient barrier islands and other shoreline features, formed at a time of much higher sea level, during the balmy mid- Pliocene epoch 3.5–3.0 million years ago.[1] Today, the base of the Orangeburg scarp stands about 85 meters (280 feet) above sea level, near where it crosses the North Carolina–South Carolina state line. When corrected for long-term regional uplift, mid-Pliocene sea level stood approximately 35 meters (115 feet) higher.[2] Other estimates vary between 10 and 30 meters (33–98 feet) higher than present.[3]

Raised coral reefs also point to past changes in sea level. Reef-building corals are colonial invertebrate animals that build large, intricately branching or rounded structures made of calcium carbonate and inhabit relatively shallow warm ocean waters. They coexist in a symbiotic relationship with blue-green algae. Some species of coral, such as *Acropora palmata*, generally grow in quite shallow water, in the Caribbean Sea and the Pacific Ocean (fig. 2.1). Therefore it has been widely used to deduce former sea level positions. Fossil *Acropora* corals, 128,000–120,000 years old, now sit 2–4 meters (6.6–13.1 feet) above modern sea level along the geologically stable coast of Western Australia.[4] They had lived during the last interglacial, or warm

Figure 2.1 *Acropora palmata* coral from the Caribbean Sea. (NOAA).

period, before the vast ice sheets of the last ice age returned and covered much of northern Europe and Canada.

Much more recently, 21,000 years ago at the height of the last ice age, *Acropora palmata* grew in shallow lagoons off the island of Barbados, in the Caribbean. Today, these fossil corals that once thrived very close to sea level are now covered by water at least 120 meters (394 feet) deep.[5] The drowned corals testify to a time at the peak of the last ice age when average sea level stood at least 120 meters lower than today.

As we saw in chapter 1, ocean height varies from place to place and over the seasons, thanks to waves, tides, and the atmosphere. Yet the ocean level remains fairly constant when averaged over the entire surface, over several years. Nonetheless, the fossil evidence clearly reveals that the average sea level in the remote past has often differed dramatically from the present. The world's average sea level has climbed nearly 20 centimeters (~ 8 inches) above that of 1900.[6] Why did sea level oscillate some 120 meters between glacial and interglacial periods? Why was sea level 10–30 meters higher 3.3 million years ago and much higher yet during the Mesozoic—the Age of Dinosaurs—than it is at present? How can we tell that today's sea level is really changing? How is sea level measured? Indeed, what is even meant by "sea level" if the ocean is so variable and capricious? The answers to these questions form the subject of this chapter.

DEFINING MEAN SEA LEVEL

Worldwide mean sea level is the consequence of diverse geological and climatological phenomena that operate over a wide range of distances and timescales (table 2.1). While waves, tides, winds, and atmospheric circulation change the ocean surface height continually, of greater interest here are slower, more enduring transformations of the shoreline, coastal ecosystems, and human habitations.

Because ocean heights are so variable, significant long-term trends appear only after sea level data from around the world have been collected and averaged yearly for at least 50 years or more. A later section describes how sea level changes are actually measured. The U.S. National Ocean and Atmospheric Administration (NOAA) defines mean sea level as the arithmetic mean of hourly water elevations recorded by tide gauges over a particular 19-year period.[7] The use of an average evens out the daily highs and lows of the tides and removes the transient effects of waves. However, the NOAA definition applies only to the mean sea level position at a particular locality. Because the elevation of the land and seafloor also vary over time independently of changes in water level (but much more slowly), a comparison of mean sea level trends from different places requires an arbitrary but fixed reference level, or datum.

A standard datum has long been the reference ellipsoid, an idealized mathematical figure, best visualized as a sphere that is flattened at the poles and bulges at the equator (see chapter 1). It mimics the general shape of the Earth and approximates mean sea level. Orbiting satellites, such as the Global Positioning System (GPS) and TOPEX/Poseidon generally refer their altitude measurements to the reference ellipsoid. Inasmuch as the Earth's gravitational field shapes the ocean surface, a more meaningful reference surface is the **geoid**. On this theoretical surface, the combined effects of the gravitational attraction relative to the Earth's center of mass and the centripetal acceleration due to the Earth's rotation are equal everywhere (fig. 2.2). This surface would coincide with mean sea level in the absence of tides, ocean currents, and winds.[8] On land, it theoretically corresponds to the water level of a network of narrow interconnected canals that are at the same level as the sea. Because the geoid approximates mean sea level more closely than does the reference ellipsoid, oceanographers and geodesists find it more

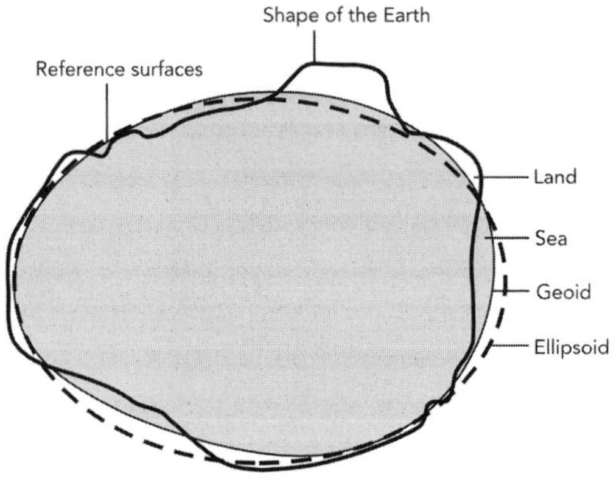

Figure 2.2 The geoid as an approximation of mean sea level. The shape of the Earth in the figure is highly distorted.

useful. Mean sea level then becomes the zero-level surface from which the elevations of mountains and other land features are ultimately measured.

In reality, the shape of the geoid is somewhat lumpy as compared to the idealized reference ellipsoid, because the mass within the Earth is unevenly distributed. Gravitational forces vary from place to place because of differences in the densities of rocks (denser rocks have more mass per unit of volume) and because of differences in the topography. Since a large mountain is more massive than a flat plain, the gravitational attraction is stronger over a mountain and the geoid is elevated there as compared to a valley. Submerged features on the ocean floor, such as seamounts, also influence the shape of the geoid. However, geoidal undulations do not always correspond to major topographic or bathymetric features, but rather stem from density differences deep inside the Earth.[9]

Although elevations obtained from satellite altimeters and GPS were originally measured relative to a reference ellipsoid, the newer devices now routinely output altitude data relative to the geoid, based on calculations from geophysical models and gravity data obtained from land surveys, aircraft, and space.

HOW SEA LEVEL CHANGES

In contrast to the transient and regional-scale processes that stir up the restless ocean, encountered in chapter 1, this chapter emphasizes more enduring ways of modifying the ocean's height. The term **eustasy** refers to a globally simultaneous change in sea level.[10] Mean global sea level can rise or fall in two fundamentally different ways: (1) by increasing or decreasing the volume or mass of ocean water and (2) by altering the volume of the ocean basins.[11] In the first case, the volume and mass of water in the ocean de-

Table 2.1 Summary of processes affecting sea level

Process	Timescale (yrs)	Magnitude (m)
Changes in volume of ocean water		
Growth or melting of ice sheets (glacial eustasy)	10,000–100,000	120–140
Growth or melting of mountain glaciers (glacial eustasy)	10–10,000	0.5
Changes in ocean temperature/salinity (steric changes)	10–10,000	10
Changes in ocean basin volume		
Plate tectonics and mountain uplift	10,000,000–100,000,000	100–500
Sedimentation (continental shelves)	10–100,000,000	100
Glacial isostasy	10,000–100,000	100–500
Hydro-isostasy	10,000–100,000	30–40
Local sea level changes		
Sediment loading and compaction (river deltas)	10–1000	<1–20
Localized faulting and tectonic deformation	1–1000	<1–3
Subsidence due to subsurface fluid extraction (gas, oil, water)	10–100	<1–2
Oceanographic and atmospheric variability	<1–10	<1–10

Sources: Lambeck (2009); Gornitz (2005); Rona (1995); NRC (1990).

creases (or increases) as glaciers advance or recede. The volume of ocean water also changes as its temperature and/or salinity varies, while mass remains constant (see chapter 1). In the second case, the shape of the ocean basins changes, depending on varying rates at which the Earth's tectonic plates separate or converge, on eruption of ocean lavas, or on accumulation of marine sediments. The shape of the ocean basins also adjusts to the changing masses of water that are transferred between ice sheets and the seas during cycles of glacial expansion or retreat. The first case may be thought of as varying the amount of the contained liquid (i.e., the volume or mass of ocean water), the second as modifying the shape and dimensions of the container (i.e., the ocean basins). Table 2.1 summarizes these processes.

Changing Ocean Water Volume

Glaciers and ice sheets affect sea level not only because their masses can distort the shape of the lithosphere, but more importantly because their growth or demise changes the volume of ocean water. Simply put, water is transferred from the ice sheets to the ocean and vice versa. If the ocean contains more water, then sea level must rise, all other factors being equal; if ice builds up, removing water from the ocean, sea level falls. The most significant changes in ocean water volume on timescales of 100,000 to tens of million years are produced by the growth and decay of glaciers and ice sheets, for example those that once covered major portions of Scandinavia and Canada. The resulting changes in sea level are known as glacial eustasy. Sea level has oscillated by approximately 120–140 meters (390–460 feet)[12] between the glacial (cold) and interglacial (warm) cycles of at least the past 800,000 years. There is still enough water locked up as ice in Greenland, Antarctica, and mountain glaciers to raise sea level by nearly 70 meters (230 feet) if all of the ice were to melt and spread out uniformly over the entire ocean.[13] Greenland holds the equivalent of 5–7 meters of sea level rise; the West Antarctic Ice Sheet contains a somewhat lesser amount. Thus large water transfers between ocean and land (as ice) can lead to an enormous difference in sea level. Such changes have actually occurred numerous times in the geological past.

Why the dramatic oscillations in sea level of the ice ages? Although a more comprehensive discussion awaits chapter 4, suffice it to say here that the worldwide swings in sea level are closely associated with climate change, and that the climate shifts of the last million years have been linked to minor changes in the Earth's orbital characteristics. Furthermore, these orbital changes have been amplified by increases or decreases in greenhouse gases such as carbon dioxide and methane in ways that are not yet completely understood. Hence the great concern today over the rising amounts of human-generated greenhouse gases and their implications for future sea level rise. These issues will be addressed further in chapter 7.

Over much shorter timescales (decades to several millennia), changes in ocean water volume due to density variations caused by temperature and/or salinity differences (i.e., **steric changes**) become major contributors to sea level. Since 1963, around a quarter of the observed 20th-century sea level rise has derived from increased ocean warming, the rest mainly coming from more rapid melting of mountain glaciers and ice sheets.[14] This upward trend is closely associated with the 20th-century climatic warming, largely attributed to anthropogenic atmospheric greenhouse gas emissions (see chapter 6).[15] However, the role of thermal expansion is likely to diminish if ice sheets melt more in coming centuries. The maximum potential sea level rise due to thermal expansion in the deep ocean is an estimated 10 meters over a 10,000-year time span.[16]

Human interventions in land water storage could also affect sea level by apportioning the volume of water retained on land versus in the ocean. For example, dams built within the last 50 years may hold back enough water to reduce sea level rise by 3 centimeters, equivalent to a sea level drop of 0.55 millimeters per year.[17] On the other hand, **groundwater mining,**[18] increased river runoff due to deforestation, wetland destruction, and urbanization (which increases the area of impermeable ground) would add water to the ocean, thus elevating sea level. Yet, the overall contributions to sea level from these diverse activities may be small or may largely counterbalance each other.[19]

Changing the Dimensions of the Ocean Basins

Rearranging the Tectonic Plates

Although others had previously noticed the curious jigsaw-puzzle fit between the coastlines of South America and Africa, German meteorologist and geophysicist Alfred Wegener (1880–1930) took a bold step further in 1912 to propose that the continents of North America, South America, Africa, Australia, Antarctica, and Eurasia were once joined together in a giant super-continent called Pangaea, which later drifted apart to form the Atlantic Ocean. He pointed out the close similarities in rock formations on both sides of the Atlantic in South America and Africa, and fossils of very similar plants and reptiles that could not possibly have floated or swum across the vast stretches of open water now separating these continents. One famous example is *Mesosaurus*, an aquatic lizard-like reptile that lived around 300 million years ago.[20] Another is the now-extinct fern *Glossopteris* of similar age, which was once widespread across the Southern Hemisphere continents. In addition to the fossils, rocks with striations or gouges produced by moving glaciers were found in many locations, demonstrating that a major ice sheet had covered large parts of the southernmost Southern Hemisphere toward the end of the Paleozoic era, 300–250 million years ago. Needless to say, contemporary geophysicists greeted Wegener's idea with great skepticism, because they could not envision by what physical mechanism this drifting could have been achieved. But little by little, more evidence accumulated from various sources.

After World War II, extensive mapping of the seafloor revealed the world-encircling pattern of mid-ocean ridges and deep trenches ringing the Pacific Ocean and elsewhere (see fig. 1.2). Unexplained at the time was the

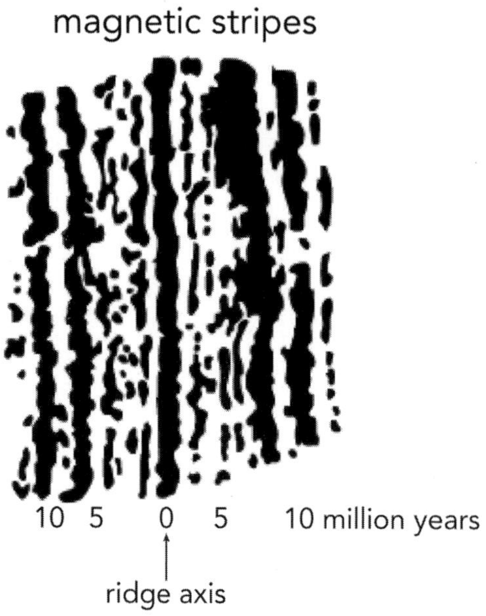

Figure 2.3 Pattern of magnetic stripes on the ocean floor along the mid-ocean ridge south of Iceland. (USGS).

high level of earthquake activity and active volcanoes associated with these physiographic landforms. Even stranger was the mirror-image "zebra-stripe" pattern of magnetized rocks on each side of the mid-ocean ridges that was discovered by magnetometers towed by research ships (fig. 2.3).

When lava congeals at the Earth's surface, grains of the mineral magnetite (an oxide of iron) line up with Earth's magnetic field, thereby acting like tiny bar magnets. Similarly, magnetite grains in sedimentary rocks lock into the orientation of the prevailing magnetic field. However, the rock record shows that the orientation of the Earth's magnetic poles has flip-flopped many times in the geologic past. That is to say, the north magnetic pole became the south pole, and vice versa (i.e., magnets that now point north would have pointed south and vice versa). The last time this happened was 780,000 years ago—the Matuyama-Brunhes geomagnetic reversal. The sequence of flip-flops, or magnetic reversals, has been well dated from samples of terrestrial rocks. Like a giant tape recorder, the successive strips of magnetized ocean rocks register the alternating changes in the orientation of the Earth's magnetic field over time.[21] It turns out that the youngest rocks occur closest to the mid-ocean ridges, while the oldest are farthest away.[22] Not only can the history of the ocean floor be deduced from this record, but so can subtle changes in the rates (and direction) of formation of the seafloor. The symmetrical magnetic stripe pattern, together with dating of progressively older rocks at increasing distance from the mid-ocean ridges, offered the strongest evidence that the continents, once conjoined, had moved apart. But how can solid rock just "drift" around?

Around 1930, the British geophysicist Arthur Holmes (1890–1965) had proposed convection as an explanation of continental drift by means of ascending columns of hot magma from the mantle that caused rock masses to separate. Convection occurs when fluids are set into motion by a sharp temperature contrast between the top and the bottom of the material. An everyday example is heating a pot of water on top of a stove. As the water is heated, the water at the bottom of the pot rises toward the surface and the cooler, surface water sinks. (As the heating continues, the water eventually begins to boil, but that is another story.)

Convection is an important process in both atmosphere and ocean. The Walker cell, described in chapter 1, is an example of atmospheric convection, in which hot, moist air rises over the "warm pool" near Indonesia and cools at high elevations. The cooler, drier upper air mass then flows toward the eastern Pacific and sinks off the coasts of Ecuador and Peru, joining the trade winds back toward the western Pacific. El Niño events disrupt this usual circulation pattern. Convection also contributes, in part, to the great ocean conveyor circulation.

Convection also occurs deep within the Earth. In the upper mantle, tongues of magma rise from a layer of partially melted, deformable rock called the **asthenosphere** (see fig. 1.2). Fresh basaltic lavas[23] erupt repeatedly along the central rift valleys at the crest of the mid-ocean ridges, slowly expanding the ocean floor. The high degree of volcanic and seismic activity along the ridge crests marks the ascending branch of mantle convection.

Building upon the concepts of convection and continental drift, other geophysicists—notably Harry H. Hess (1906–1969) and Robert S. Dietz (1914–1995), developed the more comprehensive idea of seafloor spreading in the early 1960s. As the magma from the upper mantle erupts and cools along mid-ocean ridges, magnetite crystals align themselves to the current north magnetic pole. As time passes, alternations in magnetic polarity are permanently embedded into freshly erupted lava at the ridges and older rocks gradually move apart, accounting for the "zebra-stripe" pattern of magnetization.

However, the widening ocean floor represents only half of the story. At the deep-sea trenches, cold, dense oceanic crust sinks deep into the mantle beneath continental crust or volcanic island arcs. These subduction zones mark the descending branches of mantle convection and surround the margins of the Pacific Ocean—often referred to as the "Pacific Ring of Fire" because of the numerous earthquakes and volcanic eruptions. For example, the most powerful earthquake ever recorded (magnitude 9.5 on the moment magnitude scale) struck Chile on May 22, 1960. The quake killed thousands of people and triggered a **tsunami** that crossed the Pacific, ravaging Hilo, Hawaii. More recently, a magnitude 9.0 earthquake, followed shortly thereafter by a disastrous tsunami, devastated the Sendai region on the Pacific coast of Honshu, 373 kilometers (81 miles) from Tokyo.[24] Most of the 15,000 people who died perished by drowning.

The tectonic plates represent large slabs of lithosphere—crustal and uppermost mantle rocks—that separate at the mid-ocean ridges, leisurely gliding over the asthenosphere at a snail's pace, several centimeters per year, and converge at subduction zones (see fig. 1.2). They also slide past each other along lengthy, horizontally moving faults, such as the notorious San Andreas Fault in California. These features delineate plate boundaries and are especially geologically active.

Plate tectonics slowly reconfigures the shape and depth of ocean basins over tens to hundreds of millions of years (fig. 2.4). Upwelling columns of magma erupt at the mid-ocean ridges, and as the ocean widens, the plates drift apart. Conversely, the ocean shrinks at deep-sea trenches, where oceanic lithosphere descends into the mantle along subduction zones. Marine sediments, caught in a vise between the descending plate and the more buoyant continental crust, are crumpled and crunched, ultimately rising up as mighty mountain belts such as the Andes. The collision between two continental plates creates mountain chains like the Himalayas and the Alps.

How does plate tectonics affect sea level? The remarkable continuity of magnetic striping over much of the ocean floor demonstrates that seafloor spreading has remained fairly steady over long periods of time. Yet spreading rates have varied, causing sea level to rise or fall. During periods of rapid seafloor spreading, larger volumes of younger, warmer, more buoyant basaltic crust are generated at the mid-oceanic ridges. The buoyancy of the rising magma forces the ridges to swell and rise, displacing large volumes of water that flow toward the continental margins and flood the continents. One such period was the Cretaceous (145.5–65.5 million years ago) at the culmination of the Mesozoic—the Age of Dinosaurs. The Cretaceous was mostly warm and ice-free, characterized by comparatively rapid seafloor spreading rates that inundated the continental shelves. Sea level 80 million years ago stood ~100–250 meters (328–820 feet) higher than today.[25]

Since then, as spreading rates slowed, the volume of the ridge crest decreased, and sea level slowly fell. Older ocean crust, moving away from the ridge axes, slowly cooled, became denser, and subsided. The oldest portions

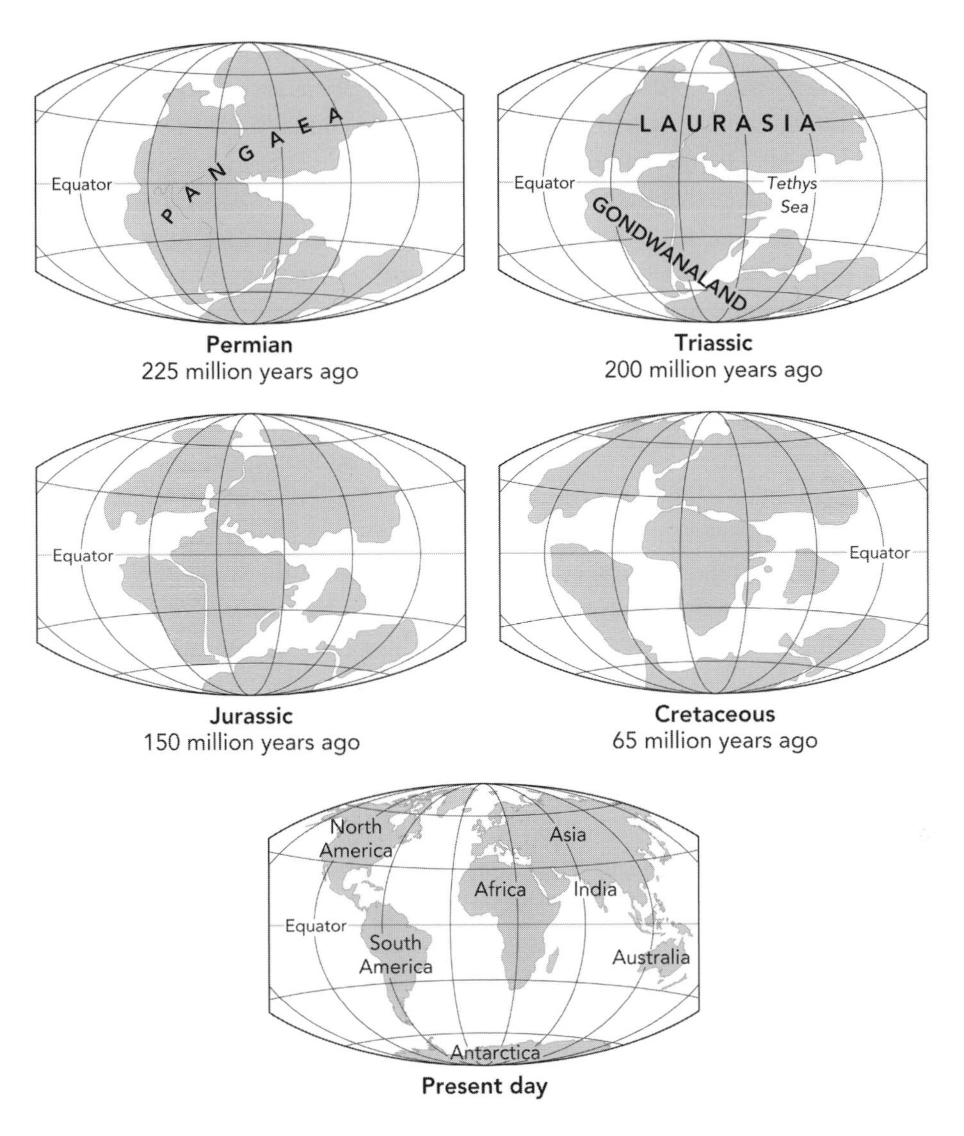

Figure 2.4 The changing positions of the tectonic plates since the breakup of Pangaea, 200 million years ago. (USGS.)

of the ocean near the continental margins are still slowly sinking. The average age of the ocean crust has increased over the last 100 million years and sea level has gradually dropped (see chapter 3).[26]

Plate tectonics, mountain building, and sea level change are interrelated in other ways as well. As two plates collide or one slides beneath the other, rocks caught in the squeeze are highly compressed and fractured and eventually thrust up into lofty mountain ranges. Shaken loose by earthquakes, rocky debris cascades down the steep slopes in rockfalls and avalanches and is washed into streams and rivers. Rivers, in turn, carve out steep valleys and carry heavy loads of suspended sediment that ultimately deposit in deltas and on continental margins. During periods of heavy offshore sediment accumulation, the extra weight of the thick sediment piles causes the continental margin to subside. As water is displaced farther inland, sea level appears to rise. But as uplift rates wane and erosion levels the relief, rates of sedimentation diminish and sea level lowers instead.

Continental collision reduces land area, since continental crust is compressed and thickened; ocean area is enlarged. If the total volume of water in the ocean stays fixed, a larger expanse of ocean would cause sea level to drop. Such a scenario likely occurred in the late Paleozoic era, when all of the continents were assembled into the super-continent, Pangaea. In addition, large portions of Pangaea were covered by a massive ice sheet at that time, which further lowered sea level. Large lava outpourings from volcanic activity at locations other than at mid-oceanic ridges also can change the volume of ocean basins.

Plate tectonics may also influence sea level indirectly through its effects on climate.[27] An example is the relationships among changes in rainfall, temperature, and erosion rates. In general, warm and wet climates accelerate the breakdown of rocks by promoting chemical reactions among water, carbon dioxide in solution (which forms a weak acid—carbonic acid), and rock minerals. These reactions not only produce new minerals (e.g., clay minerals) and dissolve others (e.g., carbonates), but also physically weaken the rock, leading to unconsolidated sediments and soils. The more loose sediment that washes into rivers, the more river-borne material is deposited offshore, making continental margins subside. Furthermore, high rainfall rates swell river currents that can then carry heavier loads. Thus, rainfall and temperature (which depend in part on elevation and continental geography resulting from plate tectonics) can change ocean volume via sediment accumulation at continental margins.

Loading and Unloading the Lithosphere

The Himalayas are the world's loftiest mountain range. Mount Everest holds the record at 29,000 feet (8,850 meters), K2 comes in second at 28,250 feet (8,611 meters), and Kanchenjunga stands at 28,170 feet (8,590 meters). At least a dozen more peaks tower above 26,000 feet (7,900 meters). These mountains soar not so much because they have been "pushed up" by colliding plates as because the collision has thickened the crust beneath them, and being less dense than the underlying mantle rock, it is therefore more buoyant. A simple analogy of the buoyancy effect is illustrated by an iceberg floating on water (fig. 2.5). The proverbial tip of the iceberg extends only around one-tenth above the water; the remaining nine-tenths remain hidden underwater. Based on Archimedes' principle, a body immersed in water sinks until it displaces its mass. Ice floats because it is less dense than water (i.e., density of ice is 0.92 gram per cubic centimeter; that of water is 1.00 gram per cubic centimeter).

Continental crust, with an average density of approximately 2.7 grams per cubic centimeter, is lighter than typical upper mantle rock, with a density of 3.3 grams per cubic centimeter. Therefore, around 20 percent of the mountain's mass projects above the surrounding plains while the remainder forms its subterranean "roots." The more massive the mountain, the higher it projects aboveground, but also the deeper its roots extend into the mantle. Although erosion of material lowers the mountain's height, it rises to compensate for the lost mass, until a new balance is established. Similarly, the deposition of sediment offshore adds an extra load to the continental shelf and slope, causing subsidence and encroachment of sea onto land.

Thanks to the softer, weaker underlying asthenosphere, the lithosphere is not perfectly rigid; it flexes under the addition or removal of loads. This process is called **isostasy**. The loads can be large piles of lava flows, massive volcanic edifices, mountain massifs, thick layers of offshore sediment, or glaciers. Thick ocean sediment deposits or massive submarine lava flows modify the shape of the ocean basin not only because they alter the underwater relief, but also because of subsidence due to the additional load. This affects sea level on a regional to global scale.

More important here, however, are those changes in surface loads produced by the waxing and waning of glaciers and ice sheets. The masses of ice that accumulate during an ice age are large enough to deform the Earth and produce substantial changes in sea level around the world.

The polar regions have witnessed the expansion and shrinkage of large ice sheets at various periods of Earth history, most recently within the last 30,000 to 20,000 years. Changes in weight from the addition or removal of large ice masses distort the shape of the lithosphere. This affects the elevation of adjacent shorelines. These changes are referred to as **glacial isostasy** (fig. 2.6). Furthermore, as the ice sheets expand or contract, glacial meltwater is either removed from (or added to) the ocean basins. This causes the total mass of ocean water to either decrease or increase. The resulting changes in water loading (i.e., **hydro-isostasy**) either elevate or depress the ocean floor. Thus, isostatic processes introduce sea level variations related to the distribution of the ice sheets.

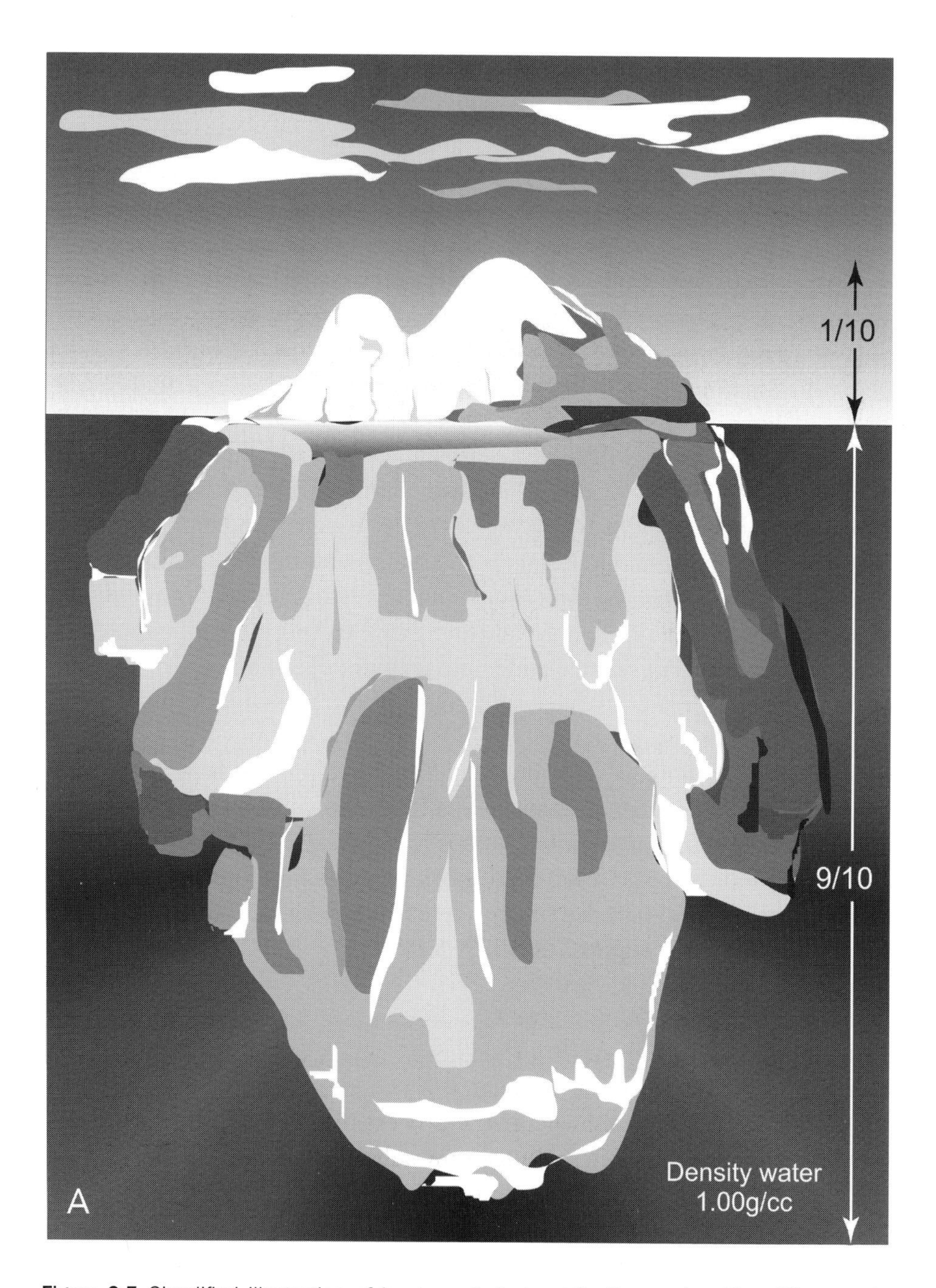

Figure 2.5 Simplified illustration of isostasy. A. Iceberg floating on ice. The difference in densities allows one-tenth of the iceberg to rise above water; the remaining nine-tenths stay submerged. B. Less-dense continents "float" on denser mantle. Continental "roots" reach depths of 40–70 kilometers (25–43 miles).

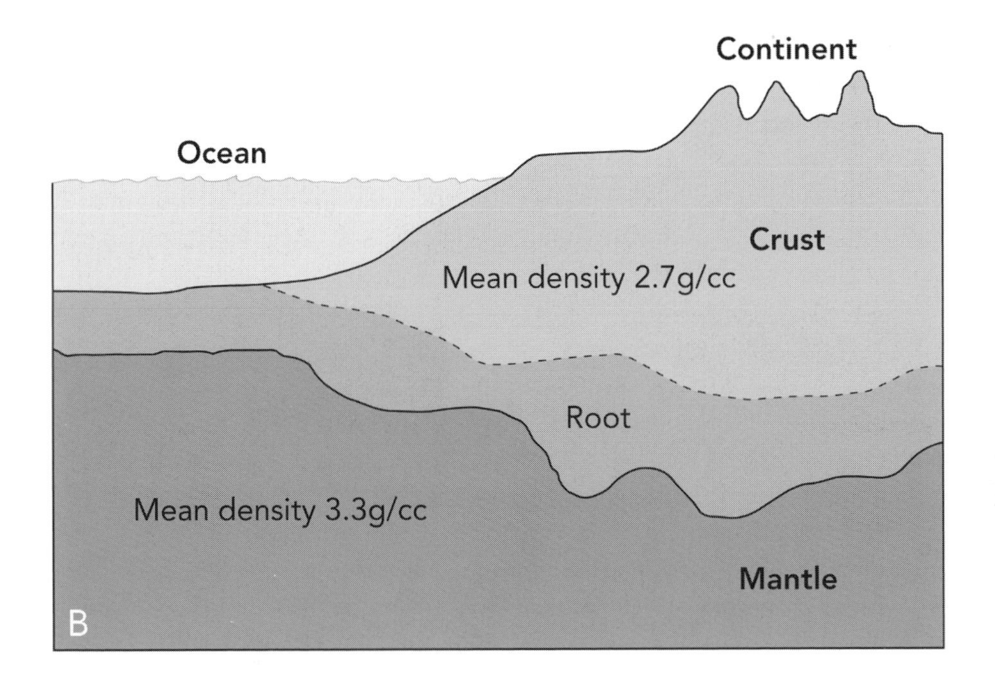

Areas once beneath ice sheets uplifted isostatically as the ice melted (fig. 2.6). Glacial rebound in Canada and Scandinavia was most rapid toward the end of the last ice age, starting around 20,000 years ago, but still continues at an ever-diminishing pace. Around the northern Baltic Sea and Hudson Bay, for example, seashells and driftwood now perched nearly 300 meters (1,000 feet) above sea level strikingly demonstrate the extent of glacial rebound. In these places, sea level still falls at rates of ~10–12 millimeters per year as the land continues to rebound. However, areas peripheral to the former ice sheets, like Atlantic City or London, warped upward as land farther north sagged under the ice load (fig. 2.6). Imagine squeezing a closed tube of toothpaste—the tube constricts where you press with your fingers, but the toothpaste thickens elsewhere. The process reversed once the ice sheets melted. The former peripheral bulge now subsides and local sea level rises! These glacial isostatic responses grow more muted away from former ice sheets, but the ocean continues to adjust to the weight of meltwater. Thus, on a number of "far field" tropical islands and continental margins, sea level appears to have dropped over the last 6,000 years, long after the ice sheets melted.

Glacial isostatic adjustment (GIA) models mathematically calculate gravitational interactions among ice sheets, land, and ocean over time and separate effects of glacial loading and unloading on sea level from the climate

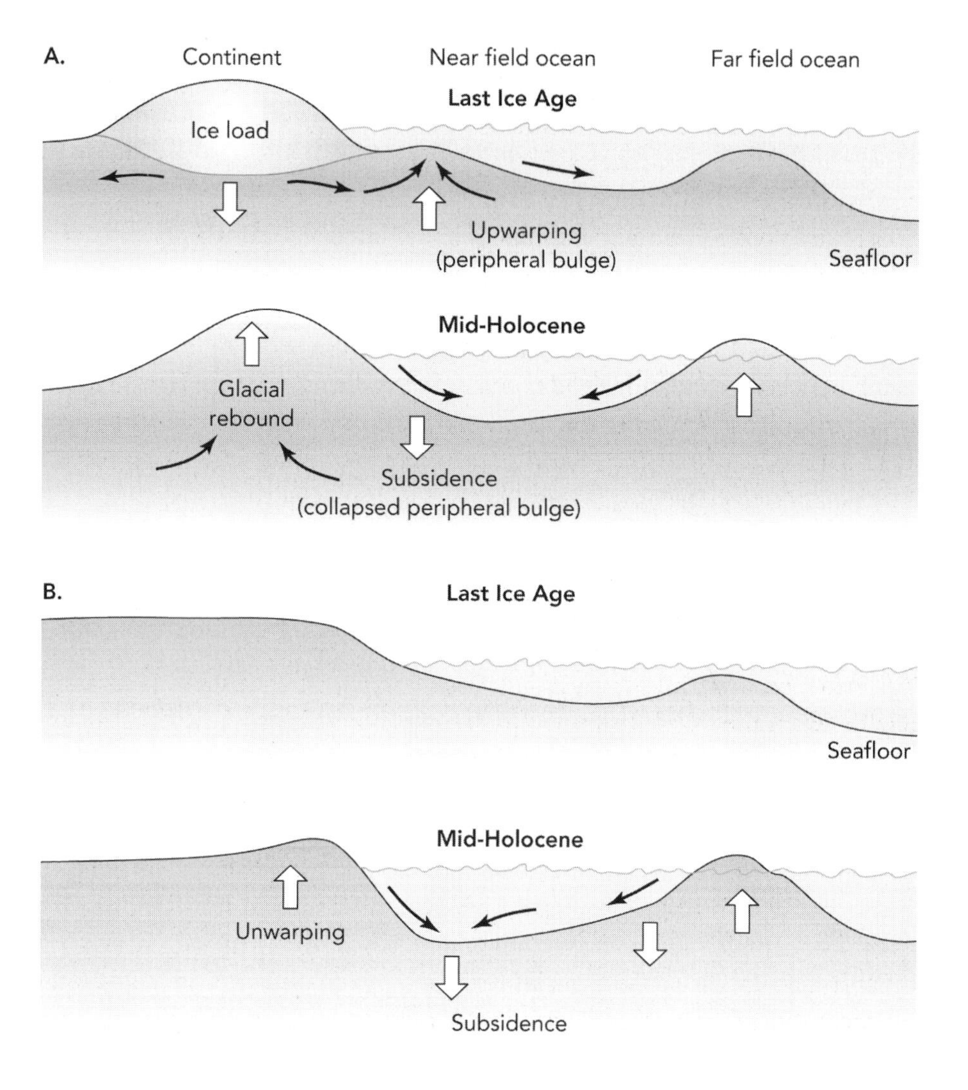

Figure 2.6 Glacial and hydro-isostatic effects. A. Glacial loading and uplifted peripheral bulge during an ice age, followed by glacial rebound and bulge collapse after the ice melts. Distant, far field islands emerge as water is "siphoned" off toward the zone of bulge collapse. B. Far from the former ice sheets, the added water loading (hydro-isostasy) on the continental shelf upwarps the continental shoreline and islands.

change signal.[28] These geophysical models compute the lithospheric reaction to changing ice and water loads, starting with the area and thickness of the ice sheet at the end of the last ice age. They also find the changes in ocean volume as glacial meltwater is added. Finally, they determine observed sea level changes at any given location over time. Input data include the maximum ice sheet thickness and area at the peak of the last ice age, the history of ice sheet retreat, and depth profiles of mantle viscosity. Mantle viscosity measures the degree of stiffness or resistance to flow within the Earth. For example, hotter, deeper rocks are softer and more yielding, and can deform more readily under a heavy load (such as a thick ice sheet). Conversely, colder rocks are more rigid and less likely to buckle under the load. Studies of earthquakes and laboratory measurements on the physical properties of rocks furnish information on the Earth's interior and its viscosity profile at depth.[29] The GIA models are tested and calibrated in order to find the closest match between calculations and independently derived past sea level records using radiocarbon-dated shells, peat, drowned tree stumps, and other near-shore fossils. The mantle viscosity parameters are also fine-tuned to provide the best fit to the observed record.

Local (Relative) Sea Level Change

The mean ocean level at any point along the coast stems from the interplay between the ocean basin shape, the volume of water it holds, localized land motions, and short-lived atmospheric or oceanographic effects (e.g., table 2.1). Their geographical and temporal variability makes it difficult to define a globally uniform sea level curve. The effects of coastal land motions must be sorted out from effects of climate change, such as changes in continental ice area or ocean water volume. Local crustal subsidence would magnify any global sea level rise, whereas uplift would offset it.

Land motions that affect sea level range from plate tectonics and coastal sedimentation to more localized fault displacements or extraction of subsurface fluids (e.g., hydrocarbons or groundwater) from porous rocks. Compaction and loading by thick sediment deposits at major river deltas cause considerable local to regional subsidence, as at the mouth of the Mississippi River, which is sinking at roughly five times the 20th-century global rate of sea level rise.[30] Over-pumping of groundwater has induced subsidence of many coastal cities, such as Houston (Texas), Bangkok (Thailand), and Venice (Italy), increasing their vulnerability to storm surges. (Recognizing this problem, Venice has halted groundwater removal since the 1970s.)

Current rates of subsidence due to aging, cooling ocean crust are essentially insignificant. Where plates converge, however, the crust can uplift several millimeters per year. For example, land is uplifting at a rate of 3–4 millimeters per year at the triple junction of the North American, Pacific, and tiny Juan de Fuca plates near Mendocino, California.

Ongoing glacial isostatic adjustments depend on location with respect to the former ice sheets. They introduce considerable modern local sea level variations. Examples include the glacial rebound of Scandinavia and Canada, or the peripheral bulge collapse of the mid-Atlantic U.S. coast, southern England, and the Netherlands. While GIA models account for these effects, they do not include consequences of tectonic activity. Therefore, most sea level studies avoid data from regions of known recent tectonic activity (as along the Pacific "Ring of Fire") and focus on data from geologically quiescent shorelines. They also do not account for subsidence due to sedimentation or subsurface fluid extraction. These must be determined independently by other means.

In summary, global sea level change is governed by: (a) changes in volume of ocean water due to glacial eustasy and steric changes, both of which are directly linked to climate, and (b) changes in volume of the ocean basins resulting from plate tectonics (table 2.1). Locally, sea level varies because of differences in sediment compaction and loading near deltas, active faulting, and changing land water storage through dam building, deforestation, or extraction of subsurface fluids. Finally, ephemeral sea level variability arises from fluctuations in ocean currents, tides, winds, and atmospheric pressure, as described in chapter 1.

MEASURING SEA LEVEL

Tide Gauges

Ever since seafarers began setting sail, the daily comings and goings of the tide have been important for navigation and shipping. The oldest system of tracking the tides was simply to mark some lines on the side of a stone pier at the harbor entrance or on a pole attached to the side of a pier and record the height of the water periodically. The oldest sea level record comes from Amsterdam, starting in 1700 and ending in 1925. The Stockholm record began in 1774; that for Brest, France, began in 1807; both continue to the present.[31] However, most sea level records start in the 20th century; a much smaller number, mainly from western Europe, date back to the mid- and late 19th century.

The traditional method for measuring sea level employs a float attached to a system of pulleys, weights, and gears that moves a pen over a paper chart.[32] As the tides push the float up and down, the strip chart records a continuous graph of changing water levels. Where the float-stilling well gauge is still in operation, it is now generally automated.

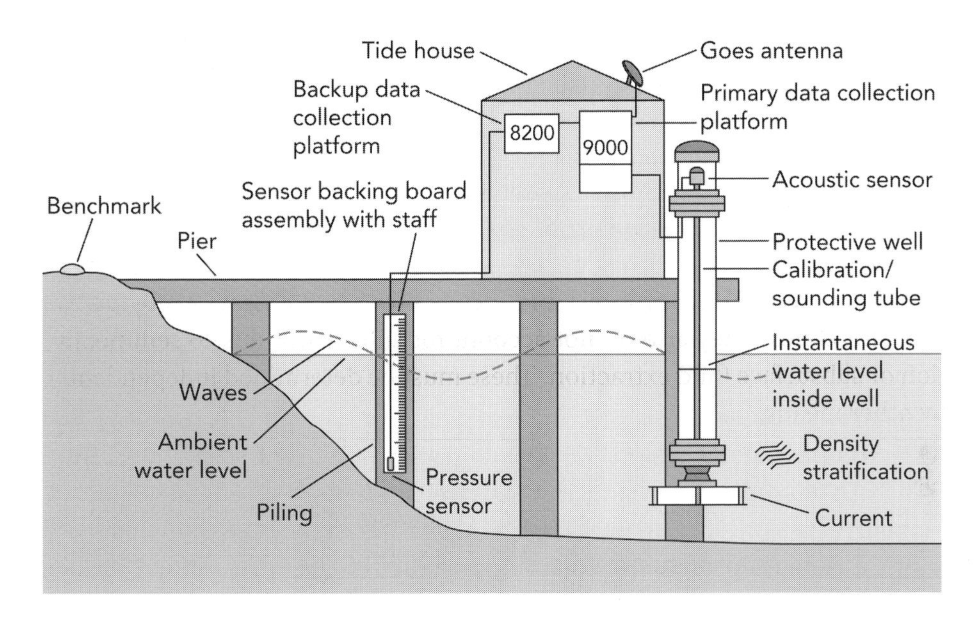

Figure 2.7 Schematic diagram of the Next Generation Water Level Measurement System (NOAA), the tide gauge currently used in the USA. (NOAA.)

By the 1990s, float gauges were gradually replaced with acoustic tide gauges, such as the Next Generation Water Level Measurement System used by NOAA (fig. 2.7). These devices measure the travel time of an emitted sound wave that is reflected off the water surface and returned to the sensor. The time elapsed changes as the water level responds to the tides. The travel time of the sound pulse is corrected for variations in velocity due to changes in temperature, humidity, and air pressure. The acoustic sensor is housed within a protective well on the side of the pier. Multiple readings are averaged frequently (e.g., at 6-minute intervals) in order to compensate for wave effects. Ancillary meteorological and oceanographic data are also collected, including air and water temperature, barometric pressure, wind speed and direction, water current speed and direction, and rainfall. Collected data are transmitted every three hours via satellite link to a central data repository. The Permanent Service for Mean Sea Level (PSMSL), located in Liverpool, UK, provides global sea level data.[33]

Tide gauges are connected by geodetic surveys to a local vertical datum and a small set of benchmarks, or fixed points, at nearby stable land sites. These in turn are referenced to a national land survey datum—a surface that parallels the geoid. For example, the current national datum in the United States is the North American Vertical Datum of 1988 (NAVD 88). Tide gauges register local (or *relative*) as well as global sea level change. Over several centuries, changes in sea level due to plate tectonics (away from plate margins) are negligible. Therefore, the global component (i.e., the eustatic or *absolute* sea level) is essentially caused by recent climate change. Local tidal and atmospheric fluctuations can be smoothed by taking monthly or annual averages or by other means. However, the effects of vertical land motions still remain. GIA models can eliminate those related to glacial isostatic adjustments, as described above, but localized tectonic displacements or land subsidence could still be present.

A Global Positioning System (GPS) station situated at or in very close proximity to the tide gauge is one means of overcoming this problem. The GPS receives signals from a worldwide network of orbiting satellites, from which the exact position and elevation relative to the geoid can be determined. A GPS unit placed near the tide gauge and its benchmarks can record vertical changes over time. Subtraction of the land-based elevation change from the mean change in sea level over a given time interval yields the absolute sea level change. Ideally, an accuracy of at least 1 millimeter in 10 years is needed in order to detect local vertical crustal movements, which are often comparable in magnitude to the rise in sea level. As of 2007, 146 GPS stations lie within 1 kilometer of a tide gauge. Because most instruments were installed within the past decade, "absolute" sea level corrections can be made at only a handful of locations at present. The GPS-tide gauge system is also used for calibrating and/or validating satellite altimeter measurements (e.g., TOPEX). For this purpose, an accuracy of ~2–3 centimeters is needed.[34]

Satellite Altimeters and Sea Level

While tide gauge measurements yield useful information about recent sea level trends, they have a number of significant limitations. For one thing, tide gauges with sufficiently long records are heavily skewed toward the Northern Hemisphere, particularly western Europe, North America, and Japan. For another, the records often contain lengthy data gaps. (At least 50 years or more of data should be collected at each station to obtain a reliable trend, because of the high degree of variability in sea surface heights due to oceanographic and atmospheric processes.) Vertical land movements other than those of glacial isostatic origin may still be present and not fully eliminated by existing geological or geophysical models. Furthermore, the tide gauges sample only sea level change at the coasts and tell us nothing about the rest of the ocean.

Satellite altimeters now provide a powerful tool for sea level studies. TOPEX/Poseidon, a joint NASA (U.S.)–CNES (France) mission that operated between 1992 and 2005, was followed by the U.S.-French Jason-1, launched on December 7, 2001, and OSTM/Jason-2, launched on June 20, 2008 (see fig. 1.5).[35] All three satellites orbit Earth at an altitude of 1,336 kilometers (830 miles), inclined to the equator at an angle of 66°. They fly over the same spot of ocean every 10 days. The satellite's position is precisely tracked by GPS and other instruments. On-board radar altimeters calculate the exact distance between the spacecraft and the ocean top based on the travel time of a microwave pulse from the spacecraft to the sea surface and back. The satellite's elevation above the ocean is determined by knowing the time and the speed of light, and correcting for atmospheric and instrumental effects. The difference between the elevation and the distance to the reference ellipsoid gives the sea surface height or sea level to within 2.5–3.3 centimeters (around 1–1.3 inches), but greater accuracy can be obtained by averaging many measurements. A trend can be established by repeating observations of the same area over many satellite cycles and averaging the observed year-to-year variations in sea surface height over all the oceans. Between 1992 and 2007, these measurements reveal an upward sea level trend of 3.4 millimeters per year.[36]

These satellites give us our first truly ocean-wide views of sea surface height fluctuations. They are therefore immensely useful not only for monitoring global sea level trends, but also for ocean circulation, or phenomena such as ENSO (see also chapter 1). However, tide gauges remain an essential part of sea level studies. The longer records of tide gauges complement those obtained from satellite altimetry and provide a historical context for interpreting the much shorter altimeter trends. Because the altimeters "see" most of the ocean,[37] they can separate out large-scale yearly variations (e.g., ENSO) from longer-lasting trends associated with climate warming. On the other hand, since the global sea level signal may be either amplified or dampened by local conditions, the relative sea level data from tide gauges enable a more localized assessment of coastal hazards and appropriate management of coastal resources. The gauges are also needed to calibrate and "ground truth" the altimetry measurements (i.e., independently verify the accuracy of the data). The tide gauge data are site-specific and offer nearly instantaneous snapshots of the local sea surface, whereas the altimeters sample larger portions of the ocean at any given time but much less frequently (returning to the same place overhead only once every 10 days). Therefore, a more comprehensive picture of short-term ocean variability and longer-term trends can be obtained from a joint system of tide gauge network combined with satellite altimetry.

Fossil and Landform Indicators of Past Sea Levels

Satellite altimeters used in sea level studies go back no more than one and a half decades; tide gauges, not much beyond 130–140 years. To extend the record beyond the instrumental period, scientists turn to "proxies" or "indicators" that are indirect measures of past sea level preserved in natural archives, such as marine fossils, sediments, peats, and submerged trees. Paleosea-level indicators are fossils or other materials that once existed or landforms created at or very close to the former shoreline.[38] In certain cases, well-dated archaeological remains may also provide clues about former shoreline positions. To be useful, the relation of the past sea level indicator to mean sea level or to some other tidal level, such as mean high water, should be known for that particular locality. Some commonly used indicators include tidal peat, wood, corals (such as *Acropora*), mollusks, and landforms shaped by the sea such as wave-cut notches or raised marine terraces. Carbon-rich materials can be dated by the radiocarbon method, which is widely used in archaeology and geology for samples up to around 55,000 years old.[39] The elevation of the dated paleosea-level indicator with respect to present sea level is carefully measured and a sea level curve is constructed by plotting the age versus elevation (or depth) over time.

Landforms

An obvious sign that sea level has changed is the occurrence of beach sands and ridges above their present position. However, they could have been deposited by storm surges or tsunamis. On the other hand, if such shoreline features of similar elevation and age recur over lengthy coastal stretches in otherwise geologically inactive regions, they may indeed reflect a regional (or even global) sea level change.

Wave-cut notches also signal a former sea level. These notches are produced by the abrasion of sand and rocky debris, carried by waves that have cut into the base of cliffs (fig. 2.8). Since it takes considerable time for loose sand and pebbles to gouge out a groove in solid rock, the notch represents prolonged wave action at the same spot.

Raised marine terraces also indicate past sea level. Marine terraces are abandoned marine shorelines, shaped by waves, that have slowly been uplifted. Thus, they occur more commonly along the tectonically active Pacific coasts. Outstanding staircases of raised terraces occur on parts of San Clemente Island off California and in the Palos Verde Hills of Los Angeles. Another well-developed example is from the Huon Peninsula, Papua New Guinea (see fig. 4.3 and chapter 4).[40]

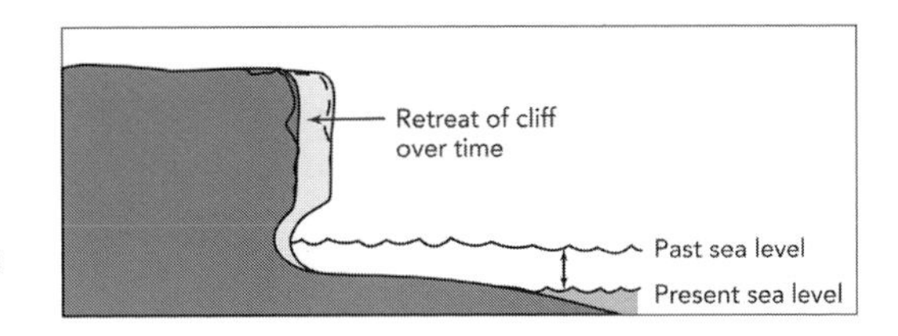

Figure 2.8 A wave-cut notch, showing location of a higher than present sea level.

The heights of past sea levels can be easily calculated. The rate of uplift (U) at a given location is determined by dividing the difference in height of a well-dated reference terrace (Href.) and sea level at that time (S) by its age (tref.), according to equation 1.

(1) $U = (Href. - S)/tref.$

Where tref. = 125,000 years, Href. = 5 ± 1 meters, S = past sea level, and U = uplift rate.

The reference terrace is generally taken as the 125,000-year-old terrace, formed during the last interglacial, when mean global sea level stood at least 4–6 meters higher than present. The terraces are usually dated using the uranium-thorium method of radioactive dating. Knowing the uplift rate, which is assumed to have remained constant on average, then the past sea level (S) is found by taking the difference between the height of any dated raised terrace (H) and the uplift rate multiplied by its age (Ut) (equation 2).

(2) $S = H - Ut$

Beachrock is another useful sea level marker. It consists of beach sediments that are primarily cemented by calcium carbonate. Beachrock forms by precipitation of calcium carbonate from evaporation of intermingling seawater and groundwater within beach deposits in the intertidal zone, in tropical and semitropical climates.

Fossils

The selection of a good fossil sea level indicator depends on knowing exactly where the organism lives relative to modern mean sea level and that its fossil counterparts have remained in situ where it once lived. While at first glance mollusks would appear to be an obvious choice, most species are problematic in that they can live over a fairly broad range of water depths, and furthermore, their shells are usually not found in place, having been moved around by the waves or surf. At best, the youngest specimen within a shelly layer in a beach deposit may denote the minimum age of the deposit, much like dating a shipwreck from the age of the youngest coin found in the treasure hoard.

Corals are widely used in paleosea-level studies. For example, *Acropora palmata* grows in very shallow water, usually within 5–7 meters (16.4–23 feet) of the surface in the Caribbean Sea and the Pacific Ocean (fig. 2.1). *Porites* forms micro-atolls atop the reef on the Great Barrier Reef of Australia and in the Indian Ocean. It lives within the intertidal zone. The range of *Montastrea* is less well constrained but can give a lower limit to past sea level for a particular location.

Submerged rooted tree stumps clearly reveal a rise in sea level. But their utility in quantifying past sea level history, as with other paleobiological materials, depends on the elevation range of the particular species. Preferable are those species that live in a relatively narrow zone near the shoreline. However, while finding in situ tree trunks buried beneath beach sand deposits clearly suggests relative sea level rise, not knowing their original distance from the former shoreline hinders a quantitative reconstruction of sea level change.

Peat deposits from salt marshes (and mangroves, in the tropics) are important indicators of past sea level rise. Salt marshes grow in the intertidal zone in sheltered lagoons behind coastal barrier islands, in major river deltas, and along the banks of river estuaries. Their upward growth generally keeps close pace with sea level rise (or fall), unless this rate is very high or the sediment supply is insufficient. Since peat is highly porous, it gradually compacts as water is squeezed out under the weight of overlying sediments. Therefore, researchers generally prefer the basal, or bottom-most, layer of peat in the sediment column, directly overlying inorganic marine sediments that predate the end of the last ice age. As sea level rose, the marsh gradually migrated inland and upward. Thus the history of sea level rise in an area can be deduced by dating a suite of basal peats extending inland from now-submerged offshore deposits to those underlying the present-day beach and the bottom of the back-barrier lagoonal salt marsh. Since errors might arise from correlating even nearby sediment cores, another approach is to use a single core that contains several different kinds of paleoenvironmental information, including past sea level indicators, such as foraminifera or diatoms.

Foraminifera (often referred to as forams) are tiny single-celled, chiefly marine, organisms that can be either free-floating or dwelling on the ocean floor or in brackish water (fig. 2.9). Most species have shells made of calcium carbonate, partitioned into chambers that are added during growth. They

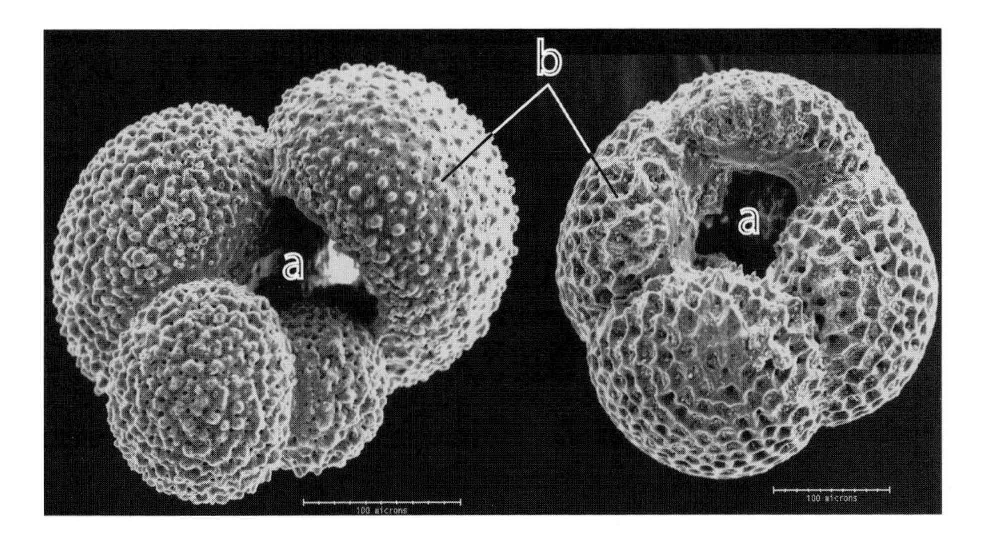

Figure 2.9 Planktonic (free-floating) foraminifera *Globigerina bulloides* (left) and *Dentoglobigerina altispira*. (H. J. Dowsett, 2009, Foraminifera, in V. Gornitz, ed., *Encyclopedia of Paleoclimatology and Ancient Environments*, fig. F2, p. 338, with kind permission from Springer Science + Business Media B.V.).

form a major component of chalk and of the limestone blocks used to build the Egyptian pyramids. Sea level studies focus on marine to brackish-water foraminifera species whose relation to the tidal cycle is known for a particular place. Brackish-water foraminifera are especially sensitive to minor changes in salinity. The intertidal zone of salt marsh environments displays a well-defined zonation of foraminifer assemblages. As the salinity increases with rising sea level, shifts in the foraminifer assemblages in the core reveal the time history of local sea level change. If similar shifts are widespread, the apparent sea level rise is likely to signal a regional or even global change.

Diatoms are minute single-celled marine or freshwater algae that can live free-floating or attached to a substrate (e.g., on rocks or sea ice). Their high sensitivity to environmental conditions, including temperature, salinity, and light levels, also makes them useful proxies in reconstructing past marsh sea levels.

The history of past sea level pieced together from most of these indicators generally extends back to the beginning of the post-glacial sea level rise. A few, like raised marine terraces, can denote sea level changes of previous glacial cycles. To peer even further beyond the veil of time, one needs to turn to different approaches. One such method employs **oxygen isotope** information preserved in the shells of tiny marine organisms.

Oxygen Isotopes

Most chemical elements have several isotopes, which differ only in atomic weight. These differences are usually expressed in terms of isotope ratios, or per mil (parts per thousand).[41] The ratio of the heavier oxygen isotope ^{18}O to the lighter, most common form, ^{16}O, is the most useful one for studying past sea levels. The ^{18}O isotope represents roughly 0.2 percent of the total oxygen in nature. A large number of geological and biological materials can be analyzed for their oxygen isotope composition, including the shells of foraminifera, corals, cave deposits such as stalagmites or stalactites, submerged tree wood, and ice, to name just a few.

A number of natural processes cause variation in the ratios of stable (i.e., non-radioactive) isotopes. Several of these are related to climate, in particular the temperature of the atmosphere and ocean, the volume of ice at the poles, and precipitation. In general, lighter isotopes of an element are more mobile and react faster chemically than heavier ones. The lighter isotopes of hydrogen and oxygen (^{16}O and ^{1}H) in water, for example, evaporate faster and therefore are preferentially incorporated into water vapor (in a process known as "isotopic fractionation"). As the vapor condenses into water in the cloud, the lighter isotopes remain in the cloud, whereas the heavier isotopes (^{18}O and ^{2}H) precipitate out as rainwater. Over successive cycles of evaporation and condensation, as air masses move from the ocean toward land and also toward the poles, rainwater becomes progressively enriched in the lighter isotopes. Thus snow falling at the poles, which eventually builds up into an ice sheet, is isotopically lighter. On the other hand, ocean water grows more depleted in the lighter O isotope as it is preferentially transferred to the ice caps.

The isotopic composition of a mineral, carbonate shell, or snow reflects that of the water from which it formed, which largely depends on temperature. The isotopic ratio of the shell of a living organism also depends on the particular species. In general, as temperature increases, the isotopic ratio becomes progressively lighter. Thus, for example, variations in the oxygen and hydrogen isotope ratios in polar ice[42] have been linked to the local atmospheric temperatures from which the snow precipitated.

Carbonate shells of bottom-dwelling or **benthic foraminifera** (which are less influenced by locally variable surface ocean temperatures than are free-floating or planktonic foraminifera) record deep-sea oxygen isotope variations that correspond to major global climate changes. But isotope ratios of benthic forams are affected by deep-ocean temperatures as well, which also vary, although less so than near the surface.

Oxygen isotope ratios oscillated significantly between glacial and interglacial periods (see chapter 4). During glacial periods when large volumes of water were locked up in polar ice sheets, oceans became relatively enriched in the heavier isotope, ^{18}O. Thus, $^{18}O/^{16}O$ ratios in forams grew heavier during glacial periods than during warmer interglacials, chiefly as a result of the greater ice sheet volume and lower sea level. For example, around 70 percent of the glacial-to-interglacial change in the oxygen isotope ratio of deep-sea foraminifera has been attributed to changes in ice volume (hence, sea level), while 30 percent results from changes in ocean bottom temperatures.[43] However, during past warm periods when polar ice sheets were small or even absent (such as the middle Cretaceous and a warm period between 40 and 50 million years ago), marine $\delta^{18}O$ values varied mainly due to water temperature changes. Other factors that determine marine oxygen isotope ratios include the particular foram species, its normal temperature range, and ocean salinity.

The Sedimentary Record

Much of the record of historical geology from Cambrian time onward has been that "the seas came in and the seas went out."

—P. B. KING[44]

The continual interaction of the ocean, lithosphere, and biosphere has changed the Earth's climate dramatically over the eons. To examine the deep past, scientists use markers of ancient climates or environments preserved in the geologic record. A number of those related to sea level have been described above. Particularly relevant to the study of ancient sea levels are sedimentary rocks that formed in coastal or near-shore settings. Most sedimentary basins, where the thickest piles of sediment accumulate, tend to occur in such environments. Therefore the stratigraphic succession of sediments offers useful insights into sea level history. The environments in which sediments deposited can be deduced from the mineral or rock composition, grain sizes, rock textures, and fossils that are present. Examples of such depositional environments include river deltas, estuaries or lagoons, beaches, and shallow, offshore deposits.

Sharp lithologic or compositional breaks in the stratigraphic column mark striking environmental, climatic, or long-term sea level shifts. This is beautifully demonstrated in the upper half of the Grand Canyon, Arizona, where the cliff-forming marine limestone strata of the Redwall Limestone

Figure 2.10 Grand Canyon. The prominent cliff near the top is the Coconino Sandstone, overlying the sloping, red Hermit Formation and Supai Group. The Redwall Limestone forms another sharp cliff halfway down the canyon. The bottom half of the photo shows the "Great Unconformity," separating pre-Cambrian from younger horizontal Paleozoic sediments. (Photo: R. Lazell.)

(~340 million years old) were overlain by intercalated reddish mudstones, sandstones, and calcareous sandstones of the Supai Group (315–285 million years old) that had deposited on a coastal plain, where the sea repeatedly advanced and retreated (fig. 2.10).[45] The Supai strata were succeeded by red beds of the ~280-million-year-old Hermit Formation—siltstones, mudstones, and fine-grained sandstones that were deposited by rivers. These rocks show cyclical alternations, presumably of climatic origin. Dried-out mudcracks and ripples at the top of the Hermit Shale hint at the onset of a much more arid climate, which culminated in the accumulation of vast sand dunes of the overlying Coconino Sandstone under full desert conditions, around 275 million years ago. The thick marine limestone sequence of the succeeding Toroweap Formation and Kaibab Limestone signals the return of the seas once more, ~273–270 million years ago.

The rock record can be used in several ways to reconstruct past sea levels. In one approach, changes in ocean basin volume over time are calculated, based on plate tectonic history, including major episodes of oceanic volcanism and deep-sea sedimentation.[46] Changes in area and depth of ocean floor are obtained by measuring the area of ocean at a given time, dating the ocean floor from its pattern of magnetic stripes, and the subsidence history of aging ocean crust. Partially subducted ocean plates can be reconstructed by assuming that the now-vanished portions were originally symmetrically distributed on both sides of the mid-ocean ridge.

Another frequently utilized (although somewhat controversial) method is that of **sequence stratigraphy,** originally developed using seismic reflection data. Seismic reflections have been widely employed in petroleum exploration since the 1960s to search for promising subsurface oil-bearing rock formations. In brief, seismic energy is induced by hammer and plate, weight drop, or explosive charge, and then the resulting seismic waves that are reflected back to the surface from rock layers of contrasting physical properties are measured. The velocity of seismic waves changes as they penetrate rock of differing densities. Major changes in seismic velocities of rock formations produce distinct reflection horizons. The seismic data are then assembled into a cross-sectional image of the buried rocks, which the trained eye views as an "X-ray" of the Earth's interior.

Exxon geologists Peter Vail, Robert Mitchum, and coworkers exploring for offshore petroleum deposits in the 1970s and early 1980s observed sharp seismic discontinuities in the rock strata, which they interpreted as **unconformities**, a term geologists use to indicate time gaps that interrupt stacks of rock layers. (A textbook example is the "Great Unconformity" of the eastern Grand Canyon, which separates tilted Precambrian sedimentary layers from horizontal Paleozoic rocks; fig. 2.10.) They called these prominent seismic reflectors **sequence boundaries**. The basic assumption is that major seismic boundaries represent stratigraphic unconformities. These delineate periods of erosion or non-deposition of sediments when the surface stood above sea level. A sequence, in this context, is defined as a set of successive strata bounded by unconformities. The **unconformity**, or **sequence boundary**, is formed during a period of falling sea level (or tectonic uplift), when the surface is exposed to erosion. Furthermore, the boundary represents a time-equivalent horizon. Periods of deposition, on the other hand, occur when either sea level is rising and/or the adjacent continental shelf is subsiding. Subsidence provides space for the sediments to accumulate in the basin and be preserved. So does a rise in sea level. Seeing similar sequences of seismic boundaries in many different localities, the Exxon geologists, subsequently joined by Bil Haq, who revised the original sea level curve, concluded that these boundaries represented periods of eustatic sea level fall and were global in extent.[47]

Concepts derived from seismic stratigraphy were subsequently applied to rocks exposed on land, using classical geological methods. A typical progression from land to the sea is illustrated by coarser sands or gravels grading into finer-grained silts, shallow-water muds and carbonates (limestones), and finally deep-water muds. The sedimentary strata may record progressive geographic shifts toward more land-like or marine environments over time, which suggest changes in relative sea level. These may arise as a result of (1) a change in global (i.e., eustatic) sea level, (2) regional changes such as subsidence of the marine basin (or uplift of the adjacent land) and/or fluctuations in the supply of sediments to the basin, or (3) a combination of these processes. Geologists interpret not only the kinds of rocks present but also how strata are truncated and overlie each other, in order to determine whether the sea was encroaching on the land or vice versa. If a similar pattern of rising and falling sea levels occurred more or less simultaneously in widely separated parts of the world, these changes are interpreted to be global, or eustatic, rather than due to local or regional factors.

Figure 2.11 illustrates these concepts schematically: Figure 2.11a (top) shows a simplified cross-section of the rock strata based on the seismic reflection data. Major sequence boundaries are delineated by thick black wavy lines, individual strata within a sequence by thinner lines. Coastal versus marine deposits are also differentiated. Figure 2.11b numbers the strata from figure 2.11a, in the three major sequences. The bottom diagram (fig. 2.11c) interprets the foregoing as changes in sea level over time. Units 1–3 in

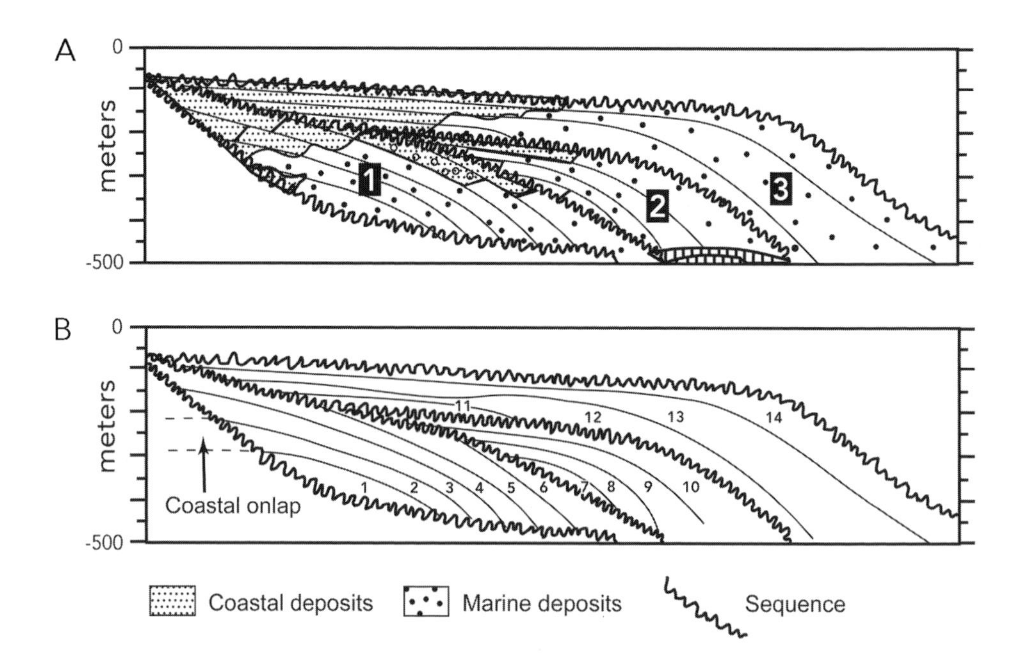

C

EUSTATIC SEA LEVEL CURVE

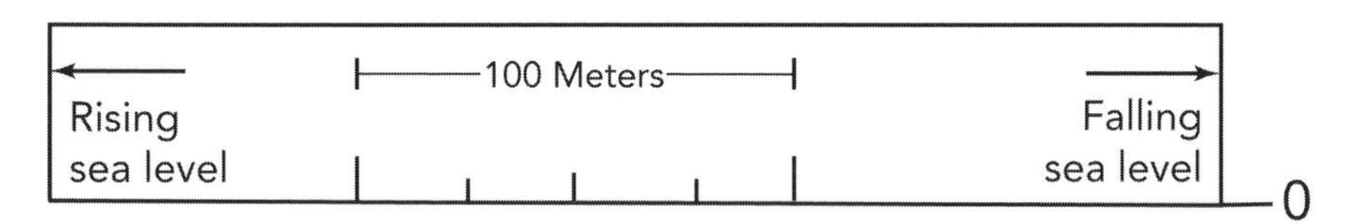

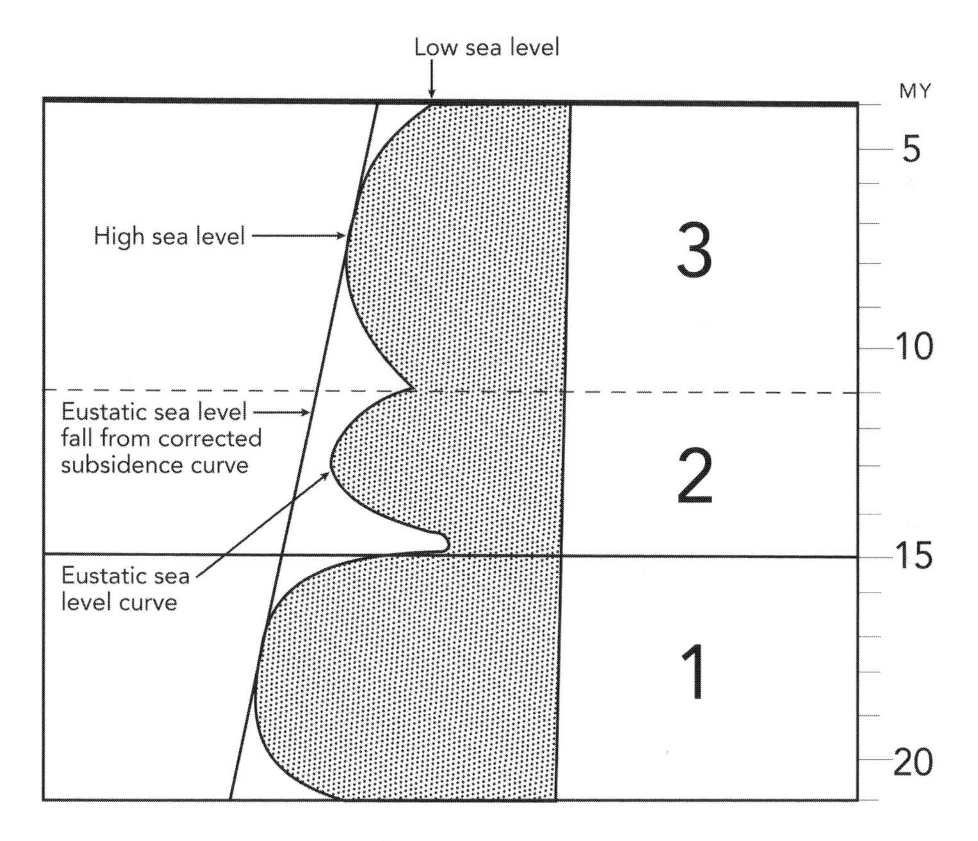

Figure 2.11 A. A hypothetical seismic reflection profile. Three major sequence boundaries are delineated by thick, wavy lines. Coastal sediments are represented by fine dots, marine sediments by larger dots. B. Sedimentary layers within the major sequences of fig. 2.11A. C. Interpretation of the sequences in terms of eustatic sea level change, after correcting for basin subsidence. The sequences indicate three cycles of rising and falling sea level. (Modified from Vail et al., 1977; Hallam, 1992; and Doyle et al., 2001.)

sequence 1 (figs. 2.11, 2.12) record a rise in sea level as marine sediments (large dots) deposit farther landward. This is followed by falling sea level, indicated by the encroachment of terrestrial sediments (small dots and circles) further seaward (units 4–6). The sequence boundary between sequence 1 and 2 represents a rapid drop in sea level. Similar cycles recur in sequences 2 and 3.

The Exxon group's sea level curves provoked intense interest, but also a host of criticism.[48] Much of the early sea level curves based on seismic stratigraphy came from unpublished, proprietary data. Subsequently, data were obtained from more widely accessible deep-sea marine cores and onshore sedimentary deposits. Another limitation is the accuracy with which sequence boundaries can be dated and shown to be truly time-synchronous.[49]

More recently, Ken Miller and colleagues from Rutgers University and elsewhere created a revised eustatic sea level curve that combined marine oxygen isotope data with continental margin and seismic stratigraphy, correcting for the effects of basin subsidence, sediment compaction and loading, using a technique called "backstripping."[50]

This chapter has briefly surveyed the major processes leading to sea level change over a broad range of timescales and has outlined the basic tools with which to recognize these changes. We are now ready to embark on a journey across time, deciphering the clues left behind by the seas that came in and went out.

During those long ages of geologic time, the sea has ebbed and flowed over the great Atlantic coastal plain. It has crept toward the distant Appalachians . . . then slowly receded . . . and on each such advance it has rained down its sediments and left the fossils of its creatures over that vast and level plain. And so the particular place of its stand today is of little moment in the history of the earth or in the nature of the beach—a hundred feet higher, or a hundred feet lower, the seas would still rise and fall unhurried over shining flats of sand, as they do today.

—RACHEL CARSON, *THE EDGE OF THE SEA*

3 Piercing the Veil of Time

Sea Levels After the Dinosaurs

THE END OF AN ERA

The world of the late Cretaceous period (80–65 million years ago) looked vastly different from the world of today. The Atlantic Ocean, which had begun to open up 180 million years ago, was still much narrower than now. A vast seaway covered much of the interior of North America. India hadn't yet collided with Asia and the Tethys Sea separated Africa from Eurasia. The atmosphere held larger quantities of carbon dioxide, and sea level, by some estimates, was as much as 170 meters (560 feet) higher.[1] The world's climate was still comfortably warm, although not quite the hothouse it had been 90 million years ago. The subtropical Arctic was inhabited by crocodilian reptiles. Ferns, cycads, and other lush vegetation grew near the poles. Dinosaurs roamed the Antarctic.[2] The world of the dinosaurs had remained fairly stable for about 150 million years, but all that came to a sudden and violent end 65 million years ago.

The agent of doom was an enormous asteroid or comet, some 10 kilometers (6 miles) across, that blazed through the sky at supersonic speed, crashing into the Yucatan Peninsula and creating a crater 180–200 kilometers (111–124 miles) across (fig. 3.1).[3] The extraterrestrial interloper generated a powerful sonic boom and heated the air to a searing intensity, producing an intense flash of light. At the moment of impact, the shock wave penetrated the ground, crushing and melting the bedrock beneath. The ground trembled and shook, setting off multiple earthquakes and a devastating tsunami that swept across the Gulf of Mexico and the Caribbean. Meanwhile, the space rock drilled deep into the limestone bedrock, gouging out a huge hole, and largely vaporizing in the process. Rock fragments were forcibly ejected to the outer edge of the atmosphere and then fell back to Earth in a hail of debris, piling up into an ejecta blanket surrounding the crater. Vaporized rock rose 10 kilometers, forming a giant fireball that spread into a mushroom cloud which resembled a mighty nuclear explosion. Within hours, the Earth was plunged into darkness as thick clouds of dust particles covered the entire sky. The dust may have been thick enough to block sunlight and interfere with plant photosynthesis for up to a year, cooling the Earth significantly. Since most of the dust, except for the very finest particles, probably rained out within months or less, any "impact winter" would have been fairly short-lived.[4] Gases, such as carbon dioxide (CO_2) and sulfur dioxide (SO_2), released during impact as the asteroid crashed into thick carbonate and sulfate deposits, may have affected climate for much longer. The warming produced by carbon dioxide could have lasted 10,000–100,000 years. A more immediate, but shorter-lived effect would have been the cooling due to sulfur dioxide, which scatters sunlight.[5] An alternate theory proposes that vast volcanic outpourings of basaltic lava in the Deccan Plateau, India, 65 million years ago, may have released copious amounts of sulfur dioxide into the atmosphere, triggering a major climatic and ecological change.

The effects on the biosphere were immediate and of far greater consequence.[6] Evidence of an ecological catastrophe comes not only from the well-known extinction of the dinosaurs and ammonites—marine mollusks related to the chambered nautilus—but from the extinction of 60–70 percent of all other species as well. The searing air temperatures following the blast triggered global-scale wildfires, leaving a layer of soot at the Cretaceous-Tertiary boundary. A "fern spike"—an abundance of fern spores—appeared in sediments directly above this. Ferns are among the first plants to re-populate an area following major ecological disasters. Ironically, we may owe our existence to nature's calamity. Mammals, small rat-sized creatures toward the end of the Mesozoic era—the Age of Dinosaurs, diversified and multiplied rapidly, filling suddenly vacant ecological niches during the following geologic era—the Cenozoic,[7] and eventually giving rise to the primates, to hominids, and ultimately to humans.

Figure 3.1 Artist's depiction of the Cretaceous-Tertiary impact event, 65 million years ago: (a) at the moment of impact of the asteroid or comet; (b) the same scene 1,000 years later, showing the 180-to-200-kilometer-wide Chicxulub crater. (Paintings by William K. Hartmann, Planetary Science Institute, Tucson.)

A SLOW DESCENT INTO THE ICEHOUSE WORLD

The extraterrestrial disaster wiped the slate clean, so to speak, by drastically altering the pathway of evolution and at least temporarily modifying the global climate. But eventually the dust literally and figuratively settled out. The close interplay of geological, climatological, and biological processes shaped the world of the next 65 million years, as described below.

Although plate tectonics and sedimentation, among other things, affect global sea level on geologic timescales, the history of sea level change, especially during the last 34 million years, is closely intertwined with climate change. Therefore, a brief review of the main steps in the lengthy transition from the "greenhouse" world of the dinosaurs to the more geologically recent "icehouse" world of waxing and waning ice sheets will help to illustrate the observed patterns of rising and falling sea level.

Long after any lingering climatic effects of the cataclysmic impact, the world remained warm, reaching a peak between 56 and 50 million years ago (fig. 3.2).[8] Subtropical ferns and alligators inhabited the polar regions.[9] Sea surface temperatures near the poles were a mild 18°C–24°C (64°F–75°F)[10] and tropical surface waters reached a sultry 35°C (95°F) or more during peak warm periods.[11] After that peak of Cenozoic warmth, the Earth's climate gradually and unevenly descended into an "icehouse" world, starting with a fairly abrupt cooling episode around 34 million years ago, at which time the first significant accumulations of ice appeared in Antarctica. Even so, small isolated glacier and ice fields may have existed well before then. For example, pebbles carried and dropped by icebergs or sea ice 45 million years ago were found embedded in marine sediments near the North Pole.[12] Somewhat younger marine sediments off the east coast of Greenland yielded similar **ice-rafted debris (IRD)**.

A major climate transition, halfway through the Cenozoic era, around 34 million years ago, marked the entry into the icehouse world. The ocean cooled by several degrees and large-scale ice masses built up in Antarctica, causing sea level to drop. The climate generally remained cool, with a short-lived thaw roughly 25–23 million years ago, succeeded by a brief glacial episode 23 million years ago.[13] The climate ameliorated 17–14 million years ago, followed by a fairly abrupt return to cold around 14 million years ago (fig. 3.2).

Around 14 million years ago, the volume of ice expanded further in Antarctica, while the surrounding Southern Ocean cooled. Fourteen-million-year-old fossil mosses, beech tree pollen, lake algae, and tiny crustaceans, as well as beetle remains accumulated in deposits left by glacial lakes in the western Olympus Range, Antarctica. Nevertheless, temperatures, although cold enough for nearby mountain glaciers to develop, were still much higher

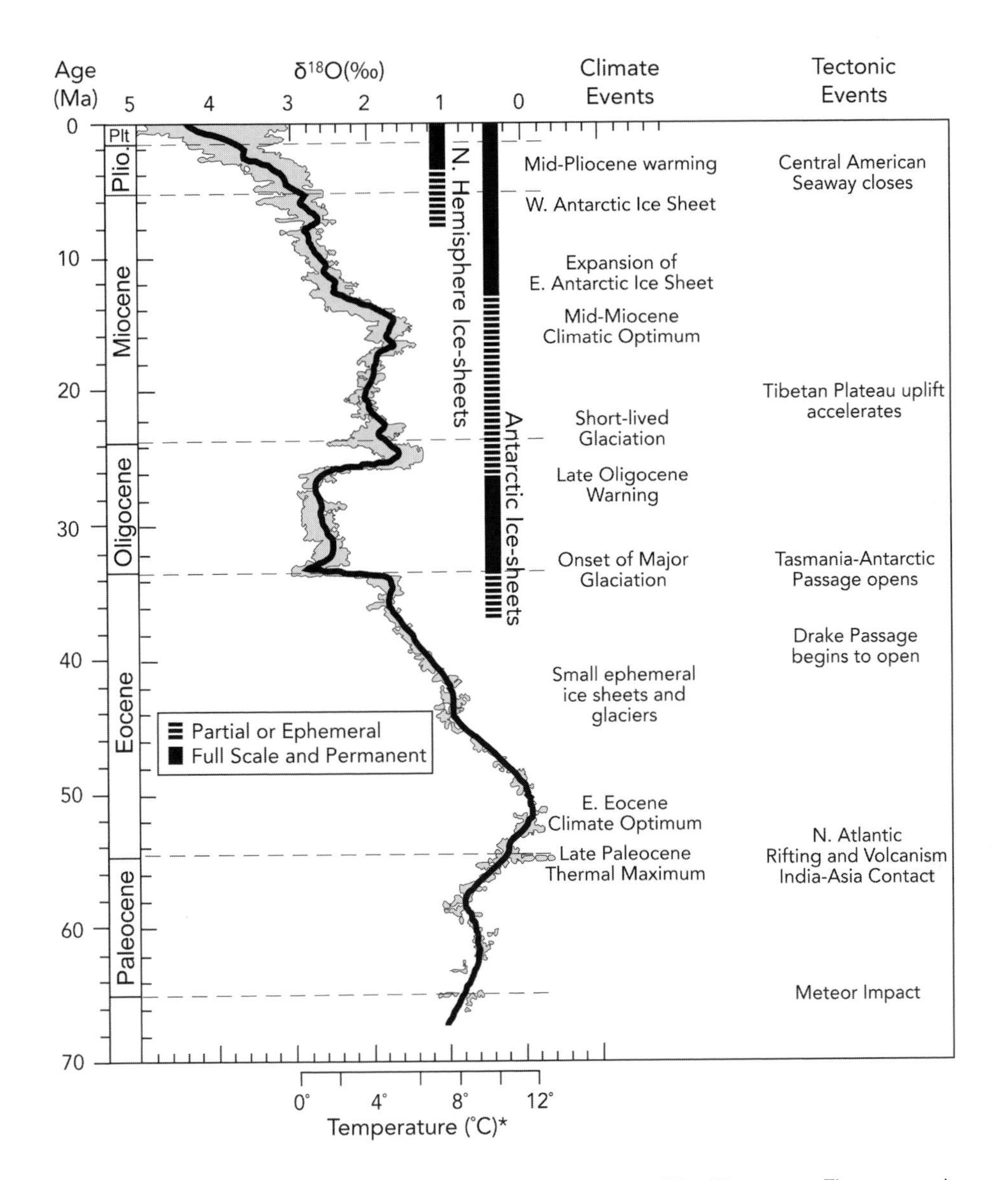

Figure 3.2 Global deep-sea oxygen isotope ratios for the last 65 million years. These records provide information about past climate and sea level changes. Major climatic and tectonic events are indicated (see text for details). (Modified from Zachos et al., 2001.)

than today. But shortly thereafter, the sediments record the beginning of a distinct cooling trend. The frigid polar desert conditions, once established, have never improved substantially in that part of Antarctica, even during brief periods when the rest of the world warmed.[14]

These findings point to the long-term stability of the East Antarctic Ice Sheet, in contrast to its neighbor, the West Antarctic Ice Sheet. Cores drilled beneath the ice cover on the Ross Ice Shelf have revealed 60 advances and retreats of the ice during the last 14 million years, attesting to a generally cold period, although interspersed by a few warmer intervals, most notably a 200,000-year ice-free period 4.2 million years ago.[15] The potential instability of the West Antarctic Ice Sheet raises concerns about future sea level in a warming world, since it holds the equivalent of at least 3 meters of sea level (see chapter 7).

The mid-Pliocene warm interlude, 3.3–3.0 million years ago, when sea levels were more than 20 meters higher than present, offers the last glimpse of formerly hotter times. The subsequent global cooling and onset of large-scale Northern Hemisphere glaciation circa 3 million years ago set the stage for the waxing and waning of the polar ice sheets during the following Quaternary period.

WHY THE WORLD COOLED

Over the 65-million-year sweep of Cenozoic history, the slow motion of the tectonic plates and their interactions have played an important role in the prolonged climatic deterioration. Some key milestones along the path of this long, drawn-out climate cooldown and global sea level fall involve the collision of the Indian plate with Eurasia, the rise of the Himalayas and the Tibetan Plateau, the closure of the Tethys Ocean, and the opening and closing of critical ocean "gateways." These last tectonic events rearranged ocean circulation patterns and the heat conveyed by currents to the poles. Decreasing atmospheric carbon dioxide levels also contributed to the cooldown.

Pangaea, the super-continent that was a single landmass 250 million years ago, initially began to break apart 180 million years ago with the opening of the North Atlantic Ocean between southeastern North America and West Africa. The South Atlantic began to open around 140 million years ago, followed by the drifting apart of Africa, Australia/Antarctica, and India. India began its slow, inexorable journey toward Eurasia, finally colliding with that landmass around 50 million years ago. The Tethys Sea, which once had separated Africa from Eurasia, began to close as the Southern Hemisphere plates drifted leisurely northward. The collision of the Indian plate with Eurasia ultimately raised the Himalayas and the Tibetan Plateau (see fig. 2.4).

This major change in the Earth's topography held significant consequences for the Earth's climate. The high average elevation of the Tibetan Plateau (close to 5 kilometers, or 16,000 feet) and its broad width strengthen the summer monsoons.[16] The rise of the plateau may have not only intensified the Asian monsoons but also modified atmospheric circulation across much of the Northern Hemisphere. However, computer climate modeling experiments suggest that the uplift-related strengthening of the Asian monsoon was insufficient to initiate the growth of large ice sheets on both hemispheres. Therefore, marine geologists and paleoclimatologists Maureen Raymo and William Ruddiman have proposed that the uplift and faulting brought fresh silicate rocks to the surface, where they were exposed to atmospheric carbon dioxide.[17] The silicate minerals in the rocks reacted chemically with the carbon dioxide, removing it from the atmosphere in the form of insoluble carbonates and silica, schematically represented by the equation

$$CaSiO_3 + CO_2 \leftrightarrow CaCO_3 + SiO_2$$

(Calcium silicate plus carbon dioxide yields calcium carbonate plus silica).

In their view, landslides, avalanches, and removal of rocky debris by fast-flowing streams and rivers unearthed fresh rock material for chemical attack. If not counterbalanced by other processes, the sequestering of atmospheric CO_2, a greenhouse gas, through chemical weathering would eventually lead to global cooling and a consequent drop in sea level. Deltas, estuaries, and continental shelves would then be subject to erosion and weathering of carbon-rich organic matter trapped in the sediments. Part of the cooling effect could be neutralized as exposed organic matter oxidized and regenerated CO_2. Another means of keeping the Earth from freezing is by volcanism. The same tectonic forces that pushed up the Himalayas produce volcanoes elsewhere, especially along subduction zones. Volcanoes release CO_2 into the atmosphere, replacing some of the gas consumed by weathering.

High atmospheric CO_2 levels had warmed the Earth during the early Cenozoic, but later as CO_2 levels began to drop, it cooled. How? Another theory suggests that the northward drift of India and closure of the Tethys Sea may also have led to cooling.[18] As India slowly migrated northward during the early Cenozoic, subduction of carbonate-rich ocean sediments beneath Eurasia would have released copious quantities of CO_2 through volcanism and **metamorphism**, keeping the Earth warm and maintaining high sea levels. Once India collided with Eurasia, 50 million years ago, not only was this "carbon dioxide subduction factory" shut off, but by that time India was positioned in the hot, humid tropical zone, where its vast expanse of basaltic lavas—the Deccan Traps—which had erupted around 65 million years ago, underwent intense tropical weathering that removed atmospheric CO_2 according to the above equation (fig. 3.3). According to the theory, the "silicate weathering machine" was in high gear! Little by little, as atmospheric CO_2

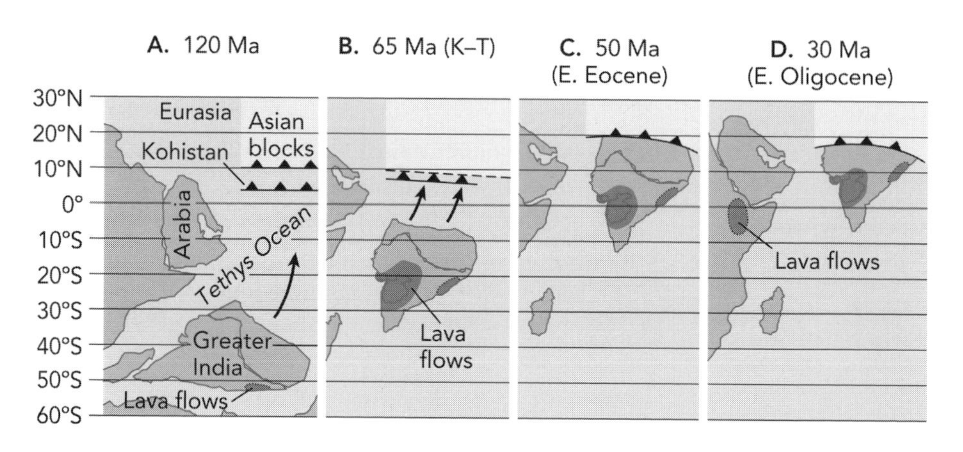

Figure 3.3 Indian Ocean plate motions (left a, b) prior to 50 million years ago—release of carbon dioxide during subduction and metamorphism of carbonate-rich oceanic sediments as India approaches Asia; and (right, c, d) after 50 million years—consumption of carbon dioxide by intense weathering of basaltic Deccan Plateau lavas within the humid tropical equatorial belt, after India collided with Asia. (Modified from Kent and Muttoni, 2008, "Equatorial Convergence of India and Early Cenozoic Climate Trends," *Proceedings of the National Academy of Sciences* 105 (42): 16065–16070. Copyright 2008 National Academy of Sciences, U.S.A.)

levels dropped, the climate cooled and the Earth was on its way to becoming an icehouse world.

The moving plates not only created lofty mountain ranges but may have led to the opening and closing of ocean "gateways" that introduced significant changes in ocean circulation, also affecting climate.[19] With the closure of Tethys and expansion of the Atlantic, ocean circulation gradually shifted from a largely east-west direction to a north-south one. In the Eastern Hemisphere, the flow of water through the Indonesian Straits grew more constricted over time. Farther south, the Drake Passage between Antarctica and South America began to open as early as 41 million years ago,[20] but slowly deepened over time as the plates drifted apart. The Tasmanian Passage between Antarctica and Australia appears to have opened somewhat later, around 34 million years ago.[21]

Some researchers believe that these gateway changes initiated the development and strengthening of the Antarctic Circumpolar Current, which isolated Antarctica from warmer waters by 34 million years ago. The formation of that powerful current and the consequent cooling of the Southern Ocean would have created favorable conditions for the accumulation of major ice sheets in Antarctica (fig. 3.4). On the other hand, ocean sediments suggest that the Antarctic Circumpolar Current strengthened much later, about 25–

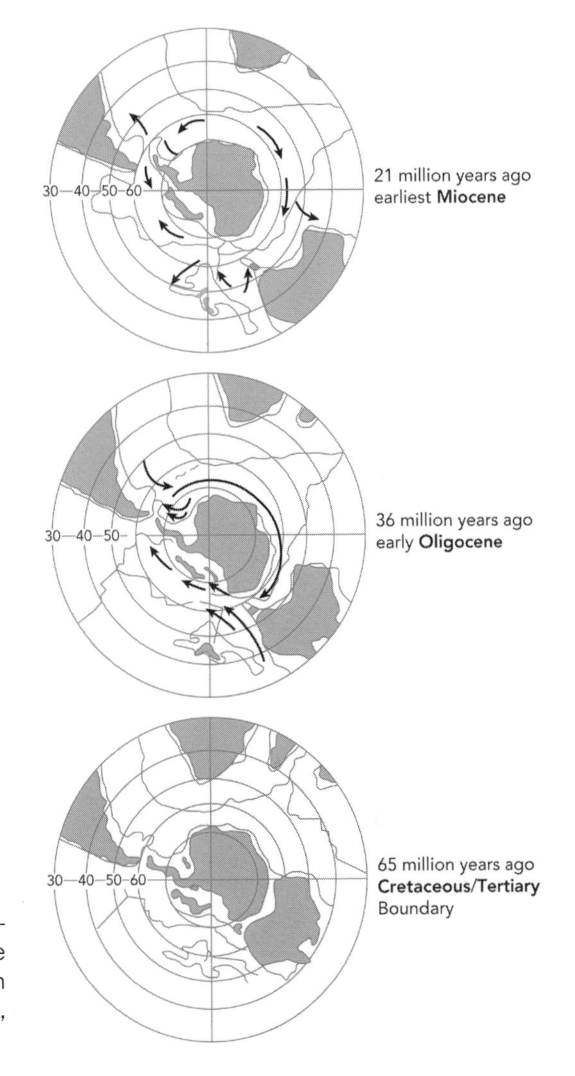

Figure 3.4 Plate motions leading to the development of the Antarctic Circumpolar Current, as the Drake and Tasmanian gateways opened. (Ocean Drilling Program ODP-Leg 189, Preliminary Report, fig. 1, 2000.)

23 million years ago, too late to trigger the initial ice buildup in Antarctica.[22] Therefore, other factors, such as declining atmospheric CO_2 levels, may have helped to cool the Southern Ocean.[23]

The Central American Seaway, separating North and South America, began to shoal around 4.6 million years ago and finally closed within the last 3 million years. This further changed ocean circulation and intensified the Gulf Stream, which by delivering more warmth and moisture to high latitudes may have fed the growth of Arctic ice sheets.[24] Instead, the gradual late Cenozoic decrease in atmospheric CO_2 concentrations may have cooled the Northern Hemisphere sufficiently to have set the stage for the onset of major glaciations there. The global climatic consequences of gateway changes remain somewhat equivocal.

ICE BUILDUP IN ANTARCTICA AND FALLING SEA LEVEL

The history of ancient sea levels is pieced together from the clues left behind by the types of sediments and their geographic distributions, climatic and environmental preferences of certain fossils, and variations in oxygen isotopes of marine foraminifera, which can be either free-floating (i.e., **planktonic**) or inhabiting the ocean floor (i.e., benthic; see fig. 2.9).

A frequently used method in studying ancient sea levels is that of sequence stratigraphy, which, as explained in chapter 2, reconstructs sea level history by interpreting the former environments in which the rock layers were deposited. The stratigraphic sea level record has been compared with oxygen isotope variations preserved in benthic foraminifera.[25] Benthic forams provide a clearer index of past global sea level and temperature changes than do their free-floating relatives, because deep-water temperatures remain more uniform than those at the surface and their isotope ratios reflect high-latitude surface conditions from which the deep water originated (e.g., see North Atlantic Deep Water formation, chapter 1). As polar ice sheets build up during glacial periods and sea level falls, ocean water becomes relatively enriched in the heavier oxygen isotope, ^{18}O. Therefore the $^{18}O/^{16}O$ ratios increase in the calcium carbonate of the foram shells that were precipitated from seawater. Conversely, during climatic thaws when polar ice sheets retreat, these isotope ratios decrease. During the very warm "greenhouse" climates of the middle to late Cretaceous and the early part of the Cenozoic era prior to 50 million years ago, when ice sheets were small or nonexistent, any variation in marine $\delta^{18}O$ values[26] arose mainly from variable water temperatures. Other factors that can affect the relation between oxygen isotope ratios in ocean water and in foraminifera include the particular species, its normal temperature range, the mean annual and seasonal water temperature, and ocean salinity.

Of greater concern in interpreting sea level history is the inherent ambiguity in the oxygen isotope data, because they vary with both temperature and ice volume. The effects of temperature can be sorted out from changes in ice volume by using independent means of estimating ocean temperatures and comparing these results to the benthic foram oxygen isotope records. The Mg/Ca ratio (and to a lesser extent, the Sr/Ca ratio) is frequently used, since these ratios change depending on the temperature of the seawater in which the forams lived. (However, changes in magnesium content of the ocean during the Cenozoic era may complicate interpretation of paleotemperature reconstructions.)

Biomarkers, like TEX_{86}, found in the marine planktonic organism Crenarchaeota, are also sensitive indicators of past sea surface temperatures. The

TEX_{86} paleothermometer is based on the relative proportion of cell membrane **lipids** within the organism, which readjusts its lipid composition in response to sea surface temperatures.[27] While local currents and climate influence ocean surface temperatures more than those on the ocean bottom, the former also respond to major global climate changes. Thus, TEX_{86} is another means to separate past ocean temperature changes from variations in ice volume or sea level.

A different approach reconstructs past sea levels from changes in the area and depth of the ocean floor over time as the tectonic plates drift apart.[28] Former plate positions and ocean spreading rates can be deduced from the pattern of magnetic stripes on the ocean floor. Subsidence of older, cooler ocean crust, as well extensive marine volcanism, affects ocean basin volume and hence global sea level. One estimate based on ocean basin dynamics suggests that sea level was 170 meters (558 feet) higher than present in the late Cretaceous, 82 million years ago. **Backstripping** of the sedimentary layers, on the other hand, shows a sea level of only 75–110 meters (246–361 feet) above present in the late Cretaceous, peaking at around 150 meters (492 feet) 53.5 million years ago, and dropping by 100 meters in the last 50 million years.[29] The much higher sea level from ocean basin dynamics has been explained by the subduction of a now-vanished portion of the eastern half of the Pacific Plate beneath North America.[30] According to this theory, the subducting slab has reached as far east as beneath the Atlantic Coast, causing subsidence of the New Jersey margin, where the stratigraphic sections had been sampled. Since the stratigraphic approach did not consider this additional subsidence, its estimate of sea level was much lower.

Different methodologies often lead to widely divergent views of Cenozoic sea level history. However, a general consensus has emerged that a significant fall in sea level accompanied the first major cooling event, 34 million years ago (fig. 3.5). Clear evidence appears at this time for the presence of a large ice sheet in Antarctica. For example, ice-rafted debris was dumped into the ocean by floating icebergs that calved off glaciers reaching sea level. One study shows that this major climate event unfolded in at least two stages[31] (table 3.1). Initially, between 34.0 and 33.8 million years ago, oxygen isotope ratios increased while shallow and deep water Mg/Ca ratios decreased, implying that both near-surface and bottom waters cooled at least 2.5°C (4.5°F) without a substantial ice buildup (or sea level fall). Later, from 33.6 to 33.5 million years ago, by contrast, geochemical proxies reveal no marked change in surface water temperatures. This implies that the observed increase in oxygen isotope ratios comes largely from the growth of an Antarctic ice sheet almost equal in size to that of today, with a corresponding drop in sea level.

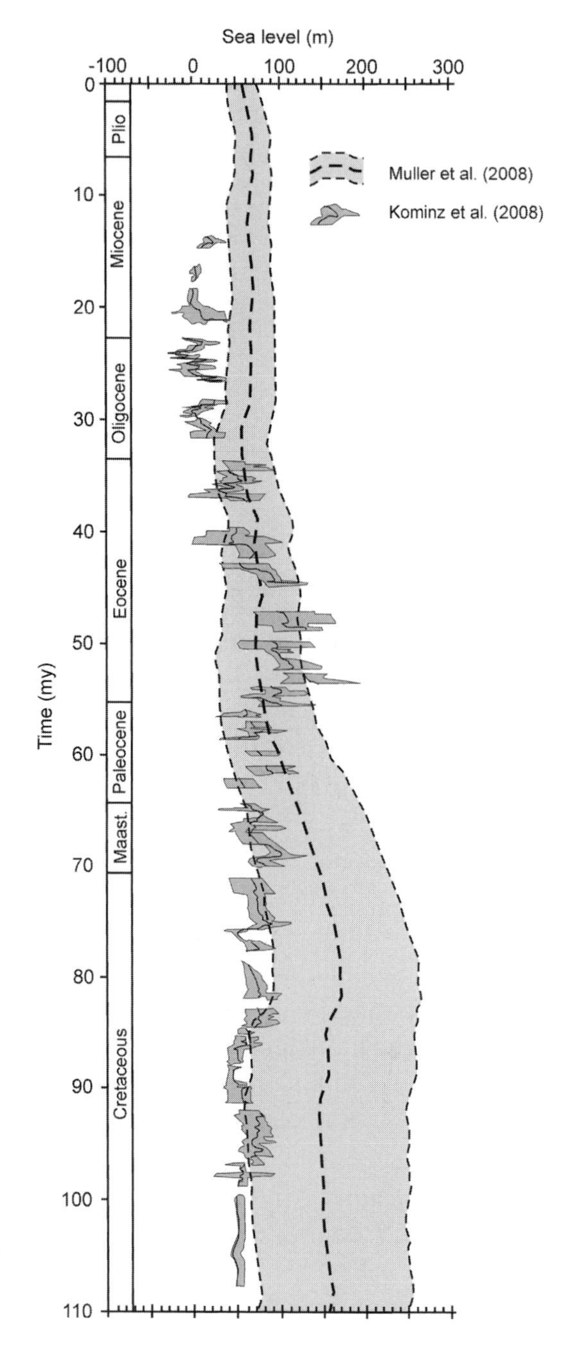

Figure 3.5 Sea level changes over the last 65 million years. (Data from Kominz et al., 2008, and Müller et al., 2008.)

Table 3.1 Summary of climate and sea level changes, 34 million years ago

Date (my)	Increase in $\delta^{18}O$		Cooling (°C)		Sea Level Fall (m)	
	K	L	K	L	K	L
33.55	1.0		2		105	
(33.6–33.5)		0.6		~0		~70
33.6	0.8		NA		>0	
33.8	0.9	0.7	2.5	2.5	30	~0
(34.0–33.8)						

K = Katz et al. (2008); L = Lear et al. (2008).
NA = no data available.
Oxygen isotope (per thousand) and temperature changes (°C) are for bottom water.

However, another study, which compares several paleoclimatic proxies within a single sediment core, divides the transition into three steps instead (table 3.1).[32] At first, 33.8 million years ago, sea level fell by 30 meters (98 feet) and temperatures cooled by 2.5°C (4.5°F). Then, 33.6 million years ago, oxygen isotope ratios increased, but Mg/Ca data are lacking for this period to differentiate between temperature and ice volume components. Lastly, 33.5 million years ago, sea level lowered by 105 meters (344 feet) and oceans cooled 2°C (3.6°F) more. An additional sea level and temperature drop may have occurred around 33 million years ago. The Antarctic ice sheet may have been 20 to 25 percent larger than present;[33] other estimates range from half to roughly twice today's ice volume. While specific details vary, it appears that the descent into the icehouse world began rather abruptly within a fairly short interval around 34 million years ago. Initial cooling left sea level relatively unchanged; subsequent stage(s) witnessed cooling oceans, falling sea level, and major Antarctic ice sheet buildup.

The discovery of ice-rafted debris off the Greenland coast that was 30–38 million years old and as old as 45 million years in the Arctic Ocean[34] has led some to speculate that ice sheets formed in the Northern Hemisphere before they did in Antarctica. However, computer climate modeling experiments suggest instead that substantial ice accumulations on Greenland and elsewhere in the Northern Hemisphere could not have developed until atmospheric CO_2 fell to approximately present-day levels,[35] cooling the Earth to the point where ice could accumulate at the poles. Independent sources of data on past atmospheric CO_2 concentrations indicate that this was unlikely to have happened before 25 million years ago. On the other hand, the computer simulations do not rule out the possibility of much older, small isolated ice caps and mountain glaciers, which could account for the presence of the Arctic-Greenland ice-rafted debris.

Significant changes in plate tectonics that affect sea level unfold over tens of millions of years and longer. Yet the stratigraphic record often preserves signs of apparent eustatic sea level changes on much shorter, million-year timescales. This evidence suggests that small polar ice sheets (although not of continental scale) waxed and waned in response to climatic fluctuations for brief periods even during the "greenhouse" world of the late Cretaceous and early Cenozoic.[36]

Following the first big step toward a colder world 34 million years ago, a short-lived cold spell occurred between 24.1 and 23.7 million years ago, at which time sea level lowered by close to 80 meters and ocean bottom temperatures chilled by 3°C.[37] Marine oxygen isotopes disclose an even greater cooling and ice sheet buildup in Antarctica around 14 million years ago (fig. 3.2).[38] Antarctic fossils also record a distinct change from a cold, alpine climate to frigid polar desert conditions at that time.[39]

However, the growth of major continental ice sheets in the Northern Hemisphere was of much more recent vintage—not more than 3 million years ago, according to many estimates. What set of circumstances plunged the north polar regions into a deep freeze long after a similar fate befell the South Pole? How did sea level change in response to the growth of Arctic ice?

INTENSIFICATION OF THE ICEHOUSE WORLD

A proposed explanation for the approximately 3-million-year onset of major Northern Hemisphere ice sheets points to drifting tectonic plates. Until at least 8 million years ago, the Isthmus of Panama remained open and water flowed freely between the Pacific and Atlantic Oceans (fig. 3.6a). As the isthmus gradually emerged, by 4.6–4.5 million years ago the gateway began to close, and it was sealed tight by 2.7 million years ago.[40] These tectonic events likely triggered a series of major shifts in ocean circulation with important climatic consequences. Caribbean water, now cut off from the Pacific Ocean, grew warmer and saltier. The Gulf Stream strengthened as it transported this warmer, more saline water toward the North Atlantic (fig. 3.6b). Approaching the Arctic, the water cooled and became dense enough to sink to the ocean bottom, thus enhancing the great ocean conveyor belt system. Along with the additional warmth, the powerful Gulf Stream brought more moisture, which fell as snow in Greenland and built up the ice sheet.

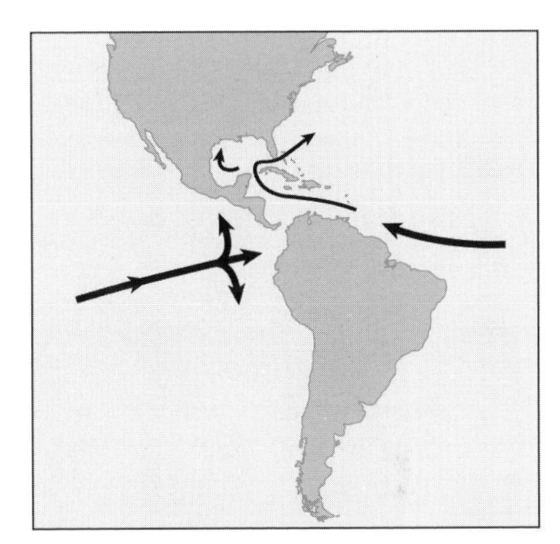

Figure 3.6a Five million years ago, a seaway separated North and South America, allowing exchange of water between the Pacific and Atlantic Oceans. (Illustration by Jack Cook, Woods Hole Oceanographic Institute.)

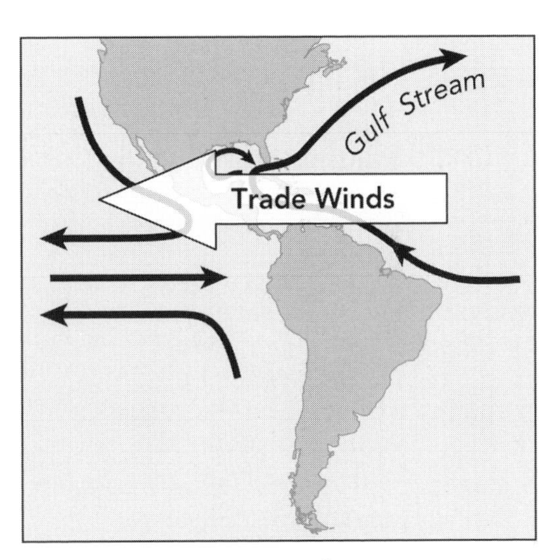

Figure 3.6b The seaway shoaled by 4.6–4.5 million years ago and closed completely by 2.7 million years ago, causing Atlantic water to grow saltier and the Gulf Stream to strengthen. (Illustration by Jack Cook, Woods Hole Oceanographic Institute.)

But this hypothesis faced a problem. The emergence of the Isthmus of Panama began long before major Northern Hemisphere ice sheets formed 2.7 million years ago. One solution was not only to transport moisture northward by an invigorated Gulf Stream, but to carry it farther aloft to Eurasia, where it fell as rain, feeding the Siberian rivers that drained into the Arctic Ocean and freshening it (fig. 3.6c). This in turn promoted growth of sea ice, which reflected sunlight and heat to space, and capped the escape of ocean

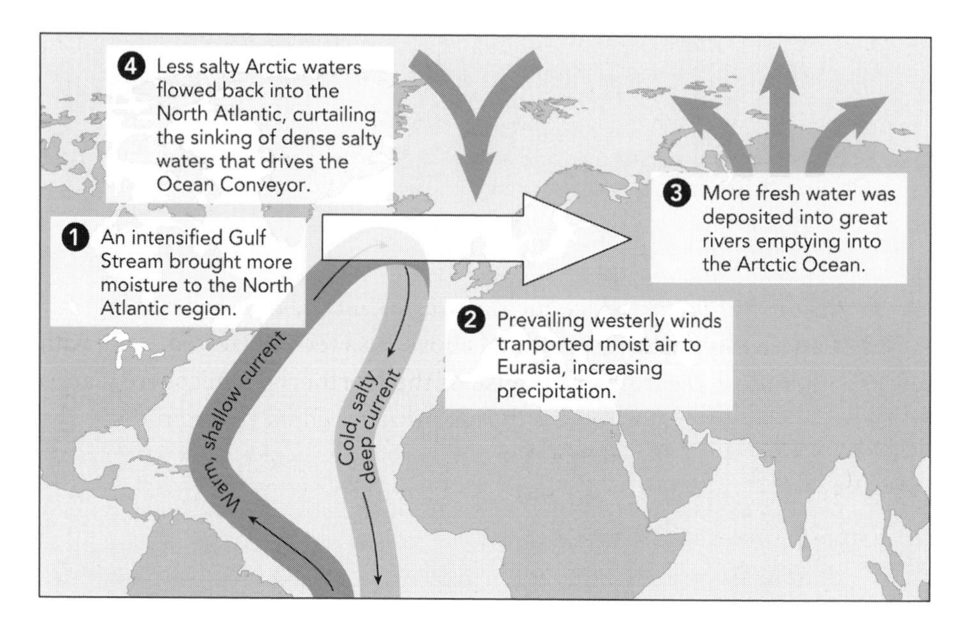

Figure 3.6c An invigorated Gulf Stream carried moisture north, and prevailing westerly winds transported moist air into northern Europe and Asia, where rainwater drained into rivers flowing into the Arctic Ocean. The influx of river water freshened Arctic Ocean water, enabling expansion of sea ice. The sea ice, being brighter than seawater, reflected more of the Sun's energy to space, cooled the region, and facilitated ice sheet expansion. (Illustration by Jack Cook, Woods Hole Oceanographic Institute.)

warmth. These processes would have cooled the Arctic regions, setting the stage for subsequent ice sheet growth.[41] Another nudge toward ice sheet expansion occurred when the tilt of the Earth's rotational axis toward the Sun was periodically reduced 3.1–2.5 million years ago. At such times, the amount of sunlight reaching the Northern Hemisphere in summer diminished and the amount of summer meltwater decreased.[42]

Information gleaned from sediment cores drilled in the North Atlantic supports a number of these proposed ideas. The amount of ice-rafted detritus dropped by drifting icebergs increased substantially after 3 million years ago. Changes in the assemblages of foraminifera point to pronounced changes in ocean circulation and climate around 2.74 million years ago.[43] Oxygen isotope ratios in the shells of deep-sea forams become heavier. This signals a major expansion in global ice volume 2.9–2.8 million years ago, equivalent to a sea level fall of 40–45 meters (131–147 feet) (fig. 3.2). Somewhat later, 2.7 million years ago, the final "climate crash" caused an even further expansion of global ice volume equivalent to another 45 meters of sea level fall, for a total lowering of ~90 meters (~295 feet) at the onset of major Northern Hemisphere glaciation. Curiously, though, sea surface temperatures were warmer during interglacial periods than at present. More heat and moisture were probably transported toward the poles by a stronger Gulf Stream–North Atlantic Current, which warmed the ocean surface. But in order to expand and maintain a large ice sheet, the winter ice growth resulting from the increased delivery of moisture to the poles would have had to exceed the greater degree of ice melting during interglacial thaws.

The "**Panama hypothesis**," outlined above, was recently tested along with several other ideas regarding the cause of the Northern Hemisphere glaciations, using computer simulations that mimic past climate conditions.[44] The computer experiments, combined with a model of ice sheet behavior, suggest that while closing the Panamanian seaway would indeed bring more snowfall to Greenland, the higher summer temperatures would instead melt much of the ice margins, reducing its volume. If, on the other hand, atmospheric CO_2 was reduced in the climate simulations, summer temperatures cooled, less ice melted, and the expansion of ice led to the equivalent of 6.3 meters of lowered sea level. The computer models, however, do not explain why atmospheric CO_2 should have dropped so suddenly 3.0–2.7 million years ago.

Another hypothesis attributes the global cooldown to tectonic changes affecting ocean flow through the Indonesian Gateway (via the Indonesian archipelago) between 3 and 4 million years ago.[45] The gradual tectonic constriction of the Indonesian Seaway initiated changes in the source of subsurface waters that stimulated cooling. Curiously, the accompanying 4°C (7°F) temperature drop in subsurface waters occurred during the otherwise fairly warm mid-Pliocene epoch.

BUCKING THE TREND: THREE CENOZOIC SEA LEVEL ABERRATIONS

The preceding discussion describes an overall declining trend in Cenozoic sea level, yet displaying considerable variability (fig. 3.5). Superimposed on this long-term trend are three unusual sea level episodes. Two of the three were associated with unusually warm periods, while the third was the consequence of tectonic upheavals. The first event appeared during the culmination of the greenhouse world around 55.5 million years ago—the Paleocene-Eocene Thermal Maximum (PETM)—named after its occurrence at the boundary between the first two geologic epochs of the Cenozoic era.[46] The second episode, circa 6 million years ago, began as the result of a set of tectonic displacements that eventually led to the near-total desiccation of the

Mediterranean Sea. When the tectonic barriers were subsequently breached, the floodwaters of the Atlantic cascaded in torrentially and rapidly refilled the Mediterranean. The third episode, 3.3–3 million years ago, was the last glimpse of a former greenhouse world before the final plunge into a cold world with permanent ice caps at both poles.

The Paleocene-Eocene Thermal Maximum (PETM)

Geoscientists suspected that something strange had happened when they discovered unusual carbon isotope compositions that appeared abruptly in 55.5-million-year-old terrestrial and ocean sediments. Other indications of an unusual event came from the deep ocean, which had suddenly warmed by as much as 5–7°C, and had also become more acidic and oxygen-deficient in many places, creating a mass extinction that wiped out nearly half of all benthic foraminifera. Mammals diversified and spread out, while other biotic changes occurred at this time as well. The acidification of the ocean caused widespread dissolution of marine carbonate sediments.

Carbon, like oxygen, has several isotopes: ^{12}C, ^{13}C, and ^{14}C. The first two isotopes are stable, with ^{12}C comprising 98.9 percent of the carbon in nature and ^{13}C making up the remaining 1.1 percent. The third isotope, ^{14}C, is radioactive, which makes it useful for dating geological and archaeological materials (see chapter 2). The ratio of the two stable carbon isotopes varies as a result of biological activity, since living organisms tend to take up the lighter ^{12}C in preference to ^{13}C. Thus, at times of high biological productivity, the $^{13}C/^{12}C$ ratio in carbonate sediments increases (or carbon gets "heavier"), because the lighter carbon has already been selectively consumed by organisms, leaving the remaining unused heavier ^{13}C carbon. Conversely, if biological activity has slowed down, or if biogenic carbon is released into the atmosphere in the form of methane (or carbon dioxide) gas, the carbon isotopes become "lighter" because the ^{12}C incorporated into living organisms now becomes available to the system.

The PETM event was characterized by a rapid decrease of 3–7 parts per million[47] in the carbon isotope ratio and a deep-sea temperature rise of more than 5°C in less than 10,000 years, with a more gradual recovery spanning 200,000 years (fig. 3.7).[48] The sudden appearance of such a large carbon isotope shift requires an enormous input of isotopically light carbon. But from where? Proposed possibilities include hydrothermal venting from North Atlantic submarine volcanism, release of deep-seated gases trapped in an isolated marine basin, or slumping of marine sediments containing gas hydrates.

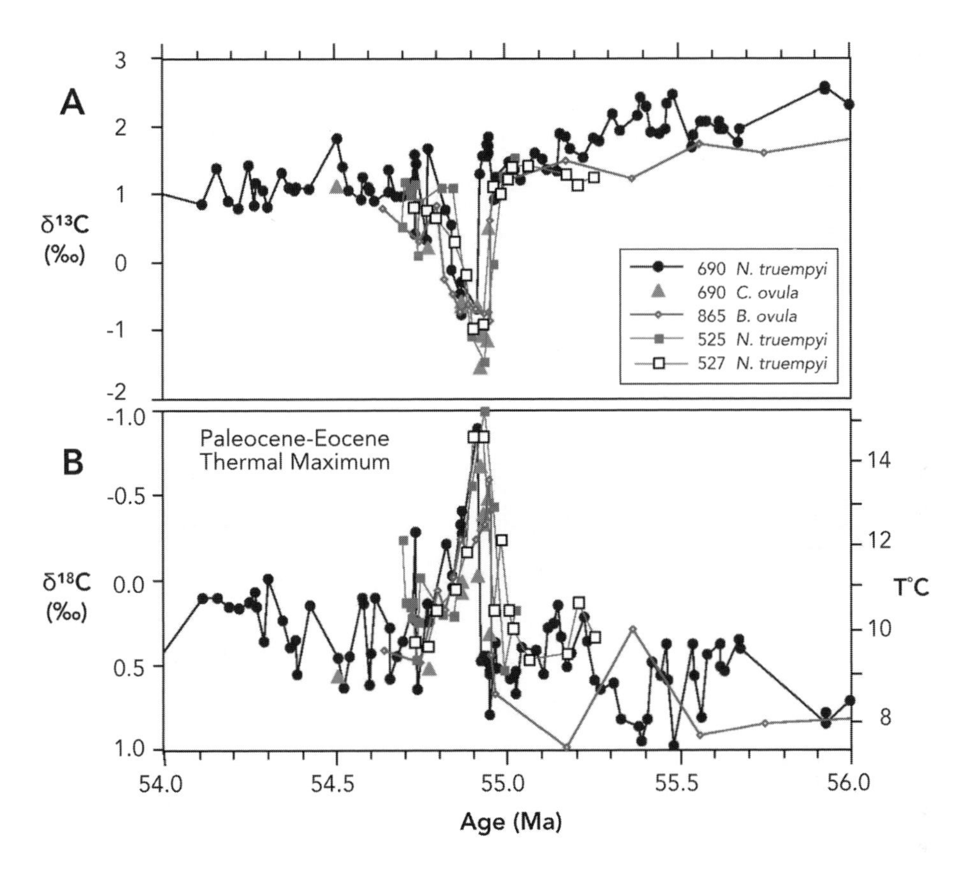

Figure 3.7 Isotope anomalies at the Paleocene-Eocene Thermal Maximum, ca. 55.5 million years ago. This event is characterized by a sharp decrease in both carbon (A) and oxygen (B) isotope ratios in deep-sea foraminifera. The low C isotope ratios imply either a drastic slowdown of biological productivity or a massive injection of methane, which rapidly oxidizes to carbon dioxide in the atmosphere. The low O isotope ratios signal a rise in temperature (and sea level). (Adapted from Zachos et al., 2001; G. A. Schmidt, 2009, "Paleocene-Eocene Thermal Maximum," in V. Gornitz, ed., *Encyclopedia of Paleoclimatology and Ancient Environments*, fig. P9, p. 697, with kind permission from Springer Science + Business Media B.V.)

One popular theory is the rapid release of methane from **gas hydrates** (also called **clathrates**). Gas hydrates are a form of ice with an open cage crystal structure that can trap gases such as methane or carbon dioxide. They are stable only at low temperatures and under pressure and occur in nature in marine sediments on the continental shelf and slope, and also buried several hundred meters deep in Arctic permafrost. As temperature rises, or pressure is relieved, hydrates decompose violently, releasing methane (fig. 3.8). The most common form of methane hydrate has the approximate formula: $8CH_4$ $46H_2O$. Most terrestrial methane is produced as a result of bacterial activity.

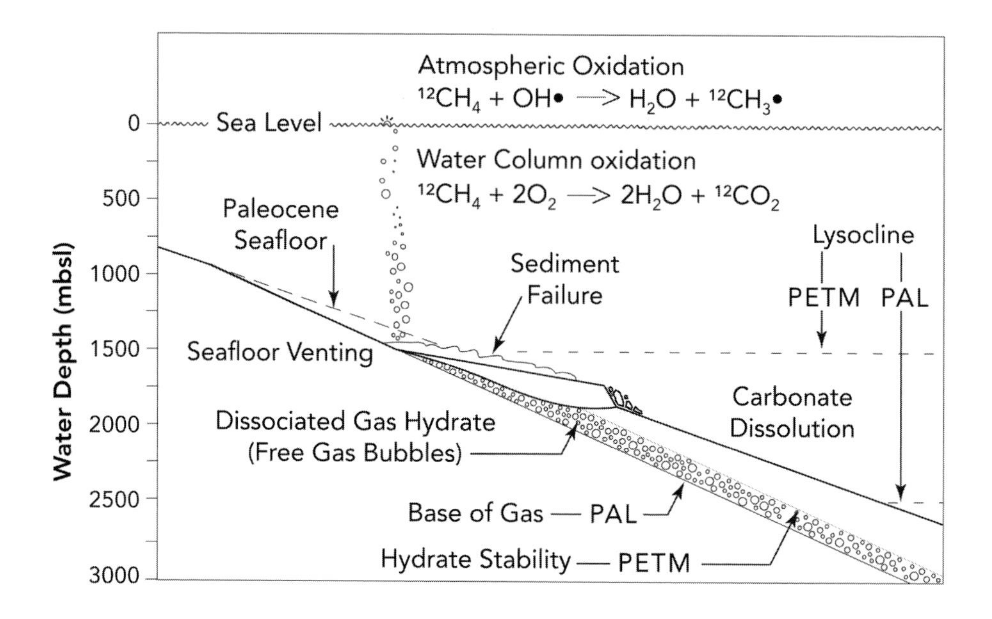

Figure 3.8 Submarine landslides triggered by the decomposition of methane hydrates in deep-sea sediments. The liberated methane gas quickly travels to the ocean surface and into the atmosphere, where it rapidly oxidizes to carbon dioxide. Since both methane and carbon dioxide are greenhouse gases, massive inputs of these gases will result in climate warming. (After G. R. Dickens and C. Forswall, 2009, "Methane Hydrates, Carbon Cycling, and Environmental Change," in V. G. Gornitz, ed. *Encyclopedia of Paleoclimatology and Ancient Environments*, fig. M21, p. 563, with kind permission from Springer Science + Business Media B.V.)

The carbon isotope ratio of bacterial methane is therefore extremely "light" (i.e., highly enriched in ^{12}C relative to ^{13}C, or in other words, depleted in ^{13}C relative to ^{12}C, hence the more negative values), so that even relatively small injections of such light methane into the atmosphere could induce a noticeable shift in isotope composition.

But what triggered the dissociation of the methane hydrate in the first place? Was it a significant ocean warming, a change in ocean circulation, or a massive submarine landslide? These processes were not necessarily unrelated. A ~5°C warming of the deep ocean could have destabilized methane hydrates in ocean sediments, provoking a rapid and forceful decomposition into the constituent methane gas and water. Such a disruption of unconsolidated sediments would have set off massive underwater avalanches (fig. 3.8). As a slurry of loose rock and mud cascaded downslope, more underlying sediments would have become dislodged, releasing additional volumes of trapped methane.

Several marine cores furnish evidence for such a scenario.[49] One core retrieved off the Atlantic coast of Florida holds chunks of mud encased in the sediment—debris from a submarine landslide—at the same stratigraphic horizon as the distinctive carbon isotope anomaly (therefore having formed simultaneously).[50] Furthermore, oxygen isotope ratios of foraminifera show gradual sea surface warming that preceded the sharp carbon isotope anomaly, favoring the idea that increasing water temperatures could have triggered the methane gas outburst. Not only was the onset of the "light" carbon anomaly abrupt, but it appeared in surface-dwelling foraminifera earlier than in their deeper-dwelling cousins.[51] This implies that the methane released by hydrate decomposition traveled rapidly to the surface of the ocean and into the atmosphere. The sudden influx of large quantities of methane into the atmosphere, which is readily oxidized to carbon dioxide (both are greenhouse gases), would have led to global warming. This in turn would have hastened ocean heating, further destabilizing the hydrates and provoking a runaway process popularly called the "clathrate gun hypothesis."[52]

One unusual feature of the PETM is the extremely warm poles. Global climate models have not been able yet to simulate such high polar temperatures without invoking unrealistically elevated CO_2 levels. Other proposed explanations involve the presence of other greenhouse gases (e.g., methane, nitrous oxide), or thicker polar stratospheric clouds, which can effectively trap heat.[53]

Sequence stratigraphic reconstructions at a number of sites around the world suggest that the PETM may have been associated with a rise in sea level. One estimate, based on changes in water depths inhabited by marine fossil populations,[54] ranges from 20 to 30 meters, but this may be too high. Small ice sheets, holding the equivalent of ~5–10 meters of sea level, may have existed even in a greenhouse world.[55] Also, the thermal expansion from a ~5°C increase in global ocean temperatures could account for up to 5 meters of sea level rise.[56] Thus sea level changes at the PETM were unlikely to have exceeded ~10 meters unless the volume of continental ice sheets had been seriously underestimated.

The PETM arouses particular interest among geoscientists because of similarities to the current rapid release of greenhouse gases into the atmosphere due to human activities. The amount of carbon (in the form of carbon dioxide and methane) expected to enter the atmosphere within the next few centuries by burning of fossil fuels and forests may attain a level comparable to that of the PETM. At the very least, the event dramatically illustrates the consequences of a sudden massive atmospheric carbon influx on the world's climate and on sea level.

The Messinian Salinity Crisis—When the Mediterranean Became a Desert

Marine geologists Kenneth Hsü, William Ryan, and Maria Bianca Cita, probing the Mediterranean seafloor from the research vessel *Glomar Challenger* in 1970, were surprised to discover a thick deposit of rock salt (halite, sodium chloride) sandwiched in between layers of sediments thousands of feet below the seafloor.[57] Minerals such as anhydrite (calcium sulfate), together with salt formed in shallow, hot briny pools, similar to the "sabkhas," or salt flats, are found along arid coastal deserts such as the United Arab Emirates on the Persian Gulf. However, questions soon arose as to whether uplift had raised the deep seafloor several thousand meters, exposing the sea to rapid evaporation and deposition of salts, followed by drastic subsidence and refilling, or whether the entire Mediterranean had somehow dried up, except for a few very deep briny pools, provoking the so-called "Messinian salinity crisis." It wouldn't take very much to dry out the Mediterranean. Today the Mediterranean loses more water by evaporation than it gains from rivers flowing into the sea, and if not for Atlantic seawater entering via the Strait of Gibraltar, it would dry up within 1,000 years. The salt would form a layer on the order of only 30 meters (98 feet) thick.[58] Instead, the salt layer was up to 1,600 meters thick. If the volume of water contained in the Mediterranean were transferred to the other oceans, global sea level would rise by only ~10 meters.

Deep-sea drilling soon established that the deep basins of the Mediterranean were already in place well before the onset of salt deposition and that the seafloor bathymetry at the time differed little from that of today. A more likely scenario was that tectonic movements began to constrict the ancestral Gibraltar Straits by around 5.96 million years ago, initiating the period of thick **evaporite**[59] deposition (fig. 3.9). But as pointed out above, a cutoff in Atlantic water influx would not have sufficed to account for the total volume of evaporites. The marine oxygen isotopes record several episodes of Antarctic ice sheet expansion and sea level fall—at ~6, ~5.8, and ~5.7 million years ago. At times, global sea level may have been tens of meters to as much as 80–100 meters (260–328 feet) lower, which reinforced the tectonically induced cutoff of Atlantic seawater.[60] At other times of higher sea level, some Atlantic seawater entered the Mediterranean.

The Mediterranean was finally completely isolated from the Atlantic around 5.6 million years ago, which led to a near-total desiccation and a period of intense erosion—the peak of the Messinian salinity crisis (fig. 3.9). Deep canyons dwarfing the Grand Canyon in size, now buried beneath thick layers of younger sediment, were cut into the ancestral Nile, Rhône, Po, and other rivers draining the Mediterranean. After a brief period of almost total

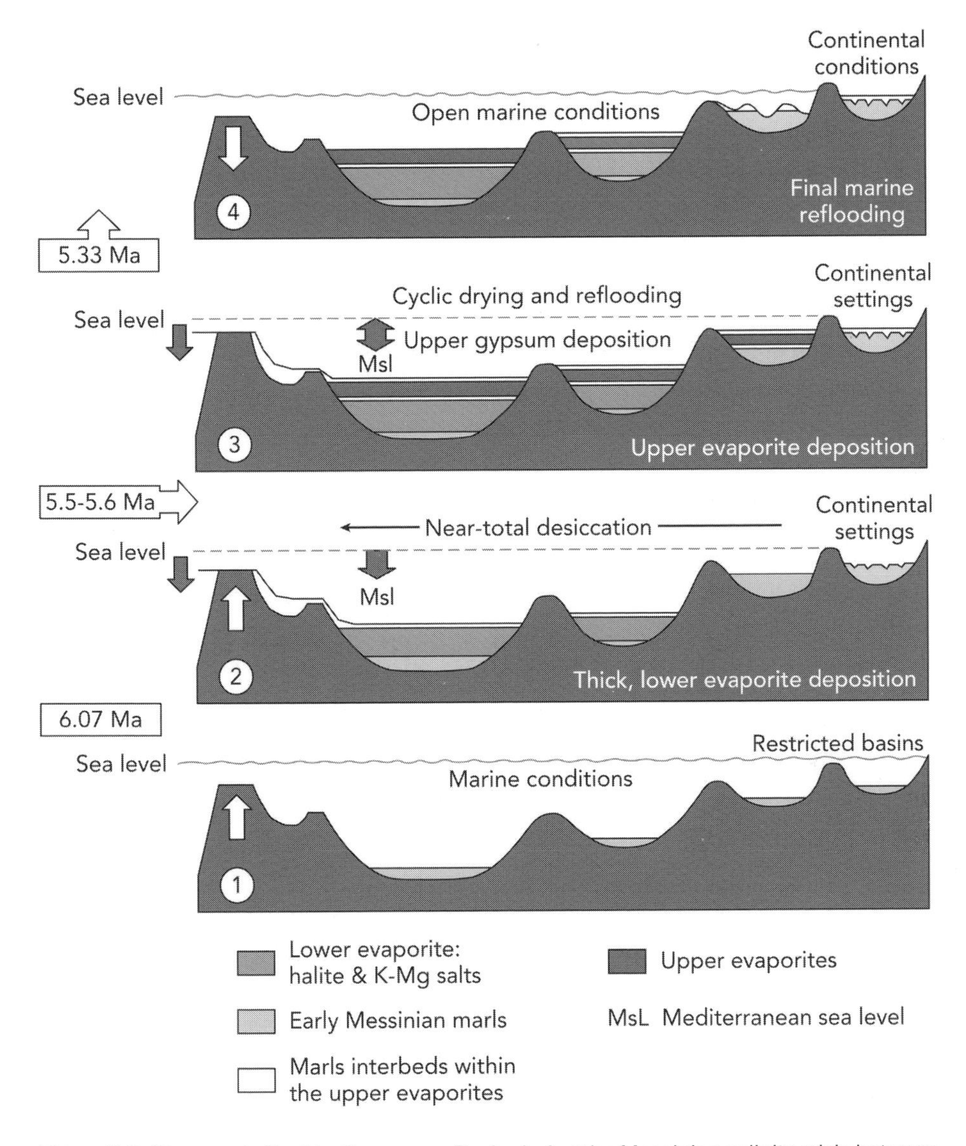

Figure 3.9 Changes in the Mediterranean Basin during the Messinian salinity crisis between 6 and 5.3 million years ago. (1) Normal marine conditions prevailed until ca. 6 million years ago. (2) Starting ~5.95 million years ago and culminating ~5.6–5.5 million years ago, tectonic movements and lowered world sea level cut the Mediterranean off from the Atlantic Ocean and the Mediterranean dried out nearly completely, leaving behind thick deposits of gypsum and rock salt. (3) After 5.6 million years, due to repeated partial opening and closing of the ancestral Strait of Gibraltar, the Mediterranean periodically re-flooded and dried out, depositing alternating layers of sediments and evaporites. (4) Finally, 5.33 million years ago, the floodgates reopened and the Mediterranean refilled rapidly. (Modified from Rouchy and Caruso, 2006.)

dryness, except for some briny pools in the deepest basins, repeated marine incursions from the west left behind a cyclic succession of sediments interlayered with gypsum, anhydrite, and salt. While the connections to the Atlantic were probably still closing, global climate ameliorated, sea level was rising, and at times may have spilled over the tectonic barriers, allowing successive thin layers of evaporites and sediments to deposit. Toward the latest stages of the Messinian salinity crisis, a wetter regional climate allowed more freshwater to enter the Mediterranean basin and shallow brackish lakes formed. Finally, the Strait of Gibraltar was breached and the Messinian salinity crisis ended abruptly, 5.33 million years ago[61] (fig. 3.9). Huge torrents of water, more powerful than Niagara Falls or Victoria Falls, cascaded over the Gibraltar spillway and catastrophically refilled the Mediterranean to the brim within a few months to two years, raising Mediterranean sea level more than 10 meters per day!

The Messinian salinity crisis may have had some widespread consequences. A transfer of all Mediterranean water to the world's oceans would have raised global sea level by up to 10 meters (33 feet). More significantly, the extensive evaporation may have removed significant quantities of salt from the world's oceans, thereby weakening North Atlantic deep-water formation and slowing the conveyor belt system, possibly leading to global cooling. Such cooling may have caused sea level to drop, intensifying Mediterranean desiccation during parts of the Messinian.[62] When the Mediterranean was almost empty, summertime temperatures at the former seafloor would have been extremely high—probably much hotter and drier than Death Valley or the Dead Sea today.

The Mid-Pliocene—Lesson for a Warmer World?

The mid-Pliocene, around 3.3–3.0 million years ago, was the final farewell to a once warmer world. It was the last prolonged period of warmth before the Earth acquired permanent ice caps at both poles and plunged into periodic ice ages. During the warmer interludes, global mean temperatures averaged 2°C to 3°C above preindustrial temperatures and sea level was much higher, resembling computer model projections of a late 21st-century world heated by anthropogenic greenhouse gases. The mid-Pliocene therefore offers a good analog of future climate change, because the ocean basins and continents had just about approached their present configurations, and most of the fossil species that provide paleoclimate information still exist, enabling direct comparisons.

Land-based and marine paleoclimate indicators show that the poles were substantially warmer, whereas the tropics differed very little from the present.[63] As a consequence, less sea ice covered the Arctic Ocean. Since ocean water reflects less light to space (i.e., absorbs more energy) than sea ice, warming ocean water and air cause more sea ice to melt and freshen ocean water. This slows down deep-water formation, thereby dampening the warming of the North Atlantic. Most atmospheric-ocean global climate models can simulate the reduced thermohaline circulation. Yet, paradoxically, paleosea surface temperature data record the greatest warming in the North Atlantic during the mid-Pliocene and a stronger ocean conveyor system.[64] Questions therefore remain about the appropriateness of the mid-Pliocene as an analog for future climate warming, in that higher greenhouse gas levels should warm all latitudes (although more so near the poles), but only higher latitudes warmed substantially during the mid-Pliocene. Furthermore, carbon dioxide levels at the time may not have been substantially higher than at present.[65]

Sea level varied considerably during the relatively mild mid-Pliocene. During the warmest intervals, sea levels reached elevations of +25–30 meters, but they rarely dropped below -25 meters.[66] These marked gyrations imply that continental ice had already built up substantially, even during this generally warm period.

Such fluctuations in sea level were not uncommon. Multiple cycles that repeat roughly every 41,000 years appear in thick shelf and shallow-water marine sediments in New Zealand.[67] This periodicity reflects the tilt cycle of the Earth's axis. According to the **astronomical theory of ice ages** originally proposed by Milutin Milankovitch (see chapter 4), the 41,000-year cycle affects the amount of the Sun's radiation that reaches the top of the atmosphere. The greater the angle between the Earth's rotational axis and the ecliptic—the plane of the Earth's orbit around the Sun—the greater the amount of summer sunlight when the Earth's axis points directly toward the Sun. The effect becomes more pronounced at higher latitudes, especially in the Northern Hemisphere, where the greater amount of summer sunlight allows for more melting of an ice sheet. Thus, the 41,000-year fluctuations in the marine sediments appear to be linked to the growth and decay of ice sheets, and hence to sea level change. During much of the mid-Pliocene, eustatic sea level fluctuated within a range of 10–30 meters. Even though this period predates the generally accepted development of continental-scale Northern Hemisphere ice sheets starting 3.0–2.7 million years ago (chapter 4), enough ice must have been present on both Greenland and Antarctica that climate oscillations left a cyclical sea level imprint in the sedimentary record.

INTO THE ICE AGES

The long stepwise descent from an early Cenozoic greenhouse world into the icehouse world toward the latter half of the era is punctuated by three major milestones at 34, 14, and 3 million years, each marked by significant changes in global climate and sea level. The first hints of the coming cold occurred around 34 million years ago, when temperatures dived 2.5°C–4.5°C (4.5°F–8.1°F) and sea levels, by some estimates, lowered by as much as 105 meters (443 feet), marking the first substantial buildup of ice in Antarctica. Then by 14 million years ago, another expansion of ice sheets in Antarctica led to an additional drop in sea level. By 3 million years ago, Greenland was deeply buried in ice and large ice sheets began to periodically cover North America and northern Europe. Sea levels alternated between low stands during the deep-freezes and high stands during the thaws. The era of the ice ages on both hemispheres had begun.

The long summer was over. For ages a tropical climate had prevailed over a great part of the earth, and animals whose home is now beneath the Equator roamed over the world from the far South to the very borders of the Arctics. . . . But their reign was over. A sudden intense winter, that was also to last for ages, fell upon our globe.

—LOUIS AGASSIZ, *GEOLOGICAL SKETCHES* (1866)

4

When the Mammoths Roamed

Sea Levels During the Ice Ages

A GLIMPSE OF THE ICE AGE WORLD

As recently as 20,000 years ago, continental-scale ice sheets blanketed most of Canada, the northern United States, and northern Europe. The future sites of New York City, Stockholm, Moscow, and other northern cities lay deeply buried under ice. The climate was bitter cold and dry. A polar desert and open tundra covered northern Europe beyond the ice sheet; **permafrost** extended south into central France. Southern Europe was mostly sparsely vegetated grassland or semiarid grassy shrubland.[1] Vast herds of large mammals, many of them now extinct, roamed the cold, windy, dusty, arid tundra steppes beyond the ice sheets. Our Paleolithic forebears depicted the animals they hunted in realistic and aesthetic images painted deep inside caves in southwestern France and Spain between 30,000 and 14,000 years ago. The ice age bestiary decorating the cave walls included mammoths, aurochs (an extinct wild ox), reindeer, bison, deer, bears, lions, and wild horses closely resembling the pony-sized, dun-colored Przewalski's horses from the steppes of central Asia, characterized by a short, dark mane and tail (fig. 4.1).

Many animals (and, likely, prehistoric hunters tracking the herds) crossed the Beringia land bridge over the Bering Strait between Asia and North America during the coldest intervals, when global sea level dropped below

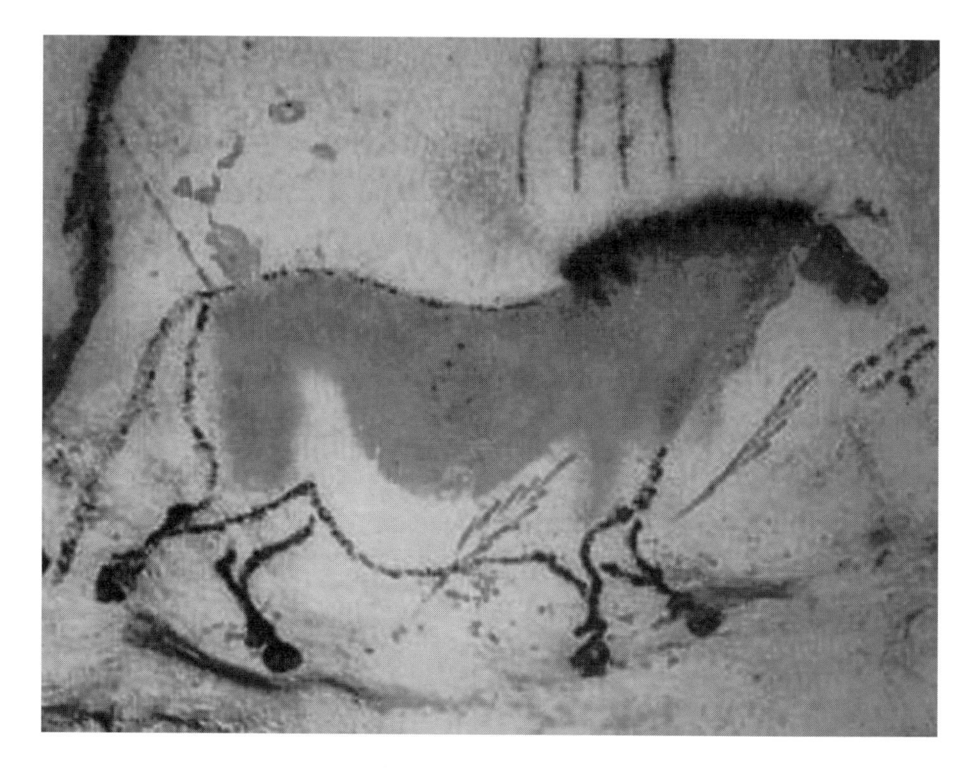

Figure 4.1 Ice age painting of a horse, Lascaux Cave, Dordogne, southwest France, ca. 17,000 years old. (Wikimedia Commons.)

100 meters (330 feet) from today. The first mammoths had entered North America from Eurasia via the Bering Strait land bridge much earlier, perhaps 1.8 million years ago. However, the roadway was a two-way street and other species, such as wild horses, occasionally traveled west in the opposite direction. The horses, originally indigenous to North America, last migrated across the Bering land bridge approximately 10,000 years ago, when sea levels were still considerably lower than present. They spread across the grasslands into Siberia, central Asia, and finally into Europe. Several thousand years later, they became extinct in North America, and no horses roamed the western prairies until Spanish conquistadores brought domesticated horses to this continent in the 16th century.

At least eight times over the last million years, sea level fell and rose as the ice sheets waxed and waned, in cycles lasting roughly 100,000 years each. Long before that, however, beginning around 3.0–2.7 million years ago, Northern Hemisphere ice had begun to expand, building up first on Greenland and then spreading repeatedly over parts of continental Canada and northern Europe during particularly cold periods, in cycles repeating approximately every 41,000 years. Whenever favorable orbital alignments dimmed the amount of northern summer sunlight, not all of the previous winter's snowfall melted and the climate grew progressively colder. Snow gradually turned to ice and built up into an extensive sheet, layer by layer, ultimately several miles thick. The transfer of so much water from the ocean into an ever-thickening ice mass led to a drop in sea level. Periodically, the situation reversed. As more sunlight reached the polar latitudes during Northern Hemisphere summer, more ice melted than froze the previous winter and sea level began to rise. But oddly, this pattern was asymmetric: the return of warmth and ice meltdowns occurred much more rapidly than the deep chills and ice accumulations.

Other rapid climate swings occurred (some within decades) in addition to those linked to periodic changes in the Earth's orbit around the Sun. While their origins are still debated, speculation centers on swift ocean circulation changes, affecting northward heat transport and ultimately, climate. The abruptness and magnitude of these changes bear cautionary warnings for our future as well. Will our growing inputs of atmospheric greenhouse gases not only raise air temperature but also initiate sudden unanticipated and unfamiliar shifts in ocean circulation and sea level change? A careful study of these long-gone events may help us prepare for such climate surprises ahead.

The history of the past behavior of the ice sheets and sea level is written in the archives of ancient coral reefs now raised high and dry by the motions of the shifting tectonic plates, in sediments at the bottom of the ocean, and in the ice sheets of Antarctica and Greenland. We now turn to these ancient tomes in order to decipher their secrets.

READING THE BOOK OF SEA LEVELS PAST

A recipe for accumulating large Northern Hemisphere ice sheets calls for mild, moist winters with heavy snowfall, yet cool enough summers to prevent all of this snow from melting. Thus, ice sheet growth is enhanced by strong, warm, salty surface ocean currents, which deliver heat and moisture to the north.

Such a scenario may have developed once the Central American Seaway was completely sealed, by 2.7 million years ago, strengthening the Gulf Stream and North Atlantic Current. More wind-borne moisture may have reached Greenland, northwestern Europe, and ultimately Eurasia, triggering feedbacks that intensified Arctic cooling. Alternatively, lowered atmospheric CO_2 may have led to cooling (chapter 3). As the climate cooled, due to these oceanic and atmospheric changes, polar ice expanded and sea level dropped.

The thick layers of seafloor sediments preserve the longest and most complete archive of past sea level changes. Tiny marine foraminifera encode the history of rising and falling sea level over the last million or so years in oxygen isotopes encased in their calcareous shells. Former beach terraces hoisted way above the present shoreline by tectonic movements also register the ups and downs of sea level during the succession of Quaternary[2] interglacials and glacials. The Quaternary climate oscillations responsible for these pronounced sea level swings also left clues in the ice sheets of Greenland and Antarctica. We now read the story told by the deep-sea sediments.

In chapter 3, we saw that the foraminiferal oxygen isotope ratios jumped upward in three pronounced steps at around 34, 14, and 3 million years ago. These steps marked worldwide cooling, sharp drops in global sea level, and ice sheet expansion, interspersed with numerous warmer interludes. The last such warm period with much higher sea levels than present ones was the mid-Pliocene, lasting from 3.3 to 3.0 million years ago.

After 2.7 million years ago, the oxygen isotope ratios of the benthic (seafloor) foraminifera grew progressively heavier, implying increasing withdrawal of ocean water and buildup of polar ice. In addition, the difference between alternating high and low temperatures expanded. By that time, Greenland had acquired a significant ice sheet and glaciers had spread over northern North America and Europe during the glacial periods. A sharp increase in the amount of ice-rafted debris at high latitudes coincided with the onset of the Northern Hemisphere glaciation. Deep-sea oxygen isotopes oscillated, repeating roughly every 41,000 years.[3] But the pattern suddenly shifted between 1 million and 800,000 years ago (fig. 4.2). The dominant isotope cycles now lasted ~100,000 years and their amplitude (difference between high and low points) widened.[4] This marked change in ocean isotope ratios has been variously called the middle Pleistocene transition, or more dramatically, the mid-Pleistocene revolution. It defines the Earth's final descent into the era of the ice ages, the last of which ended about 11,000 years ago.[5]

Staircases of raised coral reef terraces, in places such as the tectonically active Huon Peninsula of Papua New Guinea,[6] also trace the alternations in sea level, corresponding to the last one or two Quaternary glacial cycles. The Huon Peninsula is located at the southeastern end of Papua New Guinea, in an area trapped as in a vise between the converging West Pacific and Australian plates. Exposed on the slopes of the Huon Peninsula, a series of at least 20 raised coral reefs has been hoisted to heights of more than 500 meters (>1,600 feet) over the last few hundred thousand years by the steady, yet intermittent uplift (fig. 4.3).[7] These fossil coral reefs, the oldest of which is nearly 250,000 years old and now perched high and dry, once grew within a few meters of sea level.

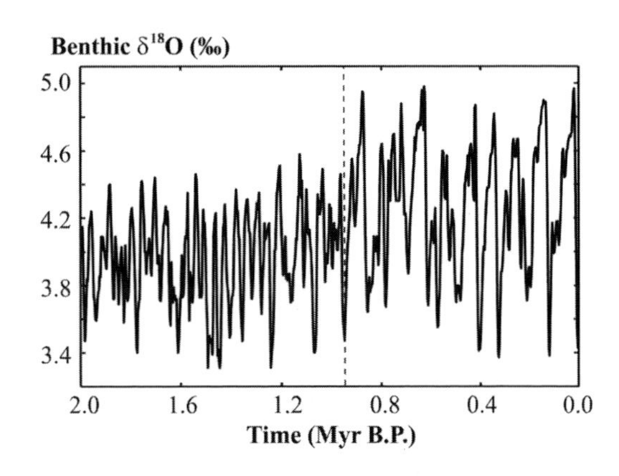

Figure 4.2 Deep-sea oxygen isotope ratios in benthic foraminifera over the last 2 million years. Higher values (heavier isotope ratios) represent colder periods characterized by lower sea levels. Note the distinct change starting around 900,000 years ago from oscillations dominated by 41,000-year cyclicity to one dominated by 100,000-year cyclicity. This change has been called the Middle Pleistocene Transition or the Mid-Pleistocene Revolution. (S. J. Marshall, 2009, "Glaciations, Quaternary," in V. Gornitz, ed., *Encyclopedia of Paleoclimatology and Ancient Environments*, fig. G45, p. 389, with kind permission from Springer Science + Business Media B.V.)

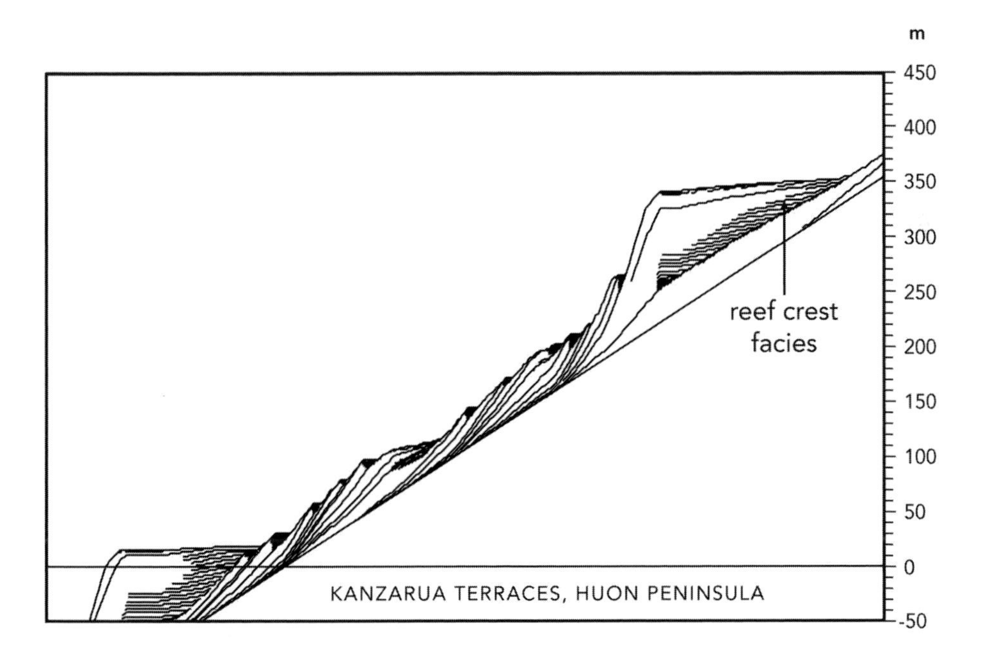

Figure 4.3 Raised coral terraces from a tectonically uplifting terrain, Huon Peninsula, Papua New Guinea. The top terrace on the right is 350 meters above modern sea level and formed at the beginning of the Last Interglacial, ~127,000 years ago. Each of the smaller steps was formed during a sea level oscillation within the last glacial cycle. (After J. Chappell, 2009, "Sea Level Change, Quaternary," in V. Gornitz, ed., *Encyclopedia of Paleoclimatology and Ancient Environments*, fig. S15, p. 895, with kind permission from Springer Science + Business Media B.V.)

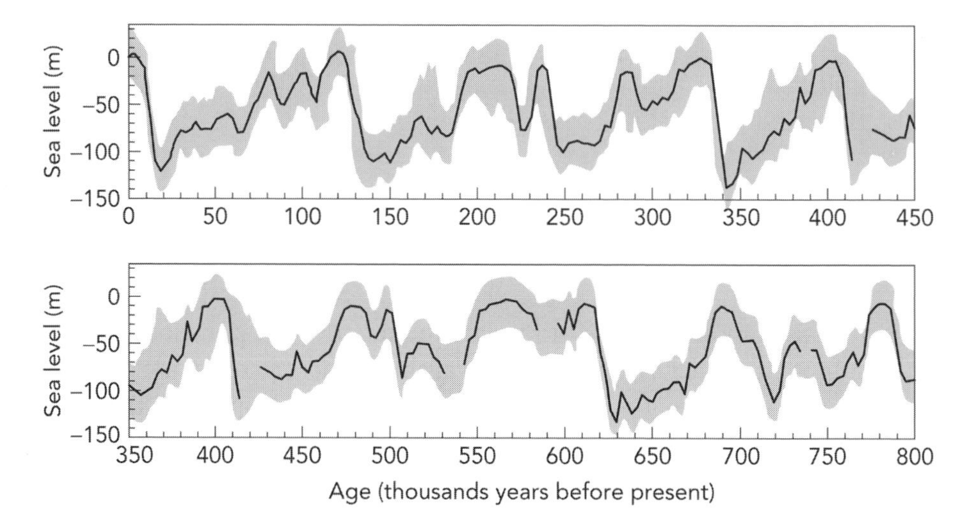

Figure 4.4 Sea level variations over the last eight glacial cycles based on marine foraminifera oxygen isotope ratios and other sea level indicators. The black curve is based on Cutler et al. (2003); the shaded area includes results from various studies and uncertainty ranges. Numbers represent Marine Isotope Stages. (Modified from Siddall et al., 2007.)

The sea level history is deduced by measuring the difference in elevation between the present height of the raised terrace, assuming that it represents a former shoreline, corrected for known uplift at that location, with a well-dated reference terrace, generally taken as the 125,000-year-old terrace from the **last interglacial**, when sea level was at least 4–6 meters (13–20 feet) higher than at present.[8] The sea level record of the raised terraces generally covers only 150,000–200,000 years, the older, more elevated terraces having been long since removed by erosion. (However, raised coral reefs on the island of Barbados, in the Caribbean, while not precisely dated, may register sea level oscillations as far back as 700,000 years.)[9] Therefore, deep-sea foraminifera still furnish the most complete sea level history stretching back many millions of years (fig. 4.4).[10]

Sea Level Gyrations

The wide swings in the foraminiferal oxygen isotopes clearly signal past climate changes, but as indicated previously, they enfold information on both ocean temperature and ice volume (i.e., sea level). During the late Quaternary, for example, the glacial to interglacial difference in oxygen isotope ratios of benthic forams is ~1.8 parts per mil—a value that includes both ice volume and temperature effects. How does one sort out the signal due to ice volume from that due to temperature? The amount of water transferred from ocean to continental ice during the last ice age lowered global sea level by ~120 meters (394 feet), an amount equivalent to ~3 percent of the ocean's total volume. This sufficed to increase the mean oxygen isotope ratio of ocean water by ~1.2 parts per mil.[11] Therefore, to a first approximation, ice volume contributes around two-thirds to the overall glacial-interglacial isotopic range. Deep ocean water cooled by 2°C–5°C (4°F–9°F) during glaciations.[12] Geochemical analyses (i.e., Mg/Ca ratios) can also be used to isolate the temperature contribution (see chapter 3). But such data are not always available. Other strategies are to select regions where deep ocean water temperatures presumably changed the least, for example at high latitudes (e.g., the Norwegian Sea) or the tropical Pacific Ocean, but even so, a minor temperature residual is likely to remain.

The sea level history inferred from the deep-sea oxygen isotopes correlates fairly well with that preserved by the uplifted coral reefs over their common time span, although differences remain in estimates of sea level elevation and exact dating (fig. 4.4). Starting roughly 140,000 years ago, sea level climbed rapidly from a low of minus 130–140 meters (426–459 feet) and reached 5–6 meters or more (16–20 feet) above present ocean level by 130,000–127,000 years ago.[13] After this interglacial highstand, sea level dropped, but not as low as before. Several more oscillations followed, in which the successive lowstands plunged deeper and deeper. Finally, around 21,000 years ago, the last glaciation reached its maximal extent, and sea level dropped by 120–130 meters (394–426 feet).[14] The last ice age ended with another rapid rise in sea level, reaching close to present levels by 7,000 years ago.

Similar patterns recurred roughly every 100,000 years over at least the last five glacial-interglacial cycles[15] in a "sawtooth" sequence (fig. 4.4). They likely prevailed over the last 800,000 years, but sea level elevations are less accurate that far back in time. Most sea level highstands generally remained below or near present levels and exceeded the present ~125,000 years ago and possibly ~400,000 years ago. Succeeding oscillations dipped lower and lower at roughly 20,000-year intervals over ~90,000 years, until attaining the deepest point in the cycle, roughly 120–140 meters (394–459 feet) below present levels. Sea level then rose rapidly from minimum to maximum, within approximately 10,000 years.

The Ice Core Time Capsules

The asymmetric sea level cycles imprinted in the ocean sediments and fossil reefs are echoed in the climate archives of ice cores from Greenland and

Antarctic. The ice cores act as time capsules, providing an amazingly detailed record of past climate history, almost layer by layer, year by year. The Greenland ice cores reach back ~120,000 years. The Antarctic ice core record, such as from the EPICA DOME C, stretches back 800,000 years! The ice seals information on past local temperature, composition of atmospheric gases, dust levels, and continental ice volume, among other climate proxies (fig. 4.5).[16] The variations in oxygen and hydrogen isotope ratios in polar ice largely depend on local air temperatures where the precipitation develops, although other factors such as season, moisture sources, topography, and clouds also affect these ratios. Despite these complexities, similar isotope patterns from different cores suggest that these isotopes are registering a common temperature signal and are therefore valid paleotemperature proxies.

Tiny air bubbles trapped in the ice enclose atmospheric greenhouse gases, such as carbon dioxide, methane, nitrous oxide. The greenhouse gases oscillated in sync with temperature, rising during the interglacials and falling during the glacials.[17] Isotopes of oxygen gas also trapped in the bubbles (as

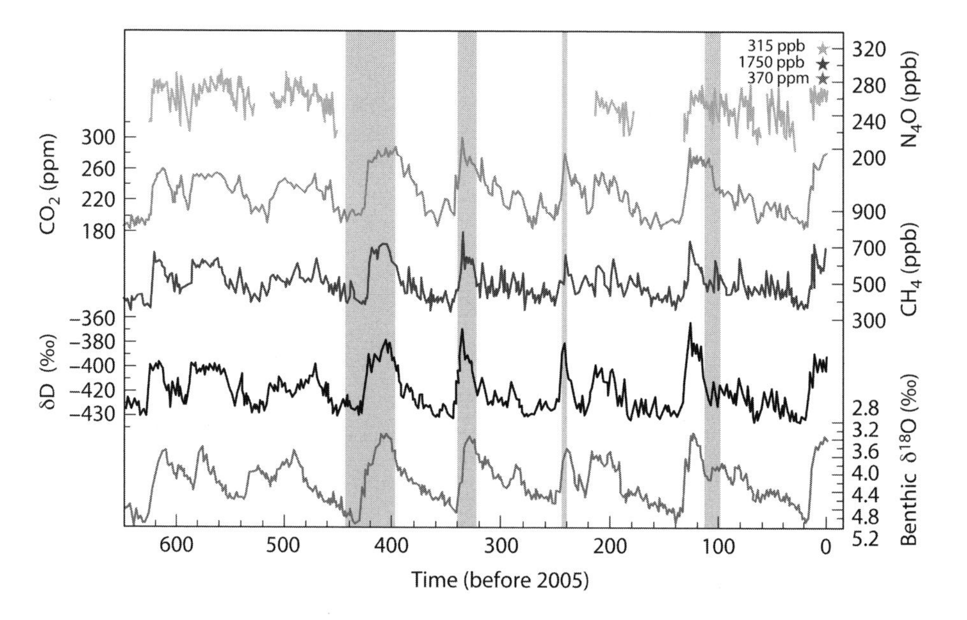

Figure 4.5 Greenhouse gas variations and deuterium variations (a proxy for local air temperature) from the EPICA ice core, Antarctica, compared with marine benthic oxygen isotope data for the last six glacial cycles. (IPCC, 2007a; Climate Change 2007: The Physical Science Basis, Working Group I Contribution to the Fourth Assessment Report of the Intergovernmental Panel on Climate Change, Figure 6.3; Cambridge University Press.)

distinct from the oxygen of the ice itself) varied along with the climate cycles. The atmospheric oxygen had originally been produced by marine photosynthesizing organisms, which presumably inherited their isotopic composition from seawater, which varies mainly with ice volume. The atmospheric oxygen ultimately found its way to the poles, where it was incorporated into the ice. The process is schematically illustrated below:

Oxygen isotope ratio (seawater) ⇒ photosynthesis (marine organisms) ⇒ oxygen isotope ratio (atmosphere) ⇒ precipitation (polar ice) ⇒ oxygen isotope ratio (ice bubbles).

In principle, the atmospheric oxygen could serve as an alternative proxy for ice volume (hence sea level).[18] However, the isotopic fractionation of molecular oxygen is affected by a number of hydrological and ecological processes. Photosynthesis and respiration of plants on land and in the ocean alter the isotopic composition of oxygen, so that atmospheric oxygen isotope ratios differ from those of seawater by 23.5 parts per thousand. This and other factors complicate the interpretation of atmospheric oxygen isotope changes derived from continental ice volume or sea level. Therefore, the deep-sea foram isotope data provide a better archive of long-term sea level change.

Dust levels in the ice also varied inversely with past temperatures. During the frigid glacial periods, many parts of the Earth grew more arid. The vegetation cover shrank and deserts expanded. Winds also gained strength and were able to lift more dust off the ground. Wind-borne dust, as during the Dust Bowl of the 1930s, swept across large swaths of North America (also central Asia) and eventually deposited on the ice in Greenland and Antarctica. The situation reversed during the wetter and warmer interglacials, when the rains returned, the vegetation recovered, and the polar ice was clear and nearly dust-free.

Major sea level oscillations recorded by oceanic benthic forams correspond remarkably well to the history of temperature and greenhouse gas changes stored in the ice cores (fig. 4.5). The 100,000-year cycles dominated both sets of archives for at least the last 800,000 years, interspersed with a shorter sawtooth pattern of diminishing temperatures, greenhouse gas concentrations, and lowered sea level, until the cycle reaches bottom, with maximal ice extent. The cycle terminates as temperatures, greenhouse gases, and sea level rise rapidly. However, the thaws and high sea level last only 10,000–30,000 years—just a small fraction of each glacial-interglacial cycle.

What do these cycles mean? What light can they shed on the causes of past climate and sea level changes? We now turn to the subject of unraveling the riddle of the cycles.

ASTRONOMICAL CYCLES AND THE ICE AGES

The French mathematician J. Adhémar (1797–1862) was among the first to propose a connection between the ice ages and the Earth's orbital cycles. He observed that the gravitational pull of the Sun and Moon acted on the Earth's equatorial bulge to cause **precession of the axis**—a wobble like that of a spinning top—which repeats every 26,000 years. But more importantly for climate change, the Earth's elliptical orbit swings around the Sun over a 22,000-year cycle, during the course of which the position of perihelion (shortest distance between the Earth and the Sun) changes with respect to the seasonal equinoxes and solstices. This cycle is known as the **precession of the equinoxes**. For example, the Northern Hemisphere winter solstice (December 21) occurs when the Earth's rotational axis points directly away from the Sun, creating the most hours of darkness. Today, this closely coincides with perihelion (January 3) (fig. 4.6a). But 11,000 years ago, the Northern Hemisphere winter solstice occurred instead at aphelion, at the time of the longest Earth-Sun distance (fig. 4.6b), while 22,000 years ago, the Earth was in today's orbital position. Adhémar reasoned that ice ages developed whenever the winter hemisphere, with its most total hours of darkness, dovetailed with the Earth's greatest distance from the Sun.[19] However, his theory predicted an ice age roughly 11,000 years ago, when actually the ice age was rapidly ending, and 22,000 years ago, when the Earth's orbit was like it is today, our planet was still in the grip of an ice age!

James Croll (1821–1890), a Scottish geologist, also emphasized the importance of changes in the shape of the Earth's orbit. The astronomer Johannes Kepler originally demonstrated that the Earth's orbit is elliptical rather than circular. However, Croll determined that the orbital eccentricity, or degree of departure from a true circle, varied over a 100,000-year period, becoming more or less circular over this period (the **eccentricity** cycle). (Eccentricity also varies over a longer, 400,000-year cycle). Although eccentricity barely affects the total amount of heat received from the Sun, Croll reasoned that even a small reduction in the amount of winter sunlight would allow more snow to accumulate, and furthermore, the snow's brightness would reflect more heat back to space, causing additional cooling (a positive feedback effect). In his view, the winter snow cover, combined with the reduction in solar radiation when the winter hemisphere was farthest from the Sun (precession), would trigger an ice age.[20] During periods of low eccentricity (or nearly circular orbit), winters would be neither unusually cold nor unusually warm. However, at times of high eccentricity, colder than normal winters develop when the winter solstice and greatest distance from the Sun coincide with the farthest orbital elongation, and conversely, much warmer winters result

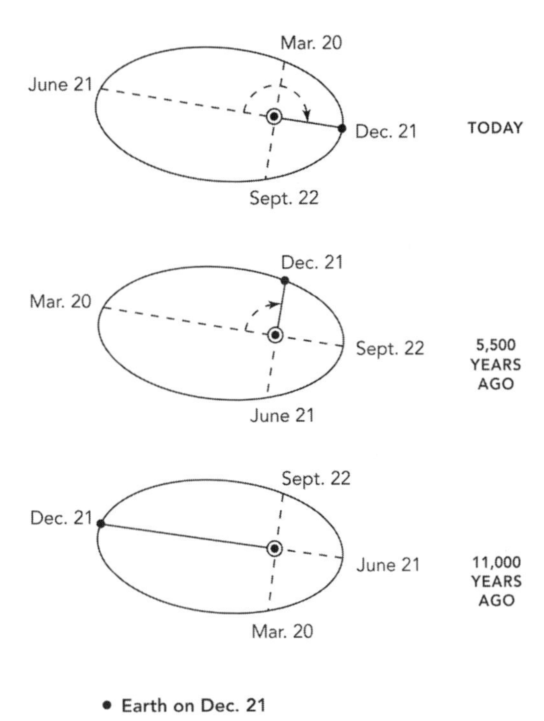

Figure 4.6 Precession of the equinoxes. Top: At present, perihelion (closest approach to Sun) nearly coincides with the Northern Hemisphere winter solstice. Bottom: However, 11,000 years ago, perihelion occurred at the summer solstice. (John Imbrie, used with permission.)

when the winter solstice lines up with perihelion at the short side of the orbit. Thus, his theory combines both eccentricity and precessional cycles.

Interestingly, Croll sensed that the subtle changes in Earth's orbital geometry were insufficient to explain the accumulating geologic evidence of former ice ages, a problem revisited by more recent scientists. He therefore suggested that changes in poleward ocean heat transport resulting from the orbital cycles may have magnified their climatic effects.[21]

The astronomical theory of the ice ages was further developed by Milutin Milankovitch (1879–1958), a Serbian geophysicist and mathematician, while in wartime captivity.[22] Milankovitch realized that the critical factor in initiating Northern Hemisphere glaciation was the reduction in the amount of sunlight during the summer months at high latitudes.[23] Arctic winters are so cold that even during an interglacial like today, snow can accumulate. On the other hand, the net amount of summer melting versus winter accumulation determines the ability of a glacier to build up and grow. Thus, in order to have an ice age, high-latitude summers must remain cold enough to prevent the previous winter's snow from totally melting. Milankovitch reasoned that the tilt of the Earth's axis (the **obliquity**)—the angle between the rotational axis and the perpendicular to the plane of the orbit (the ecliptic)—played an important role in initiating an ice age (fig. 4.7). Today, this angle is 23.5°, but

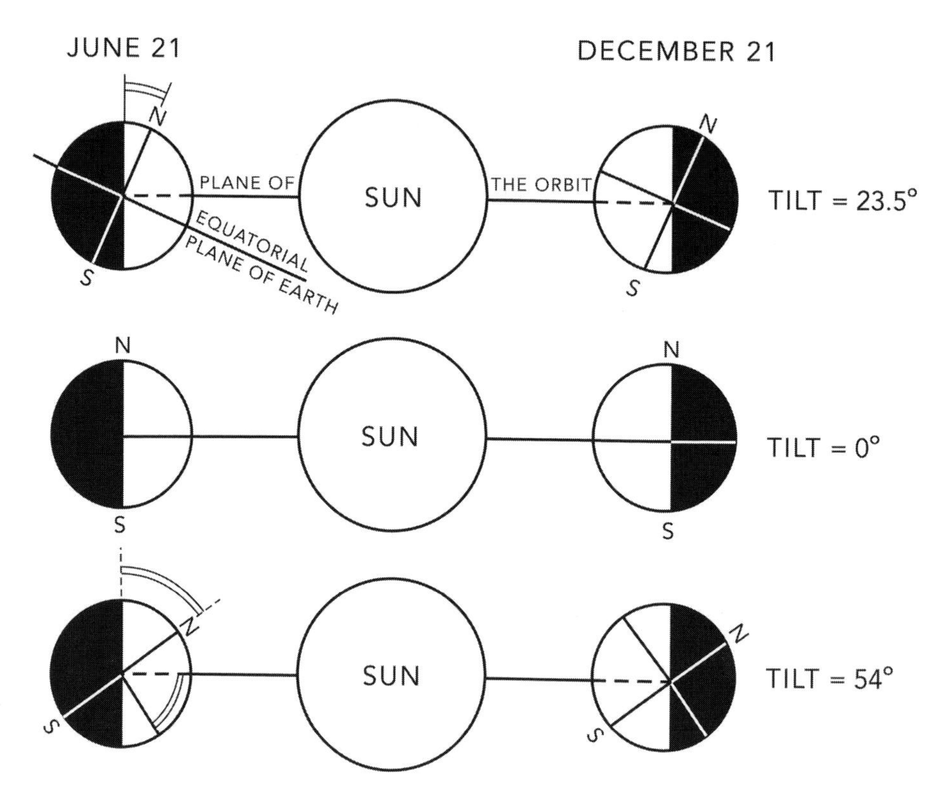

Figure 4.7 Effect of the Earth's tilt (obliquity) on the latitudinal distribution of sunlight on the Earth's surface. Top: Present-day tilt angle (23.5°). Middle: At very low tilt angles (0°), sunlight is evenly distributed geographically, favoring an ice age. Bottom: A very high tilt angle (54°) creates hot polar summers, favoring an interglacial. (John Imbrie, used with permission.)

it varies between 22° and 24.5° over a 41,000-year cycle. The greater the tilt, the more incoming solar radiation occurs at high latitudes during the summer (when the Earth's axis points more directly at the Sun) and more snow/ice can melt. Six months later, the winters get cold, but as we have seen, once water freezes, it doesn't really matter how much colder it gets. Conversely, a smaller tilt angle means cooler summers and milder winters (but never mild enough to completely melt all the snow or ice). According to Milankovitch, glaciation is triggered by a low tilt angle combined with aphelion at Northern Hemisphere summer solstice (i.e., cooler summers). The influence of obliquity is stronger at higher latitudes and therefore, the amount of solar radiation at 65°N is considered critical in starting (or, conversely, ending) a glacial cycle. The role of the Northern Hemisphere is more significant than that of the Southern Hemisphere, because of the greater land area over which ice sheets can spread or shrink. Precession, on the other hand, which governs the timing of perihelion or aphelion with respect to the seasons, exerts a stronger influence at low to mid latitudes.

Eccentricity variations produce only minor changes in the total solar radiation input. But at times when the Earth's path around the Sun is more elongated than usual, and obliquity and precession are also favorably aligned, it could give an extra nudge toward initiating an ice age. This is outlined schematically below.

How to start an ice age: Low Northern Hemisphere tilt angle (obliquity) + aphelion at June 21 (precession) + highly elongated orbit (eccentricity)

How to end an ice age: High Northern Hemisphere tilt angle + perihelion at June 21 + highly elongated orbit

For many years the astronomical theory was met with skepticism, largely because adequate geological evidence was lacking. But opinions shifted in the 1960s, as a result of deep-sea drilling into ocean sediments. In 1976 James Hays and his colleagues[24] clearly demonstrated the presence of the astronomical cycles in the oxygen isotope ratios of planktonic foraminifera and in abundances of radiolarians (tiny marine protozoans) from ocean sediments spanning a 400,000-year period. The dominant cyclical peaks occurred at periods close to 100,000, 42,000, and 23,000 years. These correspond to the eccentricity, obliquity, and precession cycles, respectively. The big surprise was the apparent dominance of the eccentricity cycle in the marine data, relative to the other two orbital components. Inasmuch as the eccentricity cycle contributes only a minor portion of the overall changes in the Sun's radiation, scientists have therefore long been puzzled by its unexpected strong influence in the paleoclimate record, which now extends to at least the last 800,000 years. This paradox has generated a debate that is still going on.

Clearly, other processes must be amplifying these orbital effects through various feedback mechanisms.[25] The initial development of a bright white reflecting ice sheet would allow more of the Sun's energy to be reflected back to space, promoting more cooling and more ice accumulation (the ice-**albedo** feedback). Once the ice sheet (particularly in North America) grew large enough to affect atmospheric circulation, storms would have been deflected farther south and the warmth carried by the Gulf Stream could no longer have penetrated as far north as before. This led only to further cooling and ice expansion.

The discovery that greenhouse gases trapped in polar ice cores vary in tune with the ice age cycles demonstrates their importance in amplifying the orbital effects. Biological processes largely control atmospheric levels of carbon dioxide and methane on timescales of millions of years or less. Even slight changes in sunlight and warmth levels would affect these processes, which generally become more active as temperatures rise and are slowed down as the Earth cools. At more favorable orbital configurations, Northern Hemisphere summers received more and more sunlight, initiating the end of an ice age. Temperatures started to rise, closely followed by increasing levels of CO_2 (and CH_4) in the atmosphere, which reinforced the warming.[26] The massive ice sheets also began to melt and disintegrate, leading to a rise in sea level and changes in ocean circulation that further affected climate.

While the exact reason for the changes in these two greenhouse gases is not fully understood, it is generally accepted that they involve biological processes. The variations in CO_2 likely come from the ocean, involving complex feedbacks between marine biological activity, changes in marine carbonate chemistry, and in ocean circulation.[27] On the other hand, atmospheric methane is chiefly generated in tropical and northern wetlands. The tropical wetlands are strongly influenced by the precessional cycle. As this cycle reaches a peak, summer warmth and rainfall, especially the monsoons, are strengthened and the wetlands expand.

A poorly understood issue is why the orbital cycles appear to have changed, roughly 800,000 years ago, from one dominated by the 41,000-year obliquity cycle to one dominated by the 100,000-year eccentricity cycle—which should have been the weakest of the three. While eccentricity alone weakly affects the total amount of incoming sunlight, it can magnify the impacts of precession and obliquity, especially when enhanced by greenhouse gas forcing. Although the cause of the switchover is still actively debated, one suggestion is that glaciers had grown sufficiently thick and sluggish during the later period, such that they could no longer melt as rapidly during the shorter obliquity cycles and therefore responded more effectively to the longer-lasting eccentricity cycles.[28] Deep-sea oxygen isotope ratios suggest that after the mid-Pleistocene revolution, the thickened ice sheets could lower sea level by at least 50–95 meters.[29]

A somewhat different view is that by 1 million years ago, North American ice sheets had expanded and reached a critical volume sufficient to destabilize them. Whenever the Sun's radiation and air temperatures reached a peak, the ice sheets rapidly thinned, sliding over a wet, lubricating layer at their base.[30] The ice sheets would then gradually build up over 100,000 years before reaching the critical threshold of instability once more. Regardless, the Earth's climate system apparently reacts in a nonlinear fashion (i.e., in complex, non-proportional ways) to even small variations in the amount of solar radiation caused by the changing orbital parameters.

Another intriguing observation is that the last four glacial cycles (since 450,000 years ago) seem to have larger amplitudes than the four previous cycles (between ~800,000 and 450,000 years ago). Maximum temperatures and atmospheric CO_2 levels registered in both the Antarctic ice cores, as well as in deep-sea sediments, are higher in the more recent period than in the older one, although minimum temperatures have remained roughly the same.[31] Thus the overall contrast between glacial and interglacial periods has widened over time. These changes may be part of a longer-term eccentricity cycle lasting 400,000 years.

Our picture of variations in sea level and climate becomes somewhat clearer as we approach the present. The changes in sea level during the last glacial cycle of the past 130,000 years have been studied more extensively than those of earlier cycles.

THE LAST GLACIAL CYCLE

A Balmy Thaw and High Waters: The Last Interglacial

Before the final plunge into the last ice age, the world was a few degrees warmer and sea level several meters higher than today, around 130,000–116,000 years ago. Trees grew as far north as the southern part of Baffin Island in the Canadian Arctic, and Scandinavia was an island. Vast areas of northwestern Europe and the West Siberian Plain were underwater. A more humid climate prevailed in the Near East then. Some scientists speculate that the earliest modern humans took advantage of this more benign climate to migrate out of Africa.[32] By 130,000–100,000 years ago, our human ancestors had already spread from Africa into the Levant, most likely via the Sinai and Negev Deserts, which were then more vegetated and watered.[33] Subsequent waves of early humans later swept over Asia, reaching Australia by 50,000–45,000 years ago and Europe shortly thereafter.[34] The peopling of the Americas was among the latest chapters in the saga of human migrations. The timing and pathways of these migrations are not known with certainty, but several portions of the possible routes (especially crossing from Malaysia via Indonesia into Australia) involved sailing over water in canoes or on rafts, a task more easily accomplished at times of low sea level. Thus, past sea level oscillations may have affected the course of human history—a theme that will be revisited in chapter 5.[35]

The previous thaw between the ice ages, when the ice briefly relaxed its frigid grip, is called the last interglacial (also known as the Eemian in

Europe or the Sangamonian in North America),[36] lasting from approximately 130,000–116,000 years ago. As orbital alignments grew more favorable by 130,000–127,000 years ago, the Northern Hemisphere basked under more intense summer sunlight and Arctic temperatures rose some 3°C–5°C above present ones.[37] The last interglacial offers a close analog for conditions resulting from anticipated anthropogenic climate warming and sea level change. Many aspects of the world then differed little from now. Continents sat in their present positions; mountain ranges were as high as they are today, although their appearance was further molded by the subsequent ice age. The major elements of atmospheric and ocean circulation were similar to that of the current world. But it was the last time before the present that temperatures were as warm and sea levels stood higher than now.

During the last interglacial, ocean waters worldwide stood at least 4–6 meters (13–20 feet), perhaps as much as 6.6–9.4 meters (21.7–30.8 feet), above the present level.[38] One data compilation, including such features as wave-cut notches, raised beach terraces, changes in rock types (e.g., see chapter 2), from numerous worldwide localities shows two sea level highstands, the first reaching +2.5 meters above present levels between ~132,000 and 125,000 years ago and the second from 124,000 to 121,000 years ago, with an average rise of 3–4 meters. Sea level may have spurted up sharply by 6 or more meters briefly between 121,000 and 119,000 years ago, at an average rate of ~3–4.5 millimeters per year, but it fell rapidly soon thereafter.[39]

Another study, based on Red Sea foraminiferal oxygen isotopes, finds that the highest sea levels may have peaked somewhat later, at ~123,000 and again at ~121,500 years ago, dropping off sharply after 119,000 years ago.[40] Sea level may have climbed as fast as 2.5 meters (8.2 feet) per century. At an average rate of sea level rise of ~1.6 meters (5.2 feet) per century, a Greenland-sized ice sheet could have vanished within four centuries.[41]

In a computer model simulation, the warmth produced by the additional sunlight under the last interglacial's more favorable orbital lineup would have enabled the Greenland Ice Sheet and western Arctic ice fields to retreat enough to raise sea level by 2.2 meters (7.2 feet) within approximately 2,000 years. After another 1,000 years into the simulation, the meltdown of Greenland would generate a total of 3.4 meters (11.2 feet) rise.[42] An increase of 3.4 meters in 3,000 years corresponds to an average rate of sea level rise of 11.3 centimeters (4.4 inches) per century—a rate roughly comparable to that of the early 20th century. In the model, sea level would have risen worldwide up to 4–6 meters, at an average rate of 40–60 centimeters (16–24 inches) per century during the time of maximum insolation between 130,000 and 129,000 years ago.[43]

The modeling study implies that melting Antarctic ice probably added to the high water as well.[44] Portions of the margins of Antarctica may have been ice-free. Diatoms, single-celled photosynthesizing algae that live in the ocean or lakes, were found in marine sediments from the Ross Embayment. So was the isotope of beryllium, ^{10}Be. Since ^{10}Be is produced by the interaction of cosmic rays with the atmosphere, its presence in marine sediments means that the water must have been exposed to the atmosphere and therefore not ice-covered. Ice core data also show that East Antarctica had warmed by 2.5° to more than 5°C at that time.

A statistical analysis of global sea level data corrected for glacial isostasy as well as gravitational and rotational effects suggests that sea levels were even higher—6.6–9.4 meters (21.7–30.8 feet), about 125,000 years ago.[45]

Since temperatures due to greenhouse gas climate warming could approach last interglacial levels by the end of this century, could the Greenland and Antarctic Ice Sheets dwindle as much as and as rapidly as before? These important questions will be addressed in subsequent chapters.

Descent into the Last Ice Age

Once the mild reprieve began to fade, by 116,000 years ago, the slow descent that culminated in the last ice age was well under way. By that time, Northern Hemisphere summer insolation had reached another minimum. Sea level dropped quickly and the northern ice sheets extended their territory. As in the earlier glacial-interglacial cycles, the initial cool episode formed part of a much longer cycle of sea level, temperature, and greenhouse gas oscillations. Succeeding sea level highs and lows plunged further and further until reaching the coldest part of the last ice age, 29,000–20,000 years ago (fig. 4.8).

The combined evidence from raised coral reefs, marine terraces, and the deep-sea oxygen isotope ratios indicates that sea level increased again, around 103,000–100,000 and 83,000–80,000 years ago, yet stayed under 20–30 meters below present levels,[46] well below the last interglacial highstand. During the cool spells, sea level dropped ~40–60 meters. After the ~80,000 highstand, sea level plunged nearly 60 meters in less than 6,000 years—a volume of water roughly equivalent to that of the modern Antarctic Ice Sheet.[47] The world's oceans swung up and down several more times, until the mean water level finally dropped 120 meters below present and a full-fledged ice age gripped the Earth (fig. 4.8).

A careful examination of the past sea level record soon revealed the presence of much shorter cycles than those linked to the Earth's orbital variations. These quasi cycles, recurring on average every 500 to 4,000 years, had puzzled paleoclimatologists for a long time. They appeared in the Greenland ice cores as well as in the ocean sediment archives. The stronger ones, in

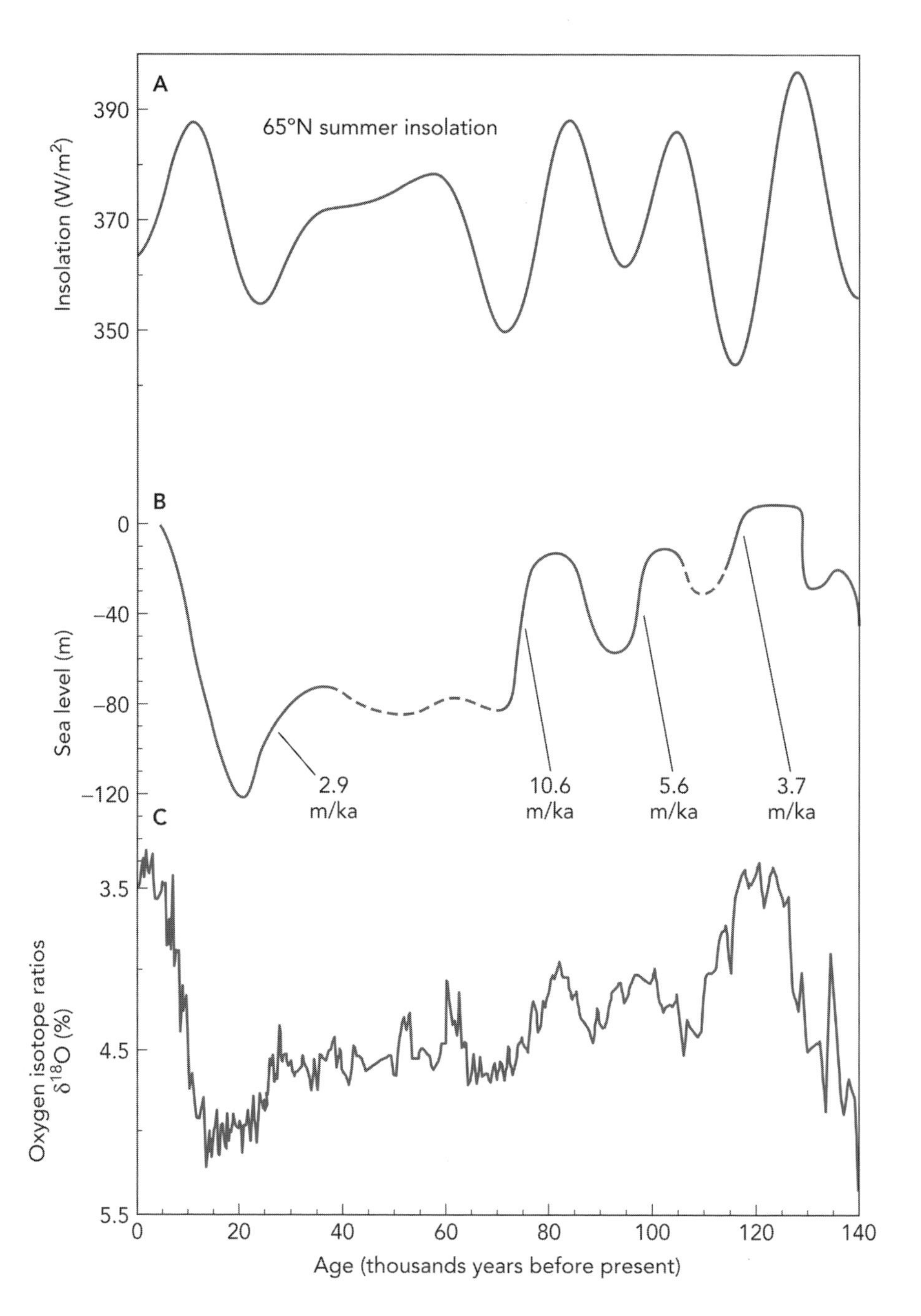

Figure 4.8 Sea level variations over the last glacial cycle: (a) summer insolation at 65 °N in watts per square meter; (b) sea level record from raised coral reefs, Papua New Guinea and Barbados. Numbers represent average rates of sea level fall in meters per thousand years for each drop; (c) benthic oxygen isotope records from Carnegie Ridge, eastern equatorial Pacific, off the Ecuador coast. (Adapted from Cutler et al., 2003, fig. 5.)

particular, seemed to be associated with ice-rafted debris (IRD) in North Atlantic sediments. These pebbles and rock fragments had been carried on drifting iceberg rafts, which then had melted and shed their loads into the ocean. The IRDs imply an expansion of northern ice sheets, hence lower sea level and colder climates. What caused these rapid climate swings—and could they recur?

Rapid Climate Swings

Six times within the last 70,000 years, huge armadas of icebergs have dumped their rocky loads into the North Atlantic Ocean. Four of these deposits have been traced to sources near the Hudson Strait, Canada, and the other two to European sources. The thick, sandy layers shed by the floating ice rafts were deposited during recurrent cold periods spaced roughly 7,000 years apart, on average, and lasting around 500 years.[48] The IRDs appeared during the coldest phases, which ended with a sudden warming. These events are known as **Heinrich events**, after their discoverer, Helmut Heinrich, a German marine geologist. The timing of Heinrich events matches remarkably well the climate variations recorded in the Greenland ice cores and also in isotopic and geochemical records of marine sediments.

The melting of the icebergs shed by collapsing ice sheets freshened the ocean water, disrupting the thermohaline circulation and raising sea level 10–20 meters within decades to a century.[49] Greenland temperatures dropped another 3°C–6°C from the already cold and icy conditions, and sea surface temperatures in the North Atlantic declined by another 1°C–2°C.[50]

What could have provoked these dramatic climate events? One theory proposes that growing ice sheets gradually reached a critical threshold, whereupon some mix of subterranean or external heat and friction at the base of the sliding ice sheet melted its underside. The thin layer of water at the base enabled the ice mass to slide much more easily, eventually destabilizing it and breaking off multitudes of icebergs as the mouth of the glacier reached the sea (fig. 4.9). This is the so-called **binge-purge theory**.

A variant hypothesis holds that large lakes built up under the ice until the trapped water burst forth in enormous outpourings. Yet another view holds that the sea level rise caused by freshwater inputs may have lifted the floating ice shelves surrounding the major ice sheets that reached the sea, thereby weakening them and breaking them apart. Without the buttressing ice shelves to restrain them, the continental ice sheets would have surged forward, creating even more icebergs. As North Atlantic seawater freshened, its reduced salinity would have hindered the sinking of North Atlantic

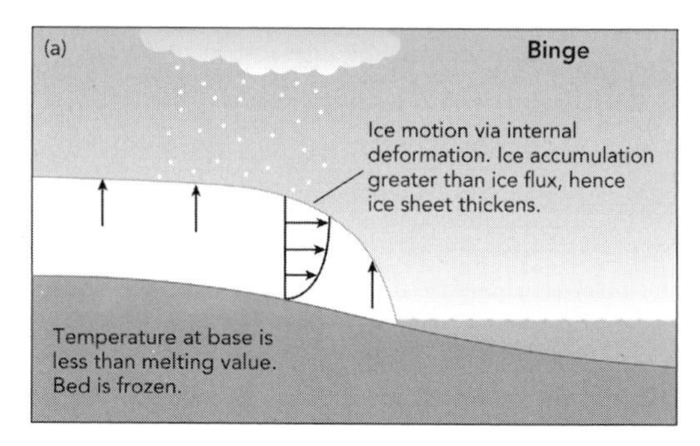

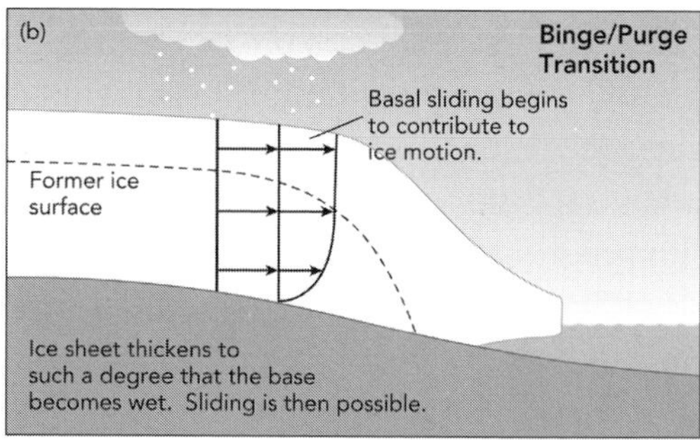

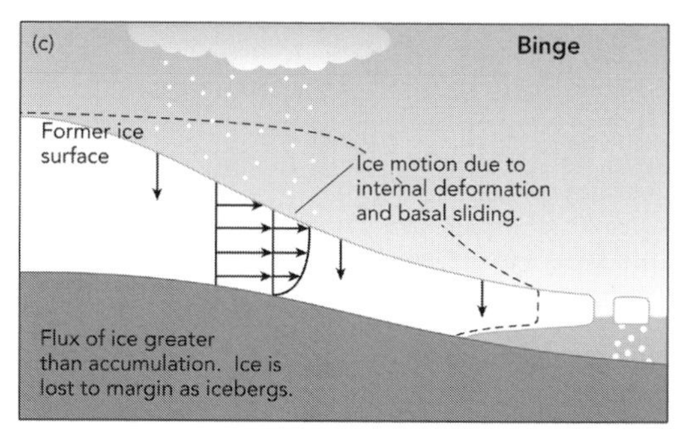

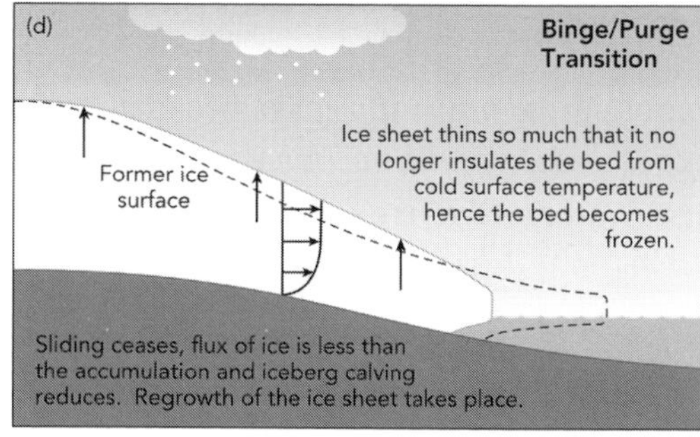

Figure 4.9 Binge-purge cycle of an ice sheet. Binge phase (top): Ice accumulation exceeds summer melting and the ice sheet thickens. Transitional phase (upper middle): The base of the ice sheet begins to melt. Purge phase (lower middle): Ice slides over a wet bottom layer and the ice sheet thins. Transition to binge phase (bottom): The ice sheet has thinned to such an extent that the base freezes to the bedrock and the accumulation phase begins anew. (M. J. Siegert, 2009, "Binge-Purge Cycles of Ice Sheet Dynamics," in V. Gornitz, ed., *Encyclopedia of Paleoclimatology and Ancient Environments*, fig. B7, p. 95, with kind permission from Springer Science + Business Media B.V.)

Deep Water (NADW), weakened the ocean conveyor system, and cooled the Northern Hemisphere (see fig. 1.4).

Even more frequent climate oscillations than the half dozen Heinrich events punctuated the last glacial cycle. These **Dansgaard-Oeschger (D-O) events**[51] are most evident in the oxygen isotope records of the Greenland ice cores that record atmospheric temperature. They often left ice-rafted debris layers in ocean sediments, although much less marked than those of Heinrich events. D-O events tended to recur roughly every 500–4,000 years, producing abrupt warmings within several decades. These events correlate with other climate proxies from different parts of the world that record high millennial variability.[52] Heinrich events occur after several D-O cycles have cooled the northern oceans to a critical threshold, and they may in fact just be super D-O events. As with Heinrich events, the D-O events may be caused by "binge-purge" cycles due to a dynamic instability of massive ice sheets, or by a seesaw-like oscillation of the deep ocean water circulation. The enormous influx of melting icebergs or freshwater discharge from glacial lakes during a Heinrich event freshens and cools the North Atlantic, causing the NADW to break down (fig. 4.10a). A tongue of even colder, denser Antarctic Bottom Water then creeps northward along the ocean depths (fig. 4.10b). Meanwhile, heat is conveyed southward across the equator, warming the Southern Hemisphere but leaving the Northern Hemisphere cold. After the iceberg invasions cease and the NADW is restored, the Antarctic Bottom Water weakens and heat now crosses the equator northward, warming the Northern Hemisphere but cooling the Southern Hemisphere (fig. 4.10b). During "normal" glacial conditions, the two bottom water currents are approximately in balance.[53] This deep ocean current seesaw (also called the **bipolar seesaw**) may explain why millennial climate oscillations in both hemispheres appear to be out of phase—i.e., one hemisphere warms up while the other one cools.

Yet another hypothesis links the apparent 1,500-year quasi cycle to changes in solar activity.[54] The most familiar solar cycle is the 11-year sunspot cycle, during which the number of visible sunspots increases from almost 0 to over 200 (more accurately, the full cycle, including the return from peak sunspot activity to its minimum, lasts ~22 years). Sunspots are dark areas on the Sun's surface associated with locally strong magnetic activity. At times of low solar activity, cosmic rays reach the Earth's surface unhindered, and cosmogenic isotopes, such as ^{10}Be or ^{14}C, become more abundant. One such period of low sunspot activity was the Maunder Minimum (1645–1715), which occurred during the coldest part of the **Little Ice Age**, leading to the suggestion that the Sun may exert an influence on the Earth's climate.

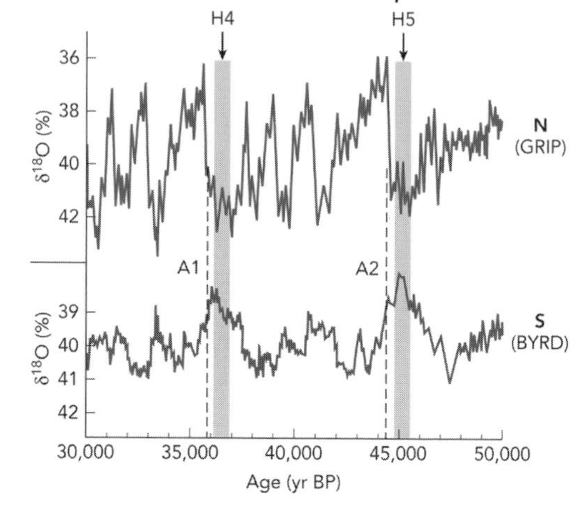

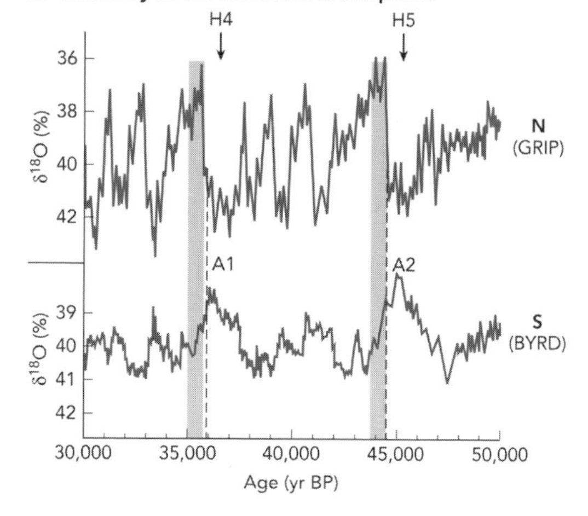

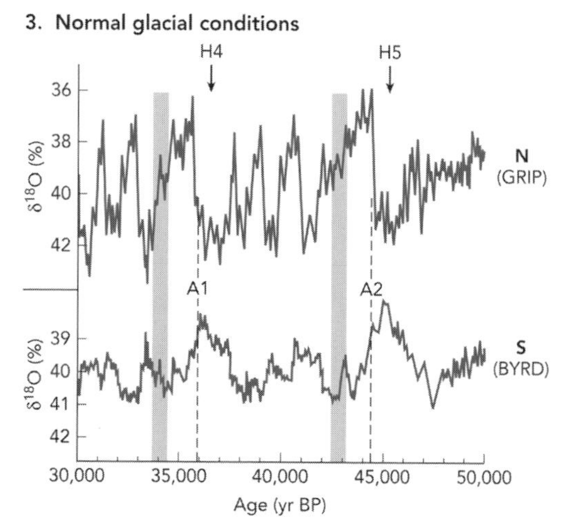

Figure 4.10a (Left): Dansgaard-Oeschger events and Heinrich events (H5 and H4) in oxygen isotope records from the GRIP, Greenland (N) and Byrd, Antarctica (S) ice cores. Vertical bars show three different time slices, starting with a Heinrich event (top), abrupt Northern Hemisphere climate recovery (middle), and intervening glacial conditions (bottom). Note the asymmetry in inter-hemispheric climate response.

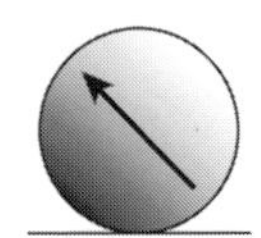

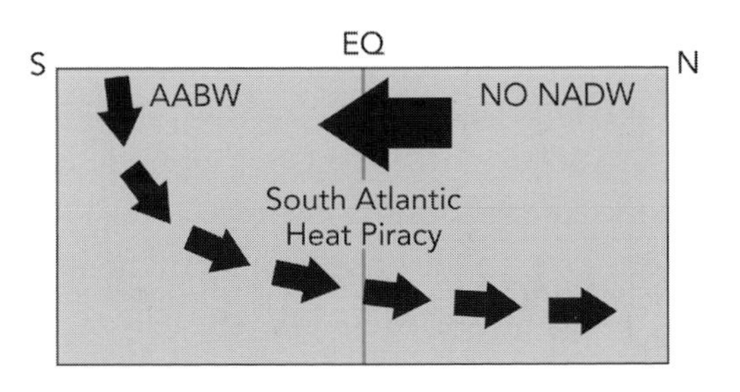

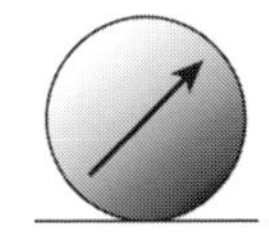

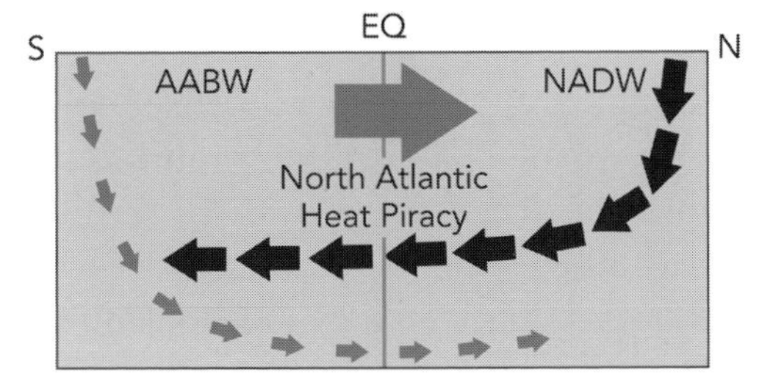

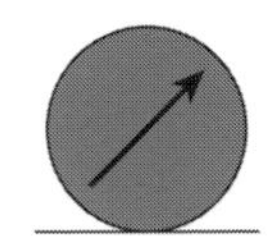

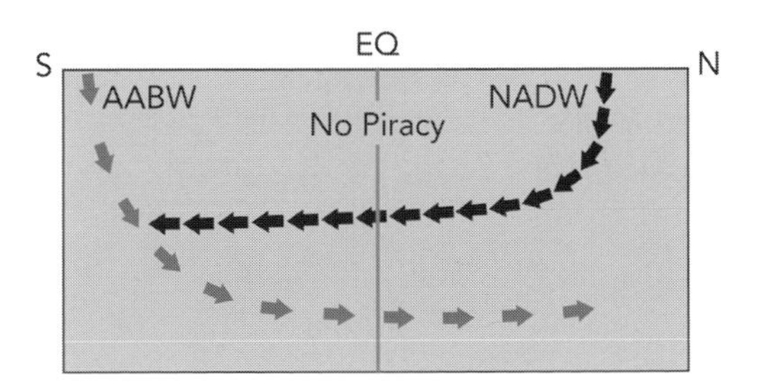

Figure 4.10b (Right): Schematic diagram illustrating the bipolar seesaw. During Heinrich event (top), suppression of NADW and heat transfer to South Atlantic. After Heinrich event (middle), restoration of NADW and heat transfer to North Atlantic. Intermediate stage (bottom)—roughly equal inter-hemispheric heat transfer. (Modified from M. Maslin, 2009, "Quaternary Climate Transitions and Cycles," in V. Gornitz, ed., *Encyclopedia of Paleoclimatology and Ancient Environments*, fig. Q8, p. 851, with kind permission from Springer Science + Business Media B.V.)

The ^{10}Be levels in Greenland ice cores and ^{14}C abundances in tree rings (markers for solar activity) correspond closely to oxygen isotope data from marine sediments over the last 12,000 years. Both data sets seem to vary over a 1,500-year cycle. Because of the alleged similarity in recurrences of the older D-O events and solar cycles, some have linked the two.[55] However, the variations in the amount of the Sun's energy due to the sunspot cycle, or the longer 1,500-year cycles, are far too weak to affect our climate, unless other amplifying factors are at work. Fluctuations in the intensity of the ENSO or in the upper atmosphere ozone layer may be potential amplifiers, at least for the shorter sunspot cycles. At any rate, the exact causes of the rapid climate changes of the last glacial cycle remain elusive.

Do these rapid climate swings also appear in the sea level record? Evidence from raised marine terraces in Papua New Guinea and oxygen isotope data from the Red Sea suggest that indeed, sea level fluctuations also correlate with these abrupt climate changes. For example, the series of raised coral terraces on the Huon Peninsula, resembling a flight of stairs, preserve information on sea level fluctuations between 60,000 and 30,000 years ago. These terraces were formed during sea level highstands, after the ocean surface had climbed by 10–15 meters. A number of the terraces correspond in age to Heinrich events marked by North Atlantic IRDs.[56] The benthic foram oxygen isotope data also display similar patterns. However, these archives show slower rises in sea level (decrease in ice volume) and deep-sea temperatures than the abrupt atmospheric warming seen in the Greenland ice cores. After the influx of icebergs had ceased, the oceans and remaining ice sheets reacted more slowly to the warming trend imprinted in the ice cores.

The Red Sea furnishes further information on rapid sea level fluctuations. It is especially sensitive to sea level variations because its only outlet to the open ocean is through the narrow Strait of Bab al Mandab in the south. The area is also hot and dry, so evaporation rates are high. Therefore, at times of low sea level, the salinity of the Red Sea increases, and conversely when sea level rises, salinity drops. Planktonic forams living in the Red Sea are quite sensitive to the local variations in salinity resulting from global sea level change. The variations in oxygen isotope ratios of the forams largely reflect changes in sea level of up to 30 meters (98 feet). Red Sea levels between 65,000 and 35,000 years ago and oxygen isotope records from both the Greenland and the Antarctic ice cores fluctuate similarly. However, the timing of sea level changes in the Red Sea matches Antarctic temperature changes more closely than it does those of Greenland.[57] The warming of Antarctica preceded that of the Northern Hemisphere, and it could have been an early contributor to global sea level rise during this period. On the other hand, the Heinrich events occurred at times of maximal Southern

Hemisphere warming (and maximal cooling in the north). This out-of-phase hemispheric temperature relationship probably stems from the bipolar seesaw mechanism described above.

The Last Glacial Maximum

New York City, like much of North America and northern Europe, lay blanketed under a massive ice sheet several miles thick at the climax of the last ice age, 23,000–19,000 years ago. Although the ice has long since vanished, traces of its former presence are still visible to the trained eye. A leisurely stroll through Central Park, a leafy oasis in the middle of bustling Manhattan, reveals a number of calling cards left by the long-gone glaciers. The sparse rocky outcrops of dark gray schist have been polished smooth, worn down by the grinding of the overriding ice sheets. Grooves or scratches (glacial striations) gouge the outcrop surfaces, created by rocks dragged along at the bottom of the ice. The outcrops on the side facing the oncoming glacier are more smoothly streamlined than on the jagged "downstream" side, where the ice plucked off chunks of rock. The resulting asymmetric mounds are known as **roches moutonnées** (Fr., sheep-like rock), because of their resemblance to the backs of sheep. The glaciers not only scoured the surface. Large boulders of exotic rocks—granite, diabase, sandstone—sit perched in precarious positions on outcroppings of the local schist bedrock (fig. 4.11). These glacial erratics were transported by the glaciers from across the Hudson River and upstate New York and dumped there once the ice melted. The ice sheet also left behind heaps of rocky debris that it had bulldozed, or sediment trapped within or on its top, as it retreated north. These ridges of poorly sorted sand and gravel now traverse the city across central Queens and Brooklyn and into neighboring Staten Island. They represent the **terminal moraine**, which marks the southernmost advance of the ice in this area.

The glaciers not only eroded or deposited material, but also changed the lay of the land, influencing local sea level to this very day! In New York City, at the southernmost edge of the long-vanished ice sheets, the land continues to sink nearly 1 millimeter (0.04 inches) per year due to the ongoing subsidence of the peripheral bulge (see chapter 2, fig. 2.6). How this land subsidence will affect the vulnerability of New York City (and other coastal cities at the periphery of former ice sheets) to future sea level rise will be explored further in chapter 9.

The Earth's uneven descent into the coldest phase of a frozen world began much earlier, however. Once the final pulse of relative warmth and higher sea level ended 70,000 years ago, the world slowly descended toward the frigid

Figure 4.11 Glacial erratic of granite on schist worn smooth by glaciers, Central Park, New York City (Photo: V. Gornitz.)

culmination of the last ice age (variously known as Wisconsinan in North America and Weichselian, or Würm, in Europe). In this frosty world, even the tropics were 2°C–3°C (4°F–5°F) colder than present, while the polar regions froze even further (e.g., average temperatures were ~9°C (16°F) lower in Antarctica and ~21°C (38°F) in Greenland).[58] The air became much colder, drier, and dustier. Tundra covered vast portions of the northern continents, grading into steppes across central Asia, China, and southern Europe, while tropical rain forests shrank and deserts expanded. In contrast, the western interior of North America was wetter than today, lake levels were higher, and scattered conifer forests grew in areas that are now shrubland or desert.

The exact timing and land covered by the ice sheets at their peak are still uncertain, but by 23,000–19,000 years ago the ice sheets had reached their greatest extent, lowering sea level by as much as 120–130 meters (394–427 feet). Data from now submerged corals in geologically stable areas, raised marine terraces at uplifted sites, and sediments from the Red Sea corroborate the timing and amount of the sea level drop.[59] At its maximum, the Laurentide Ice Sheet, which blanketed most of Canada and parts of the northern United States, formed a dome more than 4 kilometers (2.5 miles) thick, centered over Hudson Bay (fig. 4.12).[60] Melting of all this ice would

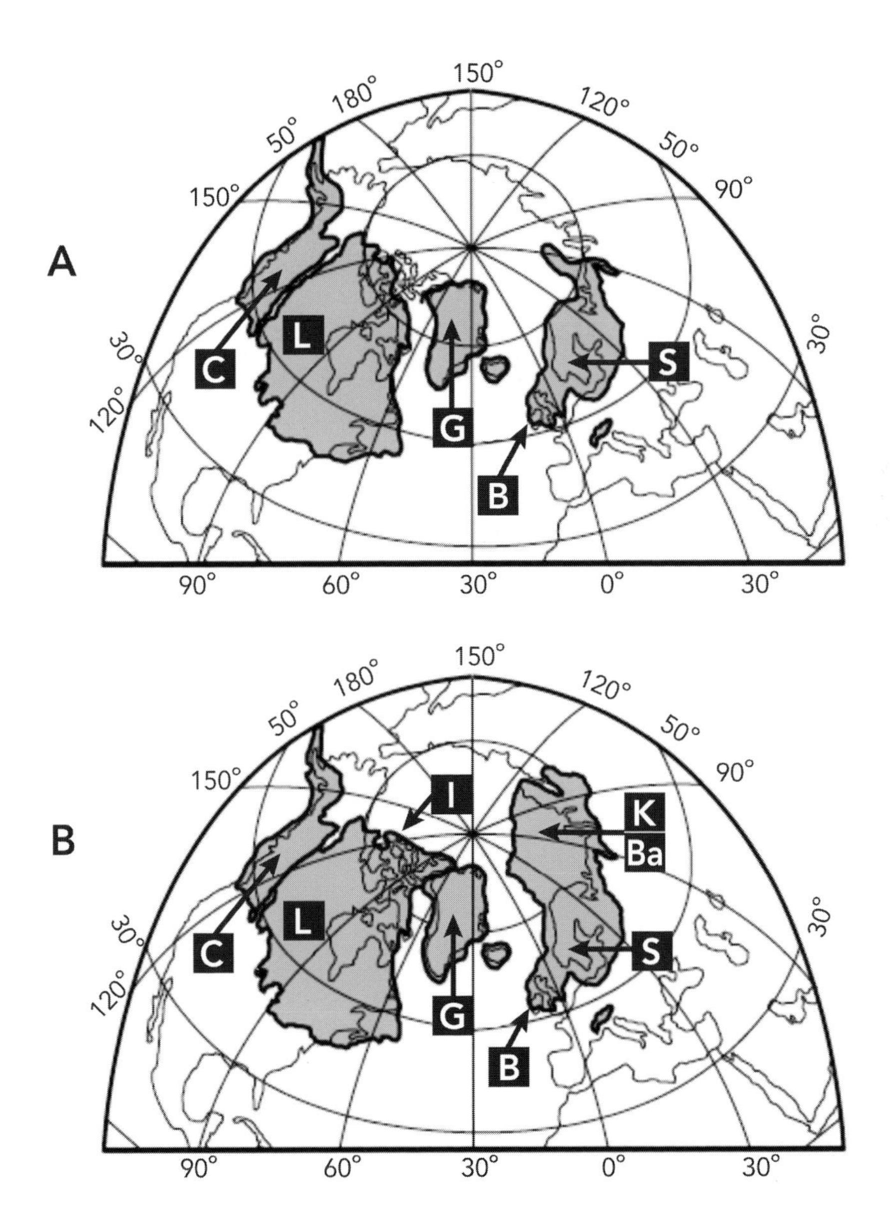

Figure 4.12 Two estimations of the extent of Northern Hemisphere ice sheets at the Last Glacial Maximum. Top: minimum reconstruction; bottom: maximum reconstruction. Ice sheets: L, Laurentide; C, Cordilleran; G, Greenland; B, British; S, Scandinavian; Ba, Barents Sea; and K, Kara Sea. (After Clark and Mix, 2002; D. Q. Bowen, 2009, "Last Glacial Maximum," in V. Gornitz, ed., *Encyclopedia of Paleoclimatology and Ancient Environments*, fig. L3, p. 494, with kind permission from Springer Science + Business Media B.V.)

elevate global sea level by 78–88 meters (256–289 feet). The Scandinavian Ice Sheet, overlying the Gulf of Bothnia, exceeded 3 kilometers (1.9 miles) in thickness.[61] Other, lesser ice sheets buried parts of the northern Cordilleras, northern Russia, Alaska, the British Isles, the Alps, and Iceland. Antarctica grew thick enough to reduce sea level by 14–18 meters (46–59 feet), as compared to its present volume, equivalent to only 61 meters (200 feet) of sea level rise if all the water were spread out evenly over the oceans. In fact, the total ice stored on all the ice sheets at the **Last Glacial Maximum**, if melted, could have raised global sea level by 178–197 meters (584–646 feet) as compared with only ~69 meters (226 feet) today.[62]

A LAST GLANCE BACK AT THE ICE AGES

Ice began to build up on Greenland by 3.0–2.7 million years ago, subsequently spreading repeatedly over parts of Canada during the coldest periods, marching to a 41,000-year beat. By a million years ago, the Earth had chilled enough that even fairly small changes in solar insolation could begin to disintegrate the bloated ice sheets within a fraction of the time it had taken them to grow in the first place. The dominant beat of the ice sheets had switched to the 100,000-year eccentricity cycle, with the 22,000-year precession and 41,000-year tilt cycles adding their overtones to the rhythm. Sea level closely followed the pattern of waxing and waning ice sheets. The past oscillations in sea level left their traces in coralline terraces now lofted high by land uplift and in the shells of tiny foraminifera at the bottom of the ocean.

In addition to the orbital cycles, other signs of even more rapid changes every few thousand years soon became apparent from ocean sediments and Greenland ice cores. Sudden climate warming and rising sea level took place, sometimes within a matter of several decades, contrasting with a much slower return to colder conditions. Although the exact cause of these rapid swings is still uncertain, growing evidence points to disruptions of the thermohaline circulation—the great conveyor belt—set off by discharges of freshwater or icebergs that then melted, reducing the salinity of the North Atlantic and preventing the formation of North Atlantic Deepwater (see chapter 1). While these changes were quite rapid by geologic standards, they were not exactly like the film *The Day After Tomorrow*, in which a greenhouse gas–induced climate warming initiated a sudden shutdown of NADW formation and a full-fledged ice age followed within a matter of days! Could they recur as rising temperatures melt more and more of Greenland, freshening northern ocean water? Later chapters will explore further the implications of future ocean responses.

After the successive ups and downs between 70,000 and 30,000 years ago, the Earth plunged into its climactic glacial cycle. At its peak, vast ice sheets covered not only Antarctica but much of North America, northern Europe and Russia. Sea level dipped 120–130 meters below present levels.

What finally loosened the grip of the last ice age? Starting 20,000 years ago, the summer solar insolation at 65°N slowly began to climb, as the Earth's axial tilt increased and the northern summer solstice swung closer and closer to perihelion. Initially, the warming was gradual, like the earliest days of spring, and the rise in sea level barely perceptible—a tiny trickle as the southernmost edges of the ice sheets began their northward retreat. By 15,000 years ago, as the northern summer sunlight gained more and more strength, the trickle had swelled into major torrents and the last big ice meltdown was well under way. Chapter 5 traces the history of the big meltdown—the global rise in sea level that followed the end of the last ice age.

Raven made a woman under the earth to have charge of the rise and fall of the tides. One time he wanted to learn about everything under the ocean and had this woman raise the water so that he could get there. He had it rise very slowly so the people had time to load their canoes and get into them. When the tide had lifted them up between the mountains they could see bears and other wild animals walking around on the still unsubmerged tops. Many of the bears swam up to them, and at that time those who had their dogs had good protection. Some people walled the tops of the mountains about and tied their canoes inside. They could not take much wood with them. . . . That was a very dangerous time. The people who survived could see trees swept up roots and all by the rush of waters, and large devilfish and creatures were carried up by it. When the tide began to fall, all the people followed it down, but the trees were gone and they had nothing to use as firewood, so they were destroyed by the cold. When Raven came back from under the earth . . . if he saw a person coming down, he would say, "Turn to a stone just where you are," and it did so. After that the sea went down so far that it was dry everywhere.

—TLINGIT LEGEND, SWANTON (1909)[1]

5 The Great Ice Meltdown and Rising Seas

THE TWILIGHT OF THE ICE AGE

At the peak of the last ice age, the thick ice sheets that blanketed North America, northern Europe, and adjacent Russia, and the tall mountain ranges stored enough water to lower sea level by at least 120 meters (394 feet) relative to the present. Before the great flood, a land bridge connected Alaska and Siberia across the Bering Strait and a now-drowned subcontinent encompassed the seas between Malaysia, Indonesia, and Borneo. One could walk from China to Japan and across the English Channel! Continental shelves lay exposed to sunlight. Since the end of the last ice age, the invading seas have drowned 5–8.5 percent of the Earth's surface. The great ice meltdown evolved over a 10,000–12,000–year time span, in gradual increases and sudden spurts.

Starting 20,000 years ago, as positions of axial tilt and eccentricity favored an interglacial, northern-latitude summer sunlight gradually intensified bit by bit and the Earth slowly began to warm. Barely perceptible at first, the warming rapidly strengthened, and persisted for several millennia before abruptly reversing and plunging into a thousand-year mini ice age. The climate then suddenly swung into the current interglacial. Ice sheets slowly began their northward retreat. The initial rise in sea level was minimal. By

16,000 years ago, the world's oceans had gained only 10 meters (33 feet) since the Last Glacial Maximum (fig. 5.1a). Four thousand years later, by 12,000 years ago, the ocean had climbed 60 meters (197 feet) above its lowest position and the great meltdown was well under way.[2] After another 4,000 years, the ocean level had climbed nearly 110 meters (361 feet) above its ice age minimum (fig. 5.1a). Three or more jumps, or **meltwater pulses**, originally described by oceanographer Richard Fairbanks, then at the Lamont-Doherty Earth Observatory in New York,[3] interrupted the overall upward trend (fig. 5.1a, b).

Why was the initial thaw interrupted by an abrupt refreezing and an equally sudden return to a more equable climate? What connection exists between these climate events, the ice sheets, and meltwater pulses? Why the uneven rise in sea level? This chapter attempts to answer these questions and others. The dramatic climate seesaws and great marine inundation concluding the last ice age were such major geological events that one cannot help but wonder whether any humans alive at the time witnessed these occurrences and, not understanding the underlying physical processes as we do today, attempted to explain them in terms they could comprehend—namely through their myths and sagas.

LEGENDS OF THE FLOOD

Diverse traditions from around the world record myths of a universal flood. Of these, the biblical tale of Noah and the Ark is perhaps the most familiar. God, angered at mankind's wickedness, decided to destroy the Earth but to spare the righteous Noah and his family. Noah was instructed to build an ark and take pairs of all animals, birds, and "creeping things," male and female. It rained for 40 days and 40 nights and floodwaters poured from the sky and the sea, until even the highest mountains were covered. All living things perished except for those aboard the ark, which landed on top of Mount Ararat. When the water finally stopped rising after 150 days, Noah sent out a raven, which flew around and quickly returned. He then sent out a dove, but it too soon came back. Some time later, he released the dove again and it returned with an olive leaf. Once again he sent it out, and since this time it did not return, Noah realized that it was finally safe to disembark. He then built an altar and made an animal sacrifice. God promised never to destroy the Earth again and produced a rainbow as a sign of peace.

Many striking parallels exist between the biblical account and the Babylonian story of Gilgamesh.[4] In the epic, Gilgamesh, king of Uruk, ruled with a cruel hand. The gods created a wild man, Enkidu, to challenge him.

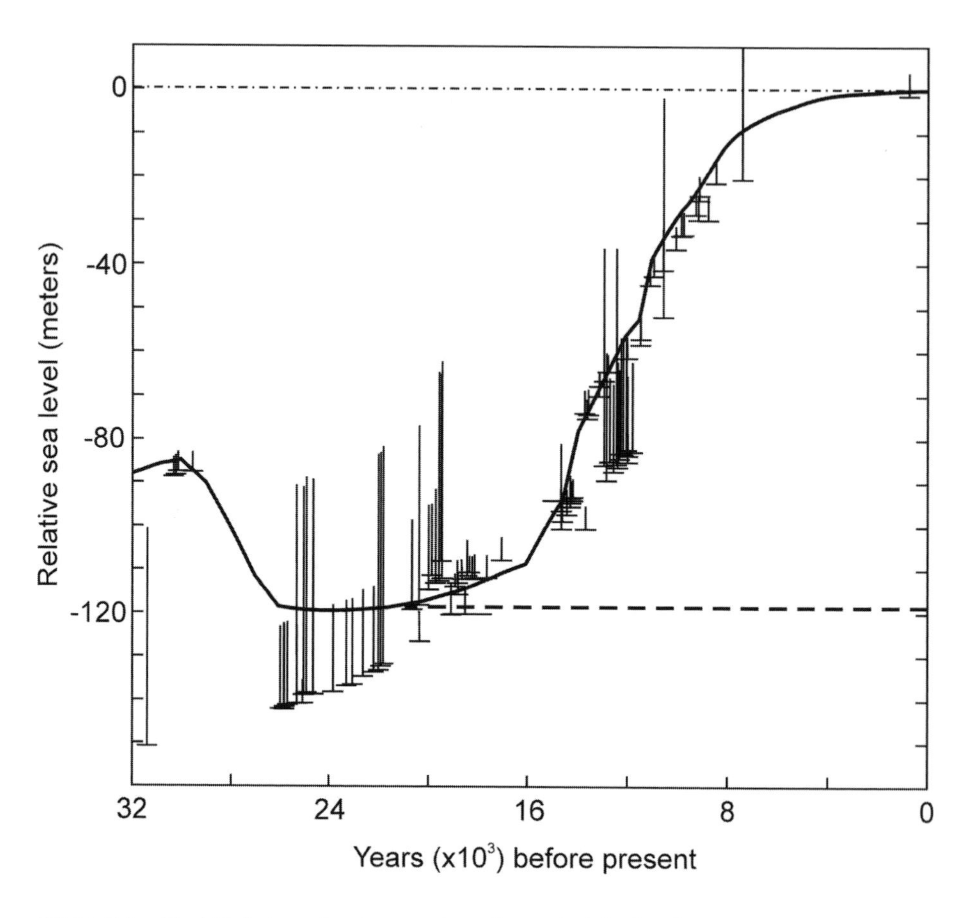

Figure 5.1a (Top) Global sea level curve since the end of the last ice age, based on a glacial isostatic adjustment model (black curve) compared with coral data from Barbados (vertical lines). Horizontal bars indicate depths of coral samples. Vertical bars represent the sea level range in which the particular coral species can live. The shortest bars (5 meters) correspond to *Acropora palmata*; the intermediate bars (20 meters), to *Monastera annularis*; the longest bars, to either *Asteroides* or *Diploria*. The last species provide only a lower limit of sea level. (Data from Peltier and Fairbanks, 2006; Gornitz, 2009, "Sea Level Change, Post-Glacial," in V. Gornitz, ed., *Encyclopedia of Paleoclimatology and Ancient Environments*, fig. S10, p. 889, with kind permission from Springer Science + Business Media B.V.)

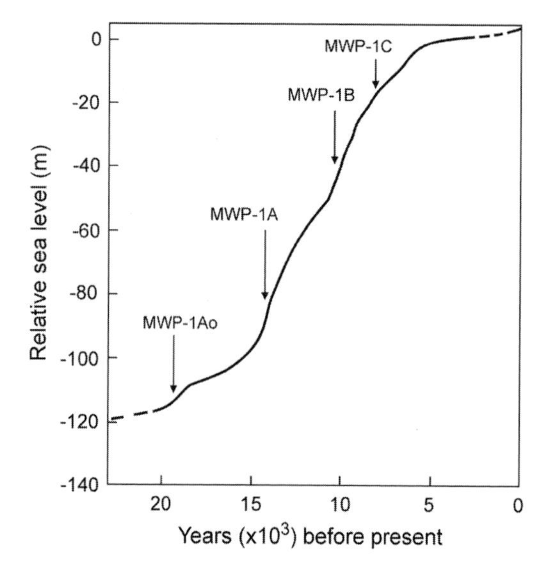

Figure 5.1b (Bottom) Generalized post-glacial sea level curve, showing the major meltwater pulses. (Data compiled from Fairbanks, 1989; Bard et al., 1990; Yokoyama et al., 2000; Hanebuth et al., 2000.)

Gilgamesh claimed the right to sleep with any bride on her wedding night. Enkidu confronted him and the two fought ferociously. Gilgamesh won, but spared Enkidu, and the two became very close friends. After many adventures together, Enkidu died and Gilgamesh was grief-stricken. He resolved to find the secret of immortality and sought out Utnapishtim, living at the end of the world; he and his wife were the only humans to survive the great flood. Although warned of the perilous journey, Gilgamesh continued until he encountered the ferryman, who took him across the Waters of Death. He eventually encountered Utnapishtim, who had received advance warning about the flood and was advised to build a boat and place his family and pairs of all living things on it. This flood lasted seven days and nights and the entire earth was covered by water. The boat came to rest on top of Mount Nimush [Nisur]. As in the Bible, birds were used to scan the terrain, but in reverse order. Utnapishtim released a dove from the boat. It flew around and returned. A few days later, a swallow was released and the same thing happened. Finally, a raven was sent out but never returned. Utnapishtim disembarked and then sacrificed a sheep. The story continues, with Gilgamesh being offered a chance at immortality, which needless to say, he failed. He thus realized that all living things must die.

In the even older Sumerian myth, the gods also decided to destroy mankind by flood, but the god Enlil warned Ziusudra beforehand. He was instructed to build a boat and carry animals and birds on it. This flood also lasted seven days and nights, and when the sun returned Ziusudra landed and sacrificed animals to the sun god. In this account, Ziusudra was granted eternal life. The Greek flood myth tells how Zeus sent a flood to destroy humanity but Prometheus advised his son Deucalion to build a chest (boat or ark?). All perished except for Deucalion and his wife. After floating in the chest for nine days and nights, they landed on Mount Parnassus. When the rains stopped, Deucalion made a sacrifice to Zeus. From India comes the story of Manu and the fish, which is part of a much longer epic of which many versions exist.[5] A common thread in all is how Manu had saved a small fish, which in gratitude warned him of a forthcoming flood that would destroy everything on Earth and to build a ship in preparation. The fish told Manu to place it in a clay jar and as it grew larger to move it to a larger vessel, and once it was fully grown Manu should return it to the sea. When the rains came and the waters kept rising, Manu tied his ship to the fish, which towed him to the highest northern mountain (in the Himalayas).

Could these myths represent dim and distant memories, later embellished with fanciful flourishes, of actual experiences of our ancient forebears who lived along the coast and fled from the rapidly rising sea at the end of the last ice age? Or are they instead very distorted and exaggerated records of real, more localized disasters? Most flood myths, although describing a seemingly universal watery catastrophe, in all likelihood reflected local events, such as an extremely high river flood, a period of exceptionally heavy and long-lasting rainfall, a particularly destructive coastal storm surge or tsunami. To most ancient peoples, their conception of the world's vastness probably extended only as far as the eye could see. Even during the most rapid phases of the great ice meltdown terminating the last glaciation, sea level did not rise much faster than 3–6 centimeters (1.2–2.4 inches) each year. While this may be quite rapid by geological standards, it is not drastic enough to cause people to build arks, head for the hills, or create flood myths.

On the other hand, over millennia, increasing global temperatures had not only thinned and shrunk the ice sheets but had weakened the remaining ice sheets enough to occasionally disintegrate them cataclysmically. Large meltwater lakes accumulating at their edges drained catastrophically as ice dams holding the water in check were overtopped and breached. At such times, huge torrents cascaded down rivers into the sea. These mega-floods contributed to the rapid jumps in sea level that punctuated the great meltdown and may have formed the basis for some of the myths. Evidence points to a sudden flooding of the Black Sea 7,600 years ago (or earlier), as the rising level of the Mediterranean Sea breached and overtopped an obstructing rock and sediment dam across the narrow Bosphorus Straits, plunging over the sill in powerful cataracts that would have dwarfed Niagara Falls.[6] The catastrophic nature of this event may have forced populations then living on the shores of the Black Sea to flee to neighboring lands, carrying memories of the rapidly rising waters with them. Their experiences may have inspired the flood stories of the Gilgamesh and the Bible. This intriguing hypothesis will be explored in greater detail later in the chapter.

The rising seas may have influenced other human migrations as well. Many mysteries surround the peopling of the Americas. The earliest migrants were big-game hunters following the vast herds of woolly mammoths, mastodons, bison, and other large mammals, many now extinct. Although it is generally agreed that the first humans to settle the New World came across the Bering land bridge from Asia toward the latter stages of the last ice age, the exact timing has aroused lively controversy. The Bering land bridge had formed during the last glacial period, once sea level had dropped ~50 meters (164 feet)[7]—as long ago as 50,000–75,000 years. After the ice began to melt, the encroaching sea reached that critical threshold by 12,000 years ago and flooded the Bering Strait, thereby closing the land bridge.[8] Therefore the nomadic hunters must have crossed the bridge well before that date.

Recent archaeological finds at Monte Verde in southern Chile, at the Debra L. Friedkin site in Texas, and at several other sites in the Americas suggest that the first people arrived in the New World before 14,600 years ago, at least 1,000 years prior to the Clovis culture, which was widespread throughout much of the United States between 13,300 and 13,000 years ago. Clovis culture was characterized by a distinctively-shaped, fluted projectile point. Prior to discoveries in South America and elsewhere, this was believed to represent the earliest firmly dated human presence in the Americas.[9] Since at the very least, several centuries must have elapsed between the time that the first arrivals crossed the Bering Strait and the time that they, or their offspring, showed up at Monte Verde, a coastal entry route was probably a more likely option at that earlier time.[10]

Once in the New World, the migrants needed to move south. Two major routes were potentially available to the wanderers (fig. 5.2). They could have headed south at times when an ice-free corridor existed between the Laurentide and Cordilleran Ice Sheets, or they could have paddled south in canoes or on rafts following the shoreline. Expanding ice blocked the interior route as early as 28,000 years ago, which remained sealed, for all practical purposes, until at least 14,000–13,000 years ago. Although the coastal route remained open before 25,000 years ago, it too became impassable during the peak of the Last Glacial Maximum, as glaciers reached all the way to the sea. But coastal deglaciation began earlier at the coast than farther inland. Segments of the outer coast became ice-free as early as 19,000–17,500 years before the present. After 16,000 years ago, deglaciation had progressed sufficiently to open up a possible sea route.[11]

Not only did the retreating ice sheets in the Pacific Northwest open up a sea passage, but they also exposed land near the coast that had been depressed under the heavy ice overburden. This land quickly rebounded once the ice was gone.[12] The glacial isostatic rebound in the Pacific Northwest was largely completed within several thousand years after deglaciation, in contrast to eastern North America, where isostatic adjustments still continue (although much slower). Ancient seafarers would have set their course depending on the position of the shoreline relative to the ice and to local sea level change. The glacial and sea level history of this region is particularly complex. Offshore islands were deglaciated and rebounded isostatically earlier than inland locations, less than 50 kilometers (31 miles) away (fig. 5.3). Within several generations, shorelines had shifted by more than 100 kilometers (62 miles).[13] Perhaps the Tlingit myth cited at the beginning of this chapter is a faint recollection of events actually experienced by their distant ancestors.

A fuller treatment of the peopling of the Americas is beyond the scope of this volume.[14] Here we have merely pointed out some potential windows of opportunity toward the end of the last ice age when entry into the New

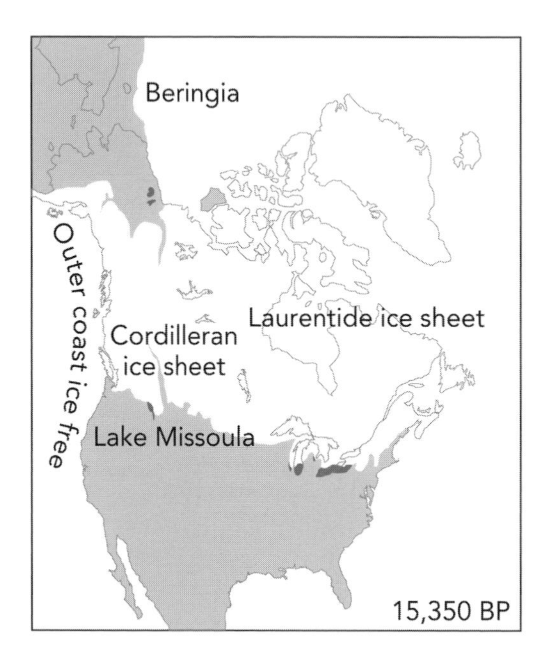

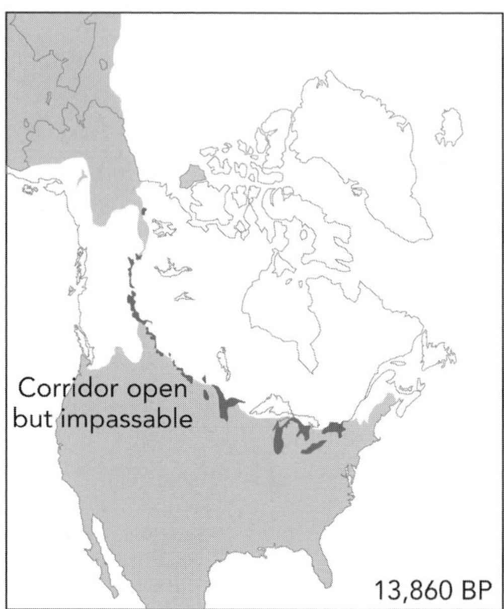

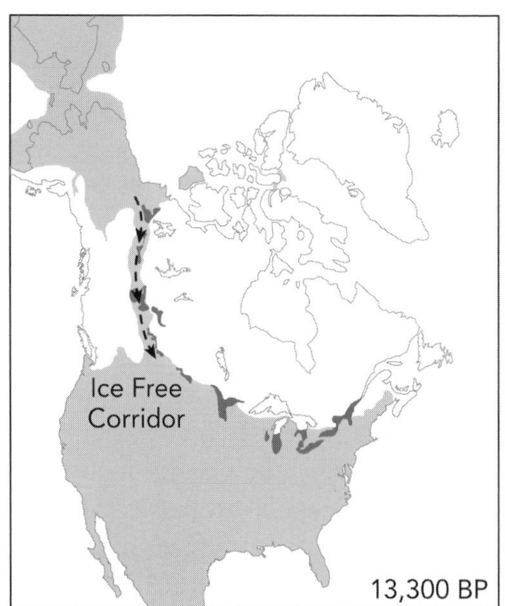

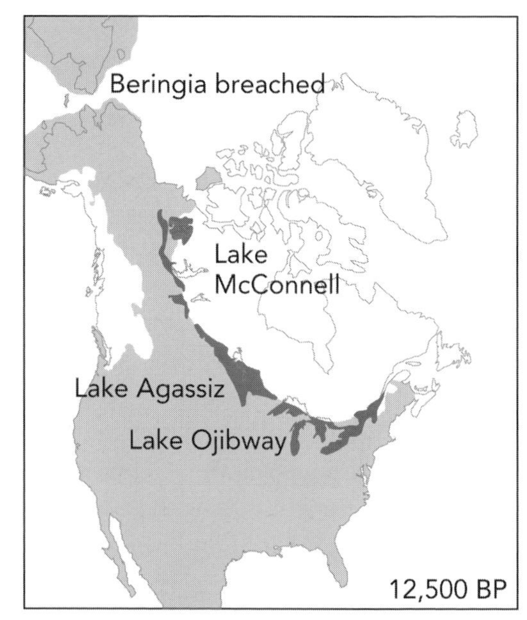

Figure 5.2 Retreat of the Cordilleran and Laurentide Ice Sheets and opening of potential ice-free migration routes. Dates are years before present. (Modified from Meltzer, 2009, fig. 6.)

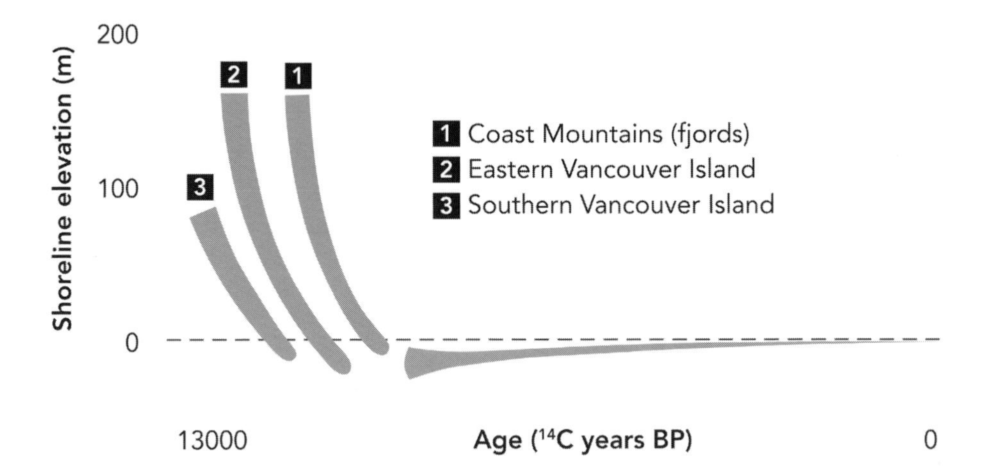

Figure 5.3 Sea level changes along the Pacific Northwest coast after the end of the last ice age. The falling sea level is due to glacial isostatic rebound (see also fig. 5.9a). Ice melted and the land rebounded first on southern Vancouver Island, followed by eastern Vancouver Island, and finally on the mainland coast. (J. J. Clague, 2009, "Cordilleran Ice Sheet," in V. Gornitz, ed., *Encyclopedia of Paleoclimatology and Ancient Environments*, fig. C73, p. 210, with kind permission from Springer Science + Business Media B.V.)

World would have been feasible, given the complex interplay between the ice sheets and world sea level, both of which were changing rapidly at the time.

MELTING ICE AND THE MARINE INCURSION

Last Gasps of the Ice Age—Wild Climate Swings

The last ice age did not end smoothly. Instead, the last deglaciation between approximately 20,000 and 8,000 years ago (also called the **Last Glacial Termination**) witnessed several pronounced shifts in global climate and sharp marine oscillations that punctuated an overall upward sea level trend. Northern summer insolation increased steadily from the Last Glacial Maximum (~21,000 years ago) and culminated 11,000 years ago, by which time the great ice meltdown was already well under way. Antarctic ice cores show that beginning 19,000–18,000 years ago, the Southern Hemisphere warmed up until shortly after 15,000 years ago. During that time, sea level already began the slow upward climb from its ice age minimum of 120 meters. Then suddenly, between 14,500 and 13,000 years ago, cold gripped southern latitudes for around 1,500 years during the Antarctic Cold Reversal before the warmth eventually returned (fig. 5.4).[15]

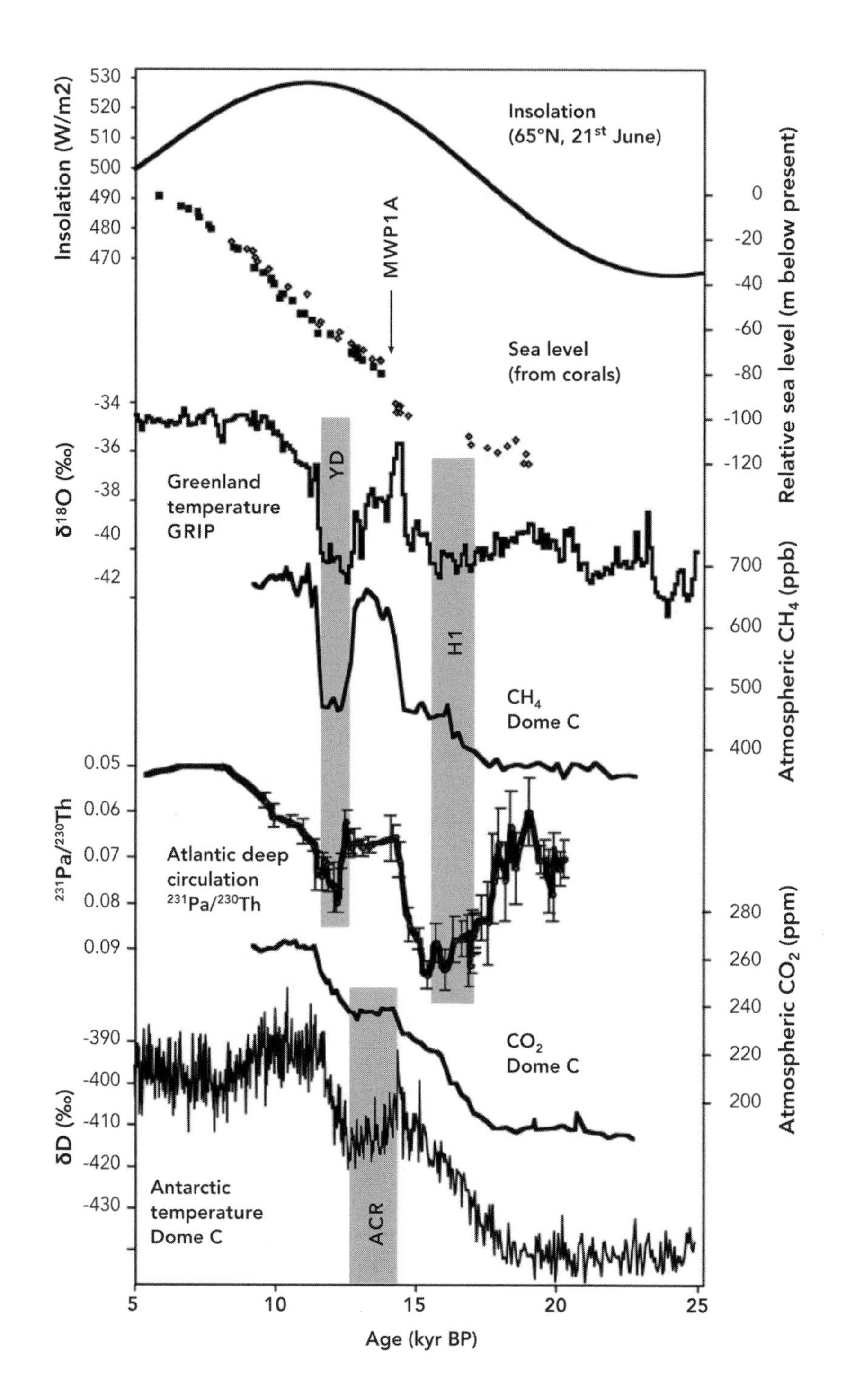

Figure 5.4 Climate variations during the Last Glacial Termination. (D. Paillard, 2009, "Last Glacial Termination," in V. Gornitz, ed., *Encyclopedia of Paleoclimatology and Ancient Environments*, fig. L5, p. 497, with kind permission from Springer Science + Business Media B.V.)

The north, meanwhile briefly breaking free of icy conditions, enjoyed a mainly warm respite between 14,700 and 12,800 years ago during the Bølling-Allerød **interstadial**), which roughly spanned the same time interval as the Antarctic cold spell (fig. 5.4).[16] The Bølling-Allerød interstadial began abruptly within three years, according to some Greenland ice core records![17] Ice sheets retreated rapidly in the north, and tundra gave way to birch and conifer forest. Sea level jumped upward sharply during this period—Meltwater pulse 1A (see below). The thaw was followed by a return to near glacial conditions starting 12,900 years ago, during the Younger Dryas (named after *Dryas octopetala*, an Arctic-Alpine flower). During the preceding warm period, this flower had shifted its range northward, following the retreating ice sheets, but it suddenly reappeared over much of Europe during the colder Younger Dryas.[18] The Younger Dryas chill took root more slowly, within two centuries, but lasted a millennium (12,900–11,700 years ago). Glaciers re-advanced and the tundra moved south once again. Air temperatures dropped sharply—by 6°C–9°C (11°F–16°F) at higher latitudes. Tundra vegetation spread south; boreal forests replaced temperate deciduous forests, and much of Africa became drier. Sea level continued to rise, but more slowly than before. Finally, by 11,700 years ago, the ice age ended once and for all, within decades with another fairly sudden warming, and shortly thereafter, sea level climbed upward again.

As in earlier abrupt climate events, the Southern Hemisphere reacted in an opposite manner. While the Younger Dryas mini ice age gripped the Northern Hemisphere tightly, Antarctic temperatures climbed steadily—the Antarctic Cold Reversal had ended. The Younger Dryas cooldown may be another manifestation of hemispheric climate asynchrony and the bipolar seesaw, in which the ocean conveyor circulation switches direction and trades hot and cold between the hemispheres, causing the abrupt climate swings.

It is widely believed that the massive influxes of freshwater from melting ice and overflowing mega-lakes during the brief Bølling-Allerød warming lowered the salinity (and density) of the North Atlantic, thus preventing the sinking of North Atlantic Deep Water (NADW). As a consequence, warm currents from the south failed to reach the northern latitudes and the entire North Atlantic region regressed to near-glacial climates. Changes in atmospheric circulation may have been involved as well. The Younger Dryas cooling affected not only the North Atlantic region but also many North American and semitropical to tropical sites. The Southern Hemisphere, on the other hand, showed little signs of cooling.

Glacial Lake Agassiz at the southwestern margin of the Laurentide Ice Sheet and the shrinking Keewatin ice dome (on the western Laurentide Ice Sheet) are believed to have been the source of this meltwater. The weight of the ice sheet had depressed the land beneath it, creating a moat along its margins. As the ice began to retreat, the moat filled with water and huge glacial lakes formed. The largest and best-known such glacial lake was Lake Agassiz, named after the Swiss paleontologist Louis Agassiz (1807–1873). At its maximum extent, the lake covered parts of North Dakota, Manitoba, and western Ontario (fig. 5.5).[19] How did the water reach the sea? A model

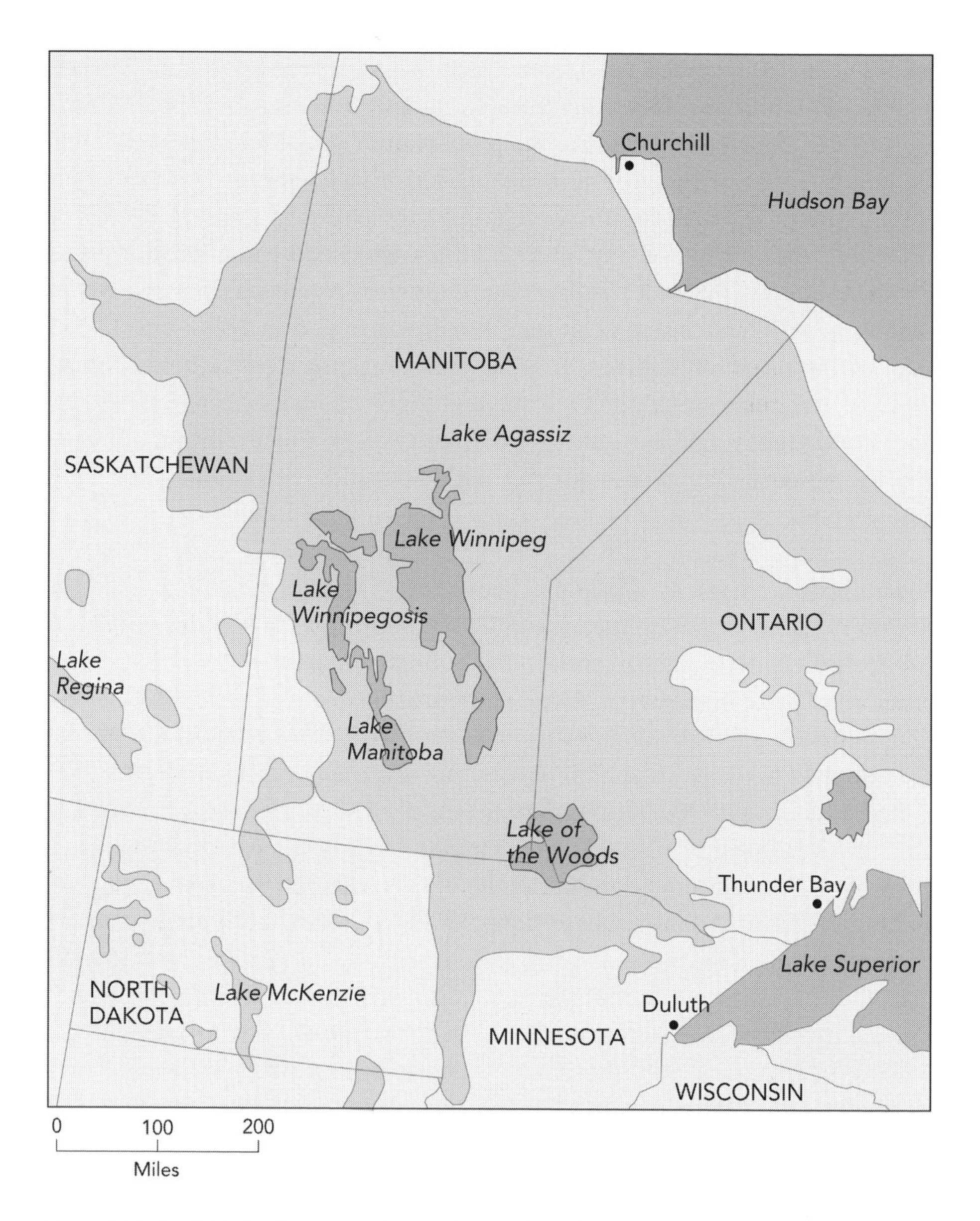

Figure 5.5 Glacial Lake Agassiz at the southern edge of the Laurentide Ice Sheet, at its maximum extent. (Modified from J. P. Bluemle, 2007, North Dakota Geological Survey.)

reconstruction of the glacial melt history finds major northward meltwater discharge 13,100–12,500 years ago, at the beginning of the Younger Dryas. The outflow entered the Arctic Ocean, via the Mackenzie River, Fram Strait, and ultimately reached the eastern North Atlantic.[20] Geomorphological data, on the other hand, suggest that ice still blocked routes to the north and east toward the St. Lawrence Seaway until the end of the Younger Dryas.[21] Sea level curves from Tahiti, New Guinea, and Barbados show a small step (under 6 meters) around 13,000 years ago near the Younger Dryas onset,[22] which may have come from this deluge.

The Younger Dryas ended as solar insolation approached a maximum. (Around 11,000 years ago, the precession of the equinoxes was optimally configured for Northern Hemisphere warming: summer solstice—June 21—occurred at perihelion, when Earth made its closest annual approach to the Sun; see fig. 4.6). Nearly 8 percent more solar radiation bathed the Northern Hemisphere during summer months at that time than today, and the additional summer warmth hastened ice sheet melting.[23] Sea level sprang upward again on the heels of the revived warm-up.

A Space Intruder?

A more controversial hypothesis for the origin of the Younger Dryas invokes an extraterrestrial intruder, that may have contributed to the Younger Dryas cooling12,900 years ago, the megafaunal extinction,[24] and sudden disappearance of the Clovis culture in North America.

A distinctive thin, dark carbon-rich layer has been spotted in many locations across North America at the boundary between the onset of the Younger Dryas, exactly 12,900 years ago, and overlying sediments. The dark layer often lies over another layer rich in magnetic microspherules and grains, iridium and nickel, carbon spherules, charcoal, soot, and fullerenes.[25] The fullerenes contain helium-3, an isotope rare on Earth but often found at known impact sites. The thin layer also yields three types of nanodiamonds that can form under the high pressures and temperatures of an impact.[26] The nanodiamonds were recovered from 16 different sites across North America and northwestern Europe—where the Younger Dryas cooling was strongest. Many of these indicators, such as iridium (Ir), also occur at known impact sites. For example, the discovery of a worldwide Ir-rich layer at the Cretaceous-Tertiary (K-T) boundary was one of the first indications of a catastrophic impact.[27] The proposed impact may have set off intense wildfires, producing soot, charcoal, and carbon spherules, as at the K-T boundary. The meteor or comet may have struck the Laurentide Ice Sheet, destabilizing it and triggering a chain of events otherwise similar to the explanation given above—namely, that the disintegrating ice sheet released a cascade of icy fresh water that disrupted normal ocean circulation, preventing the warmth of the Gulf Stream–North Atlantic Current from reaching the north. This plunged the region back into a near ice age. The incoming object produced no telltale crater, because it may have exploded in the atmosphere above the Earth, like the mysterious Tunguska event in Siberia 100 years ago. Perhaps the thick ice sheet may have shielded the rock beneath, or the meteor may have broken up into multiple smaller objects before striking the Earth (like comet Shoemaker-Levy, which struck Jupiter in July 1994). But the space object may have left a distinctive chemical signature in Greenland ice.[28]

Critics lost little time in denouncing this hypothesis.[29] Some questioned the identification of nanodiamonds; others their unique impact origin. Other than nanodiamonds, no other shocked, high-pressure minerals or impact-melted rocks were found. Furthermore, the thin carbonaceous "layer" spanned a broader time range. The megafauna extinctions in North America were not so instantaneous either. Carbon spherules occur in many non-catastrophic settings such as forest topsoils, grasslands, and swamps. Skeptics also noted that the alleged impact markers deposit in sediments from a steady rain of micrometeorites passing through the atmosphere. Yet, the so-called impact markers, including iridium, appear to peak at the Younger Dryas boundary, as at the K-T boundary. While iridium is rare in terrestrial rocks and sediments, it is much more abundant in meteorites. An unusually high enrichment in this element implies a spike in either micrometeorite infall or ocean sedimentation. In the case of the K-T event, the Alvarez father-son team ruled out the latter possibility on geologic grounds and correctly inferred that a catastrophic impact had struck the Earth 65 million years ago, with dire consequences for the then-existing fauna and flora.[30] A growing list of worldwide Ir anomalies and other accumulating evidence (and ultimately the discovery of the "smoking gun"—the buried Chicxulub crater in Mexico) confirmed their initial conclusion.

Additional data are needed in order to support the impact hypothesis for the Younger Dryas. Yet the hypothesis does deserve further careful scrutiny before being dismissed out of hand. A more plausible explanation, however, invokes a major flood produced by the melting ice sheets, as described above.

A Stepped Sea Level Rise

As the ice began to melt, sea level records from many locations reveal that that the marine incursion ending the ice age was punctuated by several periods of sharp and rapid submergence—the meltwater pulses that occurred roughly 14,000, 11,000, ~9,000, and 8,000–7,600 years ago (table 5.1; fig. 5.1a,b). Details of the exact number, timing, and magnitude of these major pulses and their relation to glacial outburst floods and other paleoclimate events are still being refined.

The post-glacial marine incursion began slowly, starting 21,000–20,000 years ago (fig. 5.1a).[31] An initial upswing (referred to here as "meltwater pulse $1A_0$") may have already begun by ~19,000 years ago, during which ocean levels may have climbed 10–15 meters (33 –49 feet) within a few hundred years (table 5.1).[32] Yet not all records indicate such an early meltwater pulse.[33]

Meltwater Pulse 1A

Meltwater pulse 1A provides the strongest indication of a rapid post-glacial sea level spurt (fig. 5.1a). The event has been variously dated between 14,600 and 13,000 years ago, largely falling within the comparatively mild Bølling-Allerød interstadial (table 5.1). Greenland temperatures leapt upward by ~4.5°C–9.9°C (8°F–18°F). Sea level jumped upward between 16 and 24 meters (52–79 feet). The meltwater pulse occurred at a time of maximal ice sheet retreat and lasted several hundred years, during which rates of sea level rise peaked at ~40–65 millimeters (1.6–2.6 inches) per year (table 5.1).

Comparison of a Greenland ice core chronology with data from a high-resolution marine sediment core drilled south of Greenland and with a sea level curve derived from drowned fossil corals on Barbados[34] reveals that meltwater pulse 1A began 14,110 years ago, five to six centuries *after* the onset of the Bølling warming that commenced 14,640 years ago.[35] More recent dating suggests that sea level rise accelerated rapidly after 14,300 years ago and peaked around 13,800 years ago (table 5.1). The time lag between onset of warming and rapid sea level rise may be significant in anticipating the speed at which large ice sheets may break apart in the future and unleash their watery stores once the heat is switched on. How long before a soft and weakened ice mass finally disintegrates catastrophically and how fast? This important question will be revisited in chapter 7. Suffice it to say here that although sea level began to rise slowly as early as 21,000–20,000 years ago, the period of greatest post-glacial marine incursion spanned 16,000–8,000 years ago.

Source of Meltwater Pulse 1A

The major meltwater pulses were sharp spikes superimposed on an overall upward sea level trend during the main phase of the great ice meltdown.

Table 5.1 Rapid periods of post-glacial sea level rise

Meltwater pulse $1A_0$			
Timing, yr	*Rate of SLR, mm/yr*	*Increase in sea level, m*	*Source(s)*
>19,000	—	10–15	Yokoyama et al. (2000)
~19,000	≥20 yrs	10 m in ≤500 yrs	Clark et al. (2004)
19,600–18,800	~12.5 (average)	10	Hanebuth et al. (2009)
Meltwater pulse 1A			
Timing, yr BP	*Rate of SLR, mm/yr*	*Increase in level, m*	*Source(s)*
~13,000	~39 (max.)	24 m in <1,000 yrs	Fairbanks (1989)
~13,500	~37		Bard et al. (1990)
13,800	~26 (max.)	—	Stanford et al. (2011)
14,200	—	13.5	Blanchon and Shaw (1996)
14,110	—	—	Stanford et al. (2006)
14,600—14,300	53.3	16 m in 300 yrs	Hanebuth et al. (2000);
Kienast et al. (2003)			
14,300–14,000	65	20	Liu and Milliman (2004)
Meltwater pulse 1B			
Timing, yr BP	*Rate of SLR, mm/yr*	*Increase in level, m*	*Source(s)*
9,500	25 (max.)		Stanford et al. (2011)
~10,500	26	~2	Fairbanks (1989)
10,900	19 (max.)		Stanford et al. (2011)
11,000	~25		Bard et al. (1990)
11,500–11,200	~40	45–58	Liu and Milliman (2004)
11,500	7.5 m		Blanchon and Shaw (1996)
Meltwater pulse 1C			
Timing, yr BP	*Rate of SLR, mm/yr*	*Increase in level, m*	*Source(s)*
~7,600	—	6.5 m	Blanchon and Shaw (1995)
8,260–7,680	34–44	—	Tooley (1989)
~8,000–8,170 to 8,250–8,420	4.9 (average)	1.2 m	Törnqvist et al. (2004)
7,600—8,200	~12	~6 m in ≤500 yr	Cronin et al. (2007)
~7,600	~10	~4.5 m	Yu et al. (2007)

Stanford, J. D., Hemingway, R., Rohling, E. J., Challenor, P. G., Medina_Elizalde M., and Lester, A. J. 2011. "Sea-Level Probability for the Last Deglaciation: A Statistical Analysis of Far-Field Records." *Global and Planetary Change* 79:193–203.

After a sluggish beginning 20,000 years ago, sea level began to speed up almost 4,000 years later, as solar insolation gradually intensified. Sea level continued to rise—although more slowly—even during the intervening Younger Dryas cold period,[36] which nevertheless occurred at a time when solar insolation was fast approaching a peak at 11,000 years ago.

Is it possible to pinpoint which disintegrating ice sheets generated these huge volumes of meltwater? We now turn our attention to possible sources for meltwater pulse 1A.

It is generally accepted that the breakup of the Laurentide Ice Sheet in North America supplied most of the meltwater for pulse 1A, due to its former vast extent. At the onset of deglaciation 20,000 years ago, meltwater flowed south down the Mississippi River to the Gulf of Mexico because other potential northern routes were still largely blocked by ice. By the time of meltwater pulse 1A, 14,500–14,000 years ago, the ice sheet had receded sufficiently for Lake Agassiz meltwater to drain south down the Mississippi River and Hudson River, east down the St. Lawrence River, and to a small extent north along the Mackenzie River toward the Arctic Ocean.[37] Melting of Eurasian ice sheets also added to the spurt in sea level.

An alternative Antarctic source has also been proposed.[38] Different meltwater sources (e.g., the Laurentide, Antarctic, or Barents Sea Ice Sheets) leave distinctive fingerprints, because they produce different geographic patterns of sea level rise. This variation arises from differences in isostatic loading of ice (see chapter 4) and also because of a diminishing gravitational attraction between the shrinking ice sheet and ocean water. A shrinking ice sheet weakens its gravitational attraction on ocean water, and water flows away from the ice edge even as mean global ocean level is increasing. In some geophysical models, substantial inputs from Antarctica most closely match observed sea level records for meltwater pulse 1A.[39] However, tiny marine algae in finely layered sediments from an East Antarctic fjord indicate that ice sheets there receded by 11,500–10,500 years ago, long *after* meltwater pulse 1A,[40] but closer to meltwater pulse 1B. Not only does the reconstructed history of ice sheet retreat favor North American (and Eurasian) rather than Antarctic, sources, but geologic studies show a quite small Antarctic contribution to meltwater pulse 1A.[41]

Meltwater Pulse 1B

The two other major meltwater pulses—1B and 1C—were not as distinct as meltwater pulse 1A (fig. 5.1b). While the encroaching oceans slowed down noticeably during the Younger Dryas, they nonetheless continued to climb upward between 14,000 and 12,000 years ago. Once the Earth thawed following the Younger Dryas, the pace accelerated again. This episode—meltwater pulse 1B—occurred 11,500–9,500 years ago. Sea level jumped upward ~15 meters (49 feet) at Barbados during meltwater pulse 1B, yet Pacific corals record no sharp increase in sea level at that time.[42] The reason for this disparity is unclear but may involve differential glacial isostatic or gravitational effects. Alternatively, the spurt at Barbados may have been overestimated. Recent dating of glacial deposits shows that the timing of meltwater pulse 1B coincides closely with outflow from the St. Lawrence and Hudson Rivers fed by waters from Lake Agassiz, and the onset of deglaciation of East Antarctica.[43]

Meltwater Pulse 1C

Around 9,300 years ago, a dam of glacial debris blocking the eastern end of Lake Superior suddenly gave way, draining the lake by ~45 meters (148 feet) and also cooling the North Atlantic region.[44] Hints of a small upsurge in sea level rise appear ~9,000–8,500 years ago in some sediment cores from Chesapeake Bay, various Canadian sites, and eastern China.[45] This event may mark a spike in sea level at the tail end of meltwater pulse 1B.

Another meltwater pulse (here called meltwater pulse 1C), weaker than earlier ones, occurred about 8,000 years ago. Although not apparent in the post-glacial eustatic sea level curve (fig. 5.1a),[46] evidence for a sudden sea level jump 8,200–7,600 years ago appears at diverse locations (table 5.1). The first signs of this meltwater pulse initially came from a gap in coral growth in the Caribbean ~7,600 years ago, when the reefs were drowned by rapidly rising sea level.[47] Sediment cores from the Mississippi River delta also record an abrupt sea level rise event ~8,500–8,000 years ago.[48] In addition to the ~9,000-year event, Chesapeake Bay reveals another episode of marsh drowning and rapid sea level spurt ~8,200–7,600 years ago.[49] During the younger episode, sea level rose approximately 6 meters within 500 years, at a mean rate of 12 millimeters per year. Evidence from southern Sweden also points to a sudden sea level rise of ~4.5 meters (14.8 feet), roughly 7,600 years ago, with rates averaging ~10 millimeters per year (table 5.1).[50] Other indications of rapid sea level rise 8,000–7,600 years ago occur off the German North Sea coast, the southern Kattegat Sea, the southwest Baltic Sea, Limfjord, northwestern Denmark, coastal Lancashire, England, and the delta of the Yellow River, China.[51]

Meltwater pulse 1C may be linked to a cold episode that occurred around 8,200 years ago, when temperatures in Greenland dropped by 5°C–8°C (9°F–14°F) and summer temperatures in Europe fell by 1°C–2°C (2°F–4°F).[52] It is widely believed that this cold spell was triggered by the catastrophic drainage of glacial Lakes Agassiz and Ojibway 8,470 years ago, during the final stages of the deglaciation of the Laurentide Ice Sheet.[53] The volume of water released by the sudden-outburst flood reached ~100,000 cubic kilometers within a few years or less. Although this enormous influx of freshwater to the world's oceans probably triggered the 8,200-year cold event,[54] its mark on global sea level was much smaller than that of the earlier meltwater pulses. This outflow increased sea level by only one meter or less (assuming an even distribution of water over the ocean),[55] although local changes could have been much higher.[56] Yet even this minor sea level change left a definite imprint in the stratigraphic record.

The Storegga Landslide

Without warning, a massive underwater avalanche of loose sediment mixed with methane gas and water suddenly cascaded down the flanks of the continental shelf and slope of the coast of Norway, 8,150 years ago (fig. 5.6). The humongous submarine landslide (one of the largest in the world) dislodged an estimated 3,000–3,200 cubic kilometers of debris, spread over 9,500 square kilometers.[57] The Storegga landslide triggered a powerful tsunami that inundated the coasts of northeastern Scotland and western Norway and dumped marine deposits up to 20 meters above sea level in the Shetland Islands. This forceful geological event is roughly the same age as the 8,200-year cold snap mentioned above, whose climatic fingerprints were clearly recorded in nearby Greenland ice cores. Is this uncanny proximity in timing just an intriguing coincidence or were these two occurrences more closely connected? Some scientists think so. They suggest that similar to the brief super-hot climate anomaly of the Paleocene-Eocene Thermal Maximum (PETM) (see chapter 3), the explosive decomposition of gas hydrates from the sediment disruption during the underwater avalanche triggered an enormous outburst of methane that abruptly ended the cold episode (see fig. 3.8).[58] However, a simple calculation soon showed that the **greenhouse effect** of the methane released by the Storegga Slide (the landslide lies in a natural gas field) was much too small to modify global temperature, although regional climate impacts could have been higher. Therefore the scientists proposed that some methane could have been converted to ozone, an even stronger greenhouse gas, although most would have probably been oxidized to carbon dioxide. Nevertheless, a little extra ozone may have been enough to turn the tide.

Alternatively, the timing of the Storegga landslide may have simply coincided with meltwater pulse 1C, which followed closely on the heels of the

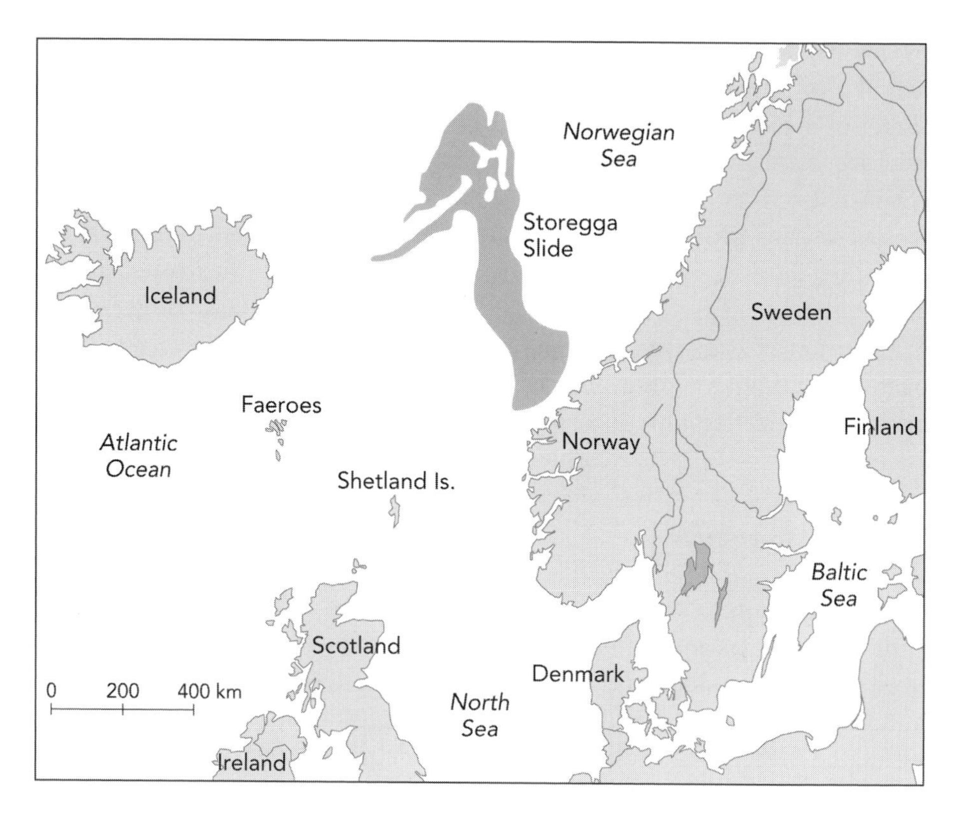

Figure 5.6 Location of the Storrega landslide off the Norwegian coast.

8,200-year cold event.[59] Although rising sea level tends to stabilize methane hydrate because of the increased weight of a thicker layer of water, warmer water penetrating ocean depths would have an opposite effect. Another potentially destabilizing force is the isostatic rebound after ice melted on nearby landmasses (i.e., Scandinavia and the British Isles). These crustal stresses could trigger earthquakes that in turn could set off underwater landslides that would disrupt the sediments overlying and trapping the methane hydrates. The removal of the rocky overburden would decompose the hydrates, freeing the caged methane gas.

Another curious confluence is the rapid inundation of the Black Sea at approximately the same time as meltwater pulse 1C. Did the Mediterranean overtop the Bosphorus sill and spill into the then lower Black Sea because of the speed-up in global sea level rise at that time? Even more intriguing is whether the catastrophic inundation of this inland sea inspired the biblical story of Noah's Flood and the even older Babylonian Gilgamesh epic. These ideas are explored further in the next section.

Inundation of the Black Sea and Noah's Flood

Building upon decades of underwater geophysical surveys in the Black Sea by teams of American, Russian, and Turkish scientists, two Columbia University marine geologists, William Ryan and Walter Pitman, hypothesized that by 7,600 years ago the sea level had risen sufficiently for the Mediterranean to breach the Bosphorus sill catastrophically, drowning a shrunken Black Sea within less than two years, in a manner curiously reminiscent of the end of the Messinian Salinity Crisis (chapter 3).[60] They further speculated that the rapidity of the inundation would have driven any coastal dwellers toward higher ground, heading in various directions toward southeastern Europe and the Middle East. Folk memories of this environmental calamity were passed down through the generations and formed the basis of myths and legends that eventually were recorded in the Bible. A number of geologic observations bolstered this claim. Submerged dunes, beaches, wave-cut shore terraces, and winding river channels surrounding the shores of the ancient Black Sea lake formed at a time when the former lake level was more than 100 meters (328 feet) below the present water level, well below the two outlet sills to the Mediterranean (i.e., the Dardanelles at 80 meters [262 feet] and the Bosphorus at 30 meters [98 feet] below present sea level). A widespread stratigraphic break was covered by a fairly uniform "mud drape" containing seashells. Beneath the sediment break was a layer formed in a semiarid nearshore landscape: mud cracks, plant roots, sand dunes, and freshwater mollusks. The overlying mud drape, by contrast, was full of marine fossils like edible mussels and cockles.

Piecing together the geologic data, Ryan and Pitman painted a picture of the Black Sea history before and during the great flood. By 14,500–14,000 years ago, the Scandinavian-Eurasian-Alpine ice sheets were in full retreat, channeling huge torrents of meltwater down the Danube, Dniester, Dnieper, Volga, and Don rivers. The water briefly pooled into a series of large lakes that ringed the southern edges of the fast-melting ice sheets and then spilled over into the Aral, Caspian, and Black seas. More meltwater followed on the heels of the Younger Dryas, raising the level of the Black Sea even higher. By then, glacial isostatic adjustments had shifted land elevations, diverting water from the upper reaches of the Dniester, Dnieper, Volga, and Don rivers toward the North Sea and Baltic Sea instead. A drier climate embraced the region and the Black Sea level dropped below its outlets to the world ocean, becoming a shrunken, brackish-water lake, a mere remnant of its former extent (fig. 5.7a). Meanwhile, rising global sea level had overtopped the Dardanelles barrier, filling the intervening Sea of Marmara and,

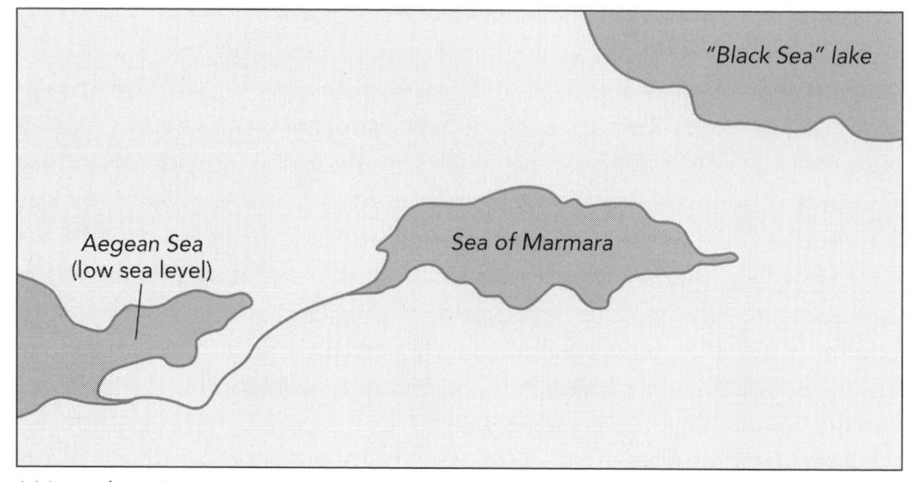

(a) Last glaciation

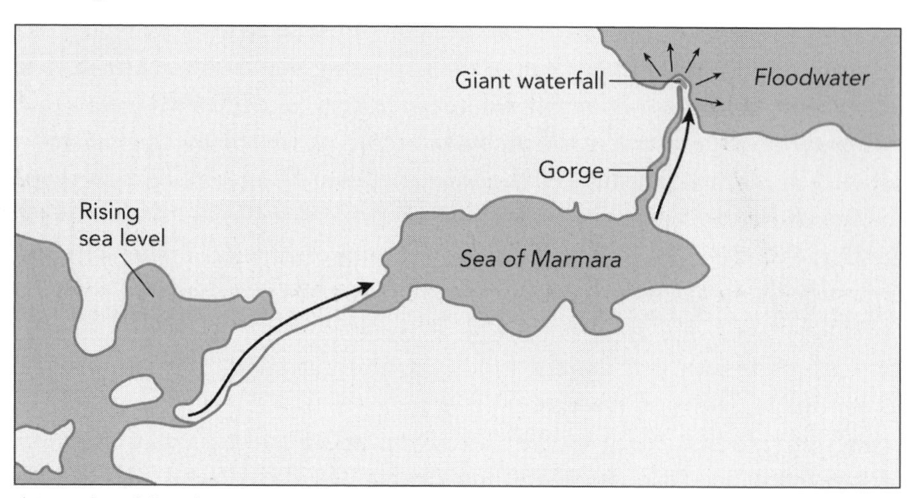

(b) Deglacial flood (7600 years ago)

Figure 5.7 A. At the end of the last ice age, the Black Sea level was at a much lower level than today. B. By 7,600 years ago, rising global sea level had over topped the Dardanelles barrier, pouring into the Sea of Marmara and across the BosphorusStrait into the Black Sea.

by 7,600 years ago, finally breaking through the ultimate barrier, a sediment dam across the Bosphorus Strait (Fig. 5.7b). Water cascaded torrentially into the Black Sea, pouring in the equivalent of two hundred times the daily flow of Niagara Falls at its peak and rapidly turning a brackish lake into a vast inland sea. The Columbia scientists also rounded out their controversial hypothesis with a wide-reaching excursion into mythology, archaeology, and even the evolution of the Indo-European languages.

The age of the proposed flood event, 7,600 years ago, falls toward the younger end of meltwater pulse 1C. At first glance, the closeness in timing suggests that global sea level was already high enough for the few extra meters from the meltwater pulse to spill Mediterranean water over the Bosphorus into the low-lying lake occupying the Black Sea basin. However, more recent dating of the marine incursion throws a damper on this appealing idea. The brackish-freshwater lake to marine salt water transition in the Black Sea was much older, ~9,500–8,900 years ago.[61] The new dates may perhaps correspond to a small meltwater pulse at around 9,000–8,900 years ago (see above). On the other hand, the Black Sea flood may be related not to any specific meltwater pulse, but merely to the post-glacial rise in sea level once it reached a high enough elevation to overtop the Bosphorus sill.

The catastrophic flood idea, while plausible, was soon attacked from many quarters. Some claim that a now-submerged 10,000–9,000-year-old delta fan at the southern exit of the Bosphorus Strait demonstrates a much older Black Sea outflow toward the Mediterranean. Delta sediments that should have been scoured by the onrushing floodwaters of the 7,600- (or 8,900–9,500) year-old catastrophic flood remained undisturbed. Instead of a catastrophic flood, a two-way flow between the Black Sea and the Mediterranean would have been established gradually.[62] Other critics point to a lack of prominent flood deposits in the Black Sea; they envision a smaller, non-catastrophic flood and gradual transition to a marine environment.[63] They also downplay the archaeological data connecting the timing of the Black Sea inundation to a spread of populations into surrounding areas.[64] Populations may have migrated in response to regional vegetation changes resulting from the 8,200-year cold event, in addition to Black Sea flooding.[65] Ryan, however, remains convinced that an abrupt flood event best fits the currently available geologic data.

Approaching the Present—Middle to Late Holocene Sea Level

By the mid-Holocene, ~7,000–6,000 years ago, melting of the ice sheets had essentially ceased and eustatic sea level approached within a few meters of modern sea level, continuing to rise very slowly (fig. 5.1a). Glacial isostatic and hydro-isostatic adjustments persisted at a diminishing rate. Therefore, local sea level was still changing. Sea level continued to drop in the formerly glaciated regions of Scandinavia and Canada (fig. 5.8a) and rise in areas of the collapsed bulge, such as the East Coast of the United States and the low countries of Europe (fig. 5.8b). In transitional zones such as Maine or the Canadian Maritimes, the early Holocene sea level fall (due to glacial

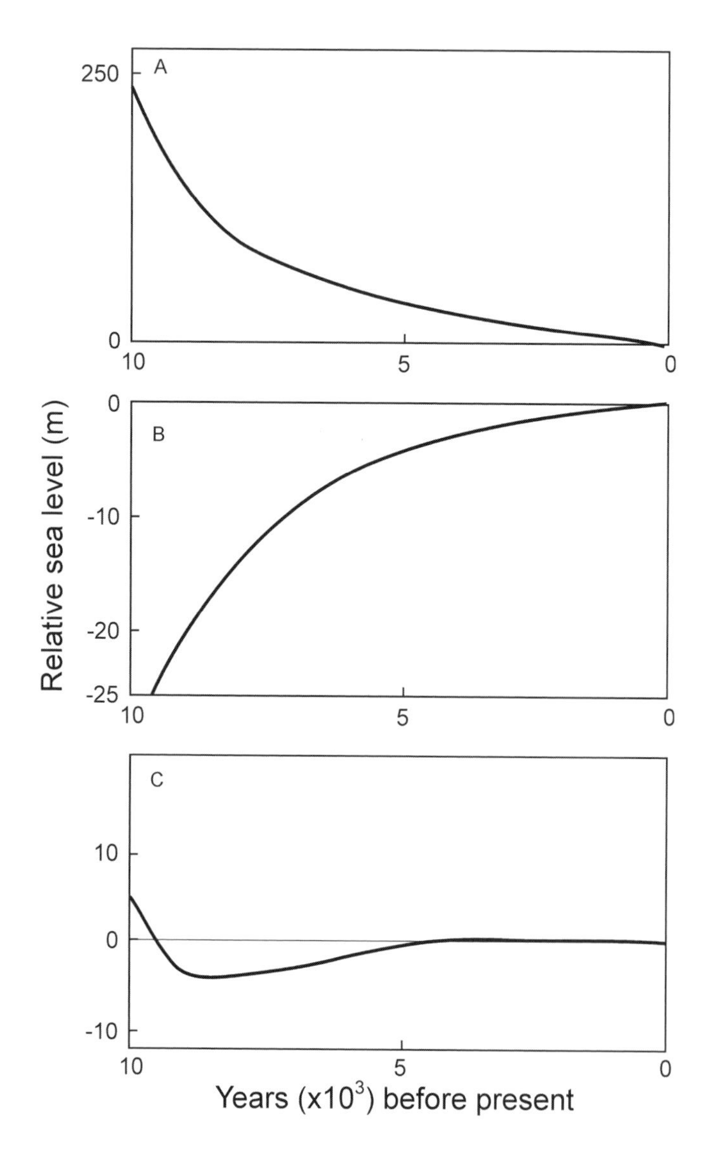

Figure 5.8 Schematic Holocene sea level curves from different zones of glacial isostatic adjustment, near former ice sheets. The vertical scale represents relative sea level change in meters (negative values are below present sea level, positive values above). These generalized curves do not represent specific localities. A. Former ice-covered regions show continuous sea level fall (land emergence), for example, Scandinavia, Canada. B. Collapsed peripheral bulge shows continuous sea level rise (land subsidence), for example, the Netherlands, Belgium, southern England, U.S. East Coast, south of New York City. C. Transitional zone shows an early Holocene sea level fall followed by sea level rise, for example, Maine, Canadian Maritimes. (Gornitz, 2009, "Sea Level Change, Post-Glacial," in V. Gornitz, ed., *Encyclopedia of Paleoclimatology and Ancient Environments*, fig. S8, p. 888, with kind permission from Springer Science + Business Media B.V.)

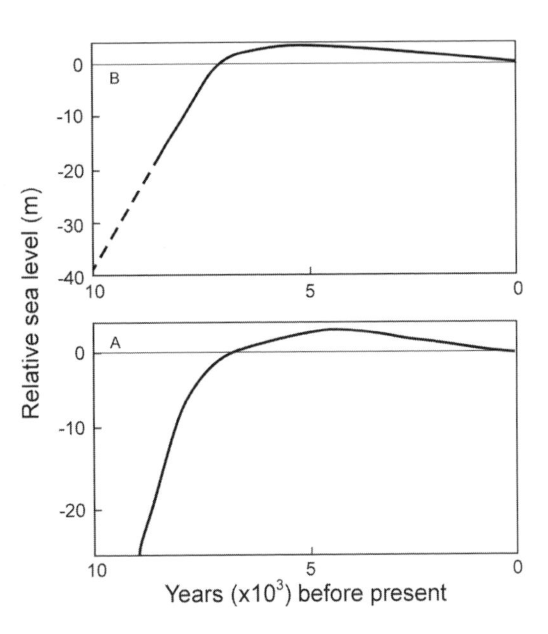

Figure 5.9 Schematic far field post-glacial sea level curves, showing mid-Holocene land emergence. These generalized curves do not represent specific localities. A. Equatorial ocean islands, for example, Cocos (Keeling) Islands, Indian Ocean; Fiji Islands, Pacific Ocean. B. Continental coastlines, for example, eastern Australia, New Zealand, Brazil, West Africa. (Gornitz, 2009, "Sea Level Change, Post-Glacial," in V. Gornitz, ed., *Encyclopedia of Paleoclimatology and Ancient Environments*, fig. S9, p. 889, with kind permission from Springer Science + Business Media B.V.)

isostatic rebound) was later overtaken by the eustatic sea level rise (fig. 5.8c). However, at many far field equatorial ocean islands and continental shorelines, sea level stood several meters higher than present during the mid-Holocene, ~6,000–4,000 years ago, and has been falling slowly there ever since (fig. 5.9).[66] Two different glacial and hydro-isostatic mechanisms are at work.[67] In the first case, water is "siphoned" away from the central equatorial ocean basins into nearshore depressions left by the collapse of peripheral bulges (see fig. 2.6). In the second, the additional weight of meltwater causes an oceanward tilting of far field continental coastlines like a seesaw ("continental levering"), causing water to flow away from the central ocean basins toward the continents (see fig. 2.6). The mid-Holocene sea level highstand has also been interpreted as eustatic in origin, presumably related to minor climate fluctuations detected in Antarctic ice cores and in several paleosea-level proxies elsewhere.

The sea level record of the last 4,000–3,000 years is not well constrained, in spite of its importance as a baseline for understanding contemporary sea level trends. In part, this arises from few accurately dated high-resolution sea level proxies, as well as difficulty in extracting a global climatic signal from low-amplitude oscillations. Minor sea level fluctuations of 1–2 meters (3–6 feet) have been reported from various sites.[68] However, these probably record local to regional climate-related processes, such as variations in river discharge, storm deposition, or coastal sedimentation, rather than true eustatic changes in sea level.

LESSONS FROM THE GREAT MELTDOWN

The great meltdown that ended the last ice age began slowly 21,000–20,000 years ago and largely played out over 12,000 years, ultimately raising sea level by some 120 meters. Several rapid spurts accompanied the overall rise in sea level: ~19,000 years ago (meltwater pulse $1A_0$); 14,600–13,500 years ago (meltwater pulse 1A); 11,500–9,500 years ago (meltwater pulse 1B); possibly 9,000–8,900 years ago, and 8,200–7,600 years ago (meltwater pulse 1C). However, some events (e.g., meltwater pulse 1A) are more firmly established than others (e.g., meltwater pulses $1A_0$, 1B, 1C), and further research is needed to confirm their exact dates and magnitudes. These rapid upswings appear to be closely linked to outbursts of glacial meltwater from rapidly disintegrating ice sheets and glacial lakes, although it has often been difficult to pinpoint specific sources. Meltwater pulse 1A was probably derived from a number of Northern Hemisphere sources, including North America and Eurasia. A minor step increase in sea level may have preceded the Younger Dryas. Some geologic data and modeling studies suggest that outflow from the Laurentide Ice Sheets via an Arctic route could have initiated this temporary return to ice age conditions. Meltwater pulse 1B, on the other hand, could be related to outflow down the St. Lawrence River and from an Antarctic source. However, a stronger case can be built for a connection between meltwater pulse 1C and an outburst of glacial meltwater from Lakes Agassiz-Ojibway, about 8,400 years ago. The influx of this cold freshwater probably triggered the 8,200-year cold event. This meltwater pulse had a weaker effect on sea level rise than the earlier ones, adding at most a few more meters to the ocean. This event essentially ended the main period of sea level rise, since most of the continental ice sheets had melted by then.

Eustatic sea level change has been relatively minor following the end of the major phase of ice sheet melting after 8,000–7,000 years ago. Crustal adjustments to glacial isostasy and hydro-isostasy have continued, although at diminishing rates, leading to variations in sea level change at different localities. The mid- to late Holocene emergence (i.e., apparent sea level fall) of many tropical ocean islands and continental shorelines in the far field has generally been attributed to ocean "siphoning" and "continental levering." Evidence for low-amplitude Holocene sea level oscillations has been reported from a number of localities, but it remains to be determined whether these represent true eustatic (i.e., global) events or regional to local variability.

What have we learned from our excursion into the ice ages and the great ice meltdown? Are we likely to experience a sudden polar ice sheet collapse, as has happened many times before? A study of past sea level changes can provide us with potential scenarios of higher rates of sea level rise from future ice melt than those projected by standard global climate models. However, such analog approaches have their limitations. For one thing, although earlier climates may have resembled that expected in the near future, the driving forces in past climate and sea level changes may have differed significantly from those of the present. For example, the warming of the previous interglacial as well as at the onset of the current interglacial were caused by high northern latitude summer insolation, as predicted by the Milankovitch theory, whereas today's global warming is related to increasing atmospheric greenhouse gas emissions. As a result, the hemispheric and seasonal distribution of warming may differ from that of the past. Another important difference is that the rapid meltwater pulses caused by catastrophic disintegration of massive continental ice sheets (like the Laurentide) during the last deglaciation will not likely play a role in the immediate future (although some believe that the remaining Greenland and West Antarctic ice sheets could potentially destabilize even after a temperature increase of a only few degrees—see chapter 7). Furthermore, the ability to determine accurate dates and rates of sea level rise during the meltwater pulses is limited by a lack of high temporal resolution paleosea-level proxies that could help narrow down the timing of the events and their magnitudes.

Before we probe further into the potential future implications of these former geological events, we still need to examine recent and current trends in ocean behavior. Chapter 6 investigates the sea level changes that have taken place within the last few thousand years and the apparent speedup that has occurred within the last 150 years.

The geological record documents that the rate of rise [of sea level] observed in the 20th century (and apparently accelerating today), is anomalous and far exceeds the natural, pre-anthropogenic rate of rise of 0.7 ± 0.3 mm/yr.

—K. G. MILLER ET AL. (2009)

6 The Modern Speedup of Sea Level Rise

VANISHING GLACIERS

The village of Gletsch in south-central Switzerland was once a way station on the old post coach route, which opened in 1866, over the Furka and St. Gotthard Passes across the Alps into Italy. A 19th-century topographic map hanging in the historic Hotel du Rhone lobby depicts the tongue of the glacier several hundred yards from the hotel. The proximity of the glacier to the hotel also appears on old drawings and postcards dating from about 1860 to 1870 (fig. 6.1). The Rhone Glacier, together with many other alpine glaciers, had advanced during the Little Ice Age, a cold period that gripped Europe and other regions between the 15th and early 19th centuries. After the mid-19th century, a warming trend set in and the glacier began its retreat up the valley. By the mid-1970s, when I visited the site, the glacier snout had receded seven-tenths of a mile up the valley, a short drive from the hotel, and several recessional moraines[1] marked the stages of its retreat. At that time, tourists could walk through a grotto carved into the ice. Since then, the glacier has receded 660 feet farther up the mountainside.[2] Tourists now climb a zigzagging path along the mountainside to reach the ice cave, which is recut frequently as the glacier edge creeps farther upslope. Overall, between 1880 and 2008, the Rhone Glacier has shrunk by 1.3 kilometers (0.81 miles).

Figure 6.1 Rhone glacier as seen from Gletsch, Switzerland, in 1870 and in 2008. (Photo: Dominic Buettner/*The New York Times/Redux*.)

At current melting rates, the glaciers still capping the majestic snowy white peaks of the Swiss Alps may become mere ice patches by the end of this century.

The Athabasca Glacier in the Canadian Rockies has shrunk by approximately 1.5 kilometers (0.93 miles) since the mid-19th century.[3] By 1986, when I toured the area, the glacier was near the Icefields Parkway, linking the Banff and Jasper National Parks, hence making it a unique tourist destination. Large "ice buses" still drive visitors around on the surface of the glacier. However, after a temporary pause between1950 and 1980, the retreat has resumed and the glacier's tongue now sits a few hundred meters farther upslope. The glaciers of Alaska, the Andes, the Himalayas, and especially the tropics tell a similar story.

The shrinking mountain glaciers are the most visible signs of global warming. Lonnie Thompson, a glaciologist at Ohio State University who studies the paleoclimate record stored in mountain glacier ice, has measured the dramatic recession of the snows of Mount Kilimanjaro. The ice on Mount

Kilimanjaro covered an area of ~11.1 square kilometers (~4.3 square miles) in 1912, but in recent years occupies only 2.4 square kilometers (0.94 square miles). Thompson estimates that at current rates the perennial ice will disappear within the next 15 to 20 years. This would mark the first time in 11,000 years that Kilimanjaro would be ice-free.[4] He is concerned that the current thaw is destroying a unique, irreplaceable archive of past climates that may provide us with crucial information about future climate changes. Already, he has had to collect ice cores at higher and higher elevations in the Andes, because the warming has obscured the annual growth layers near the surface at lower elevations.

SIGNS OF A WARMING PLANET

Soaring Temperatures

In addition to vanishing mountain glaciers, many other signs point to a heating planet. Between 1906 and 2005, global temperatures climbed by 0.74°C (1.33°F).[5] The temperature trend nearly doubled in the second half of the 20th century and the pace of temperature rise appears to have been accelerating in recent decades. Although the increase so far may seem minor, to put this into a longer time perspective, Northern Hemisphere temperatures of the past 50 years were likely the highest within the last millennium. The last time northern regions were warmer for prolonged periods than they are at present was 125,000 years ago, during the last interglacial, when the sea stood 4–6 meters or more above present levels (see chapter 4).

Northern Canada, Alaska, and central Asia have warmed the most year-round, while only a few scattered locations have cooled. Western North America, northern Europe, and central Asia have experienced the most Northern Hemisphere winter warming. Springs in Europe, northern, central, and eastern Asia have also grown milder. Other seasons are heating up as well.[6] Along with higher air temperatures, heat has penetrated into the oceans down to 3,000 meters (9,800 feet).[7] As discussed below, the expansion of warming ocean water is a major cause of recent sea level rise.

Effects on Land, Sea, and Ice

Noticeable changes have been spotted in vegetation cover, water, and ice. For example, accelerated glacier ice melting and earlier spring thaws in the mountains have increased stream runoff and generated an earlier peak discharge in spring and early summer. This in turn dries out streams during the hotter, drier summer months, reducing the water supply for downstream urban centers or agricultural irrigation. Other potential hazards facing mountainous regions include the growing risk of avalanches and flash floods as lakes formed by glacier meltwater drain catastrophically. These glacier lake outburst floods could affect people living in the high mountain valleys of the Himalayas, Andes, and Alps.

Ice melt from receding mountain glaciers and thinning Greenland and Antarctic ice sheets is elevating ocean levels. Recent trends from the two polar ice sheets raise concern.[8] Lower sections of the Greenland Ice Sheet are thinning and glaciers are disgorging ice into the ocean more rapidly, adding 0.4–0.7 millimeters (0.02–0.03 inches) per year to the sea within the last decade.[9] Proverbially slow glaciers that reach the sea have also sped up as floating ice shelves that otherwise would stem their advance break apart.

The Antarctic Peninsula, jutting out farther north, has warmed more than the rest of the continent and at least 10 large fringing ice shelves have broken up since the mid-1990s. In January 1995, the 1,994-square-kilometer (770-square-mile) Larsen A Ice Shelf suddenly disintegrated. In 2002, the Larsen B Ice Shelf lost roughly 3,250 square kilometers (1,254 square miles) in a 35-day period. It had remained stable since the end of the last ice age, 12,000 years ago. The Wilkins Ice Shelf, parts of which broke apart in 2008, continued to crumble in April 2009 (fig. 6.2).

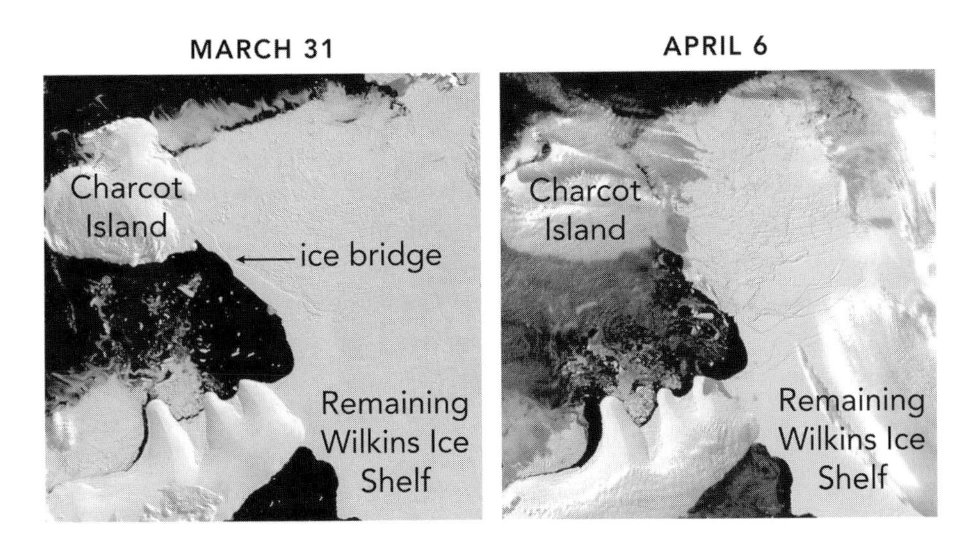

Figure 6.2 Breakup of the Wilkins Ice Shelf, Antarctic Peninsula, March 31 and early April 2009 (NASA.)

This bodes ill for the future of the West Antarctic Ice Sheet (WAIS), which is potentially unstable. Parts of the WAIS are also thinning.[10] Together, melting of Greenland and Antarctica has added ~0.7 millimeters per year (0.03 inches per year) to the sea since the 1990s.[11] While these figures may seem puny, Greenland and the WAIS hold enough ice to raise global sea level by ~6–7 meters (20–23 feet) and 3 meters (10 feet), respectively, if they melted completely!

Satellites see the area of Arctic sea ice diminish: the decline averages 2.7 percent per decade since 1978, with larger summer losses of 7.4 percent per decade.[12] The Southern Hemisphere, by contrast, shows no consistent downward trend in sea ice area or thickness yet. While floating sea ice does not affect sea level (since it has already displaced its weight in water), its disappearance would hasten climate warming and speed up land ice melting through a series of positive feedbacks. When sea ice melts during the summer, the darker water on the surface absorbs more of the sun's incoming energy, which increases the surface water temperature and hastens ice melting. Less sea ice exposes more open ocean water, which absorbs additional solar energy. Sea surface temperatures rise, inducing more sea ice melting (the ice-albedo feedback). Sea ice plays another important climate role as an insulator, sealing in the heat contained in ocean water. Once the ice cover is removed, this stored heat can escape to the atmosphere, warming it up. The enhanced polar warmth stimulates greater ice melt on Greenland, and unlike a shrinking sea ice cover, that does lead to sea level rise. Another feedback stems from the freshening seawater as the ice cover contracts. As explained earlier, this could slow down the NADW and cool the North Atlantic, partly offsetting the warming trend. However, as global temperatures continue to rise, the regional cooling (or reduced warming) will be overshadowed by the global buildup of heat.

Another mark of a warming Earth is the decreasing snow ice cover in the Northern Hemisphere, particularly in spring. Northern snow area has declined in spring and summer since the early 1920s, and even more so since the late 1970s, despite the warmer winters[13] (winters still remain cold enough for snow to fall and accumulate). Since 1988, the average yearly snow cover declined 5 percent. Between 1922 and 2005, the spring snow cover fell 7.5 percent. The decreasing snow cover creates a feedback similar to the ice-albedo feedback, since snow (particularly fresh snow), like ice, has a much higher albedo (i.e., reflectivity) than forests or tundra.

Permanently frozen ground (permafrost) in the Arctic is thawing and as a consequence, the area of wetlands and small lakes is expanding. The soil destabilizes as it thaws, undermining building foundations and roads in Alaska, northern Canada, and Siberia. Thawing permafrost in Alaska is not only buckling roads and damaging buildings but also causing black spruce to tilt at odd angles ("drunken forests"). Once-frozen soil organic matter is rapidly decomposed by anaerobic bacteria, releasing methane and some carbon dioxide. A reduced sea ice cover has also worsened coastal erosion in the Arctic. Sea ice and ice shelves that form in winter protect the coast from storm damage, particularly wherever soft, unconsolidated sediments line the shore. As the Arctic warms, the ice shelves remain for shorter periods, exposing the coast to greater storm-induced erosion.

The biosphere has also begun to respond to the rising temperatures.[14] The growing season has lengthened in many places. Plants unfold their leaves and flowers earlier in the spring, fruits ripen a few days sooner in the fall, and deciduous trees shed their leaves later in the season. Some plant species have migrated toward higher latitudes and altitudes. Alaskan temperatures have soared more than the global average, exacerbating spruce beetle infestations of spruce trees. A potential threat to health and to agriculture is the invasion by tropical and subtropical insect pests as temperatures ameliorate in formerly cooler regions. Migrating birds fly north earlier in the spring and head south later in the fall. Some bird species no longer bother migrating. In the New York City metropolitan area, for example, many Canada geese have become year-round residents, creating a public nuisance as they foul city parks and suburban backyards. Many bird species have begun laying their eggs earlier.

Increasing ocean temperatures have been implicated in more-frequent incidents of coral bleaching (whitening of coral due to the death of symbiotic algae). A severe bleaching outbreak in 1998, associated with a strong El Niño event, killed an estimated 16 percent of the world's coral reefs, especially in the western Pacific and Indian Oceans.[15] As atmospheric carbon dioxide mounts, more CO_2 penetrates into the ocean, acidifying the water.[16] Corals, composed of calcium carbonate, are more soluble in acidic water, which can impede healthy growth. Therefore, coral reefs, which not only harbor a rich, diverse marine ecosystem but also shield tropical ocean islands and beaches against the full fury of major cyclones, may grow more vulnerable to global warming. The ocean warming and possible changes in circulation may also account for the decrease in oxygen concentration observed in various locations.

THE GREENHOUSE EFFECT

As shown in previous chapters, the Earth's climate, sea level, and atmospheric carbon dioxide concentrations have varied widely in the geologic past. Therefore, why should we blame human activities for the observed recent changes in climate? Might not the rising temperatures and atmospheric greenhouse gases just follow a natural cycle—a return of warmth at the end of the Little Ice Age, which succeeded the **Medieval Warm Period**, around 1,000 years ago? Skeptics point to these natural past fluctuations as evidence that climate has always varied throughout Earth's history, so why should this be any different? Natural variability may in fact explain part of the observed changes, but accumulating evidence points to a human role.

Scientists, initially under the direction of Charles Keeling and, since 2005, his son Ralph Keeling, have measured CO_2 concentrations atop Mauna Loa, an extinct volcano on the island of Hawaii. The results show a steady increase in CO_2 from 315 parts per million in 1958, when the time series began, to 387 parts per million in 2009 (fig. 6.3). The upward curve is punctuated by a rhythmic, seasonal rise and fall due to vegetation, mostly located in the

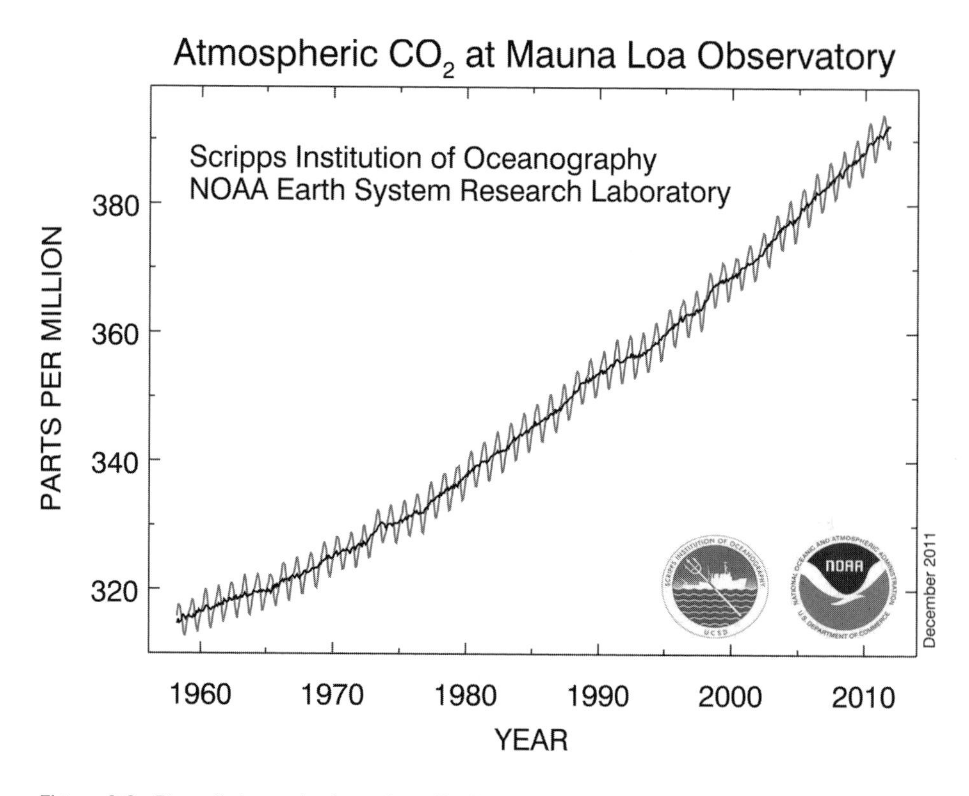

Figure 6.3 Rise of atmospheric carbon dioxide at Mauna Loa Observatory, Hawaii, between 1959 and the present. The waviness in the curve is caused by the seasonal variations in CO_2 consumption by plants, predominantly in the Northern Hemisphere. (NOAA; Scripps Institution of Oceanography.)

Northern Hemisphere, which has a much larger land area than the Southern Hemisphere.[17]

Gas bubbles trapped in polar ice sheets extend the CO_2 record well beyond the instrumental period (see chapter 4). Prior to a.d. 1800 and the onset of the Industrial Revolution, CO_2 levels had remained fairly steady at around 280 parts per million for close to a millennium, but they started climbing upward slowly during the 19th century since the beginning of industrialization. However, they rose sharply since the 1950s. Other greenhouse gases like methane and nitrous oxide also show steep increases since the mid-20th century.[18]

How do these gases affect the Earth's climate? Their heat-trapping ability has been likened to that of a garden greenhouse. The glass structure admits visible and ultraviolet radiation from the sun, which the plants need as an energy source for photosynthesis. However, the glass prevents heat energy in the form of infrared radiation from escaping back to the atmosphere, so the interior of the greenhouse warms up.[19] Similarly, ultraviolet and visible light reach the top of the Earth's atmosphere. Roughly one-third of this incoming radiation is reflected directly back to space by clouds or bright, reflective surfaces like snow or ice, while the rest is absorbed by either the atmosphere or the Earth's surface, warming it (fig. 6.4). To restore energy balance, the Earth re-emits some of this energy at longer wavelengths—in the infrared—since its surface is much cooler than that of the Sun. Atmospheric water vapor, carbon dioxide, and other trace greenhouse gases absorb a portion of the radiation. Some infrared radiation passes through the atmosphere and is lost to space. Thanks to naturally occurring greenhouse gases, primarily water vapor and to a lesser extent CO_2, the Earth's temperature is 33°C (59°F) higher than it would have been in the absence of these gases. This is known as the natural greenhouse effect. However, as indicated above, greenhouse gases apart from water vapor soared exponentially during the 20th century. Natural processes cannot explain this sudden, rapid upswing. On the other hand, the steep rise in greenhouse gases closely matches increases in fossil fuel combustion and deforestation (major sources of carbon dioxide, CO_2), rice cultivation and livestock (sources of anthropogenic methane, CH_4), and industry (nitrous oxide, N_2O).

CLIMATE CHANGE—THE LAST MILLENNIUM

That our planet is heating up is now evident. Yet could this not be just another sharp natural climate swing, like those of the geologic past? While this possibility cannot be entirely excluded, it is increasingly unlikely. Comparison

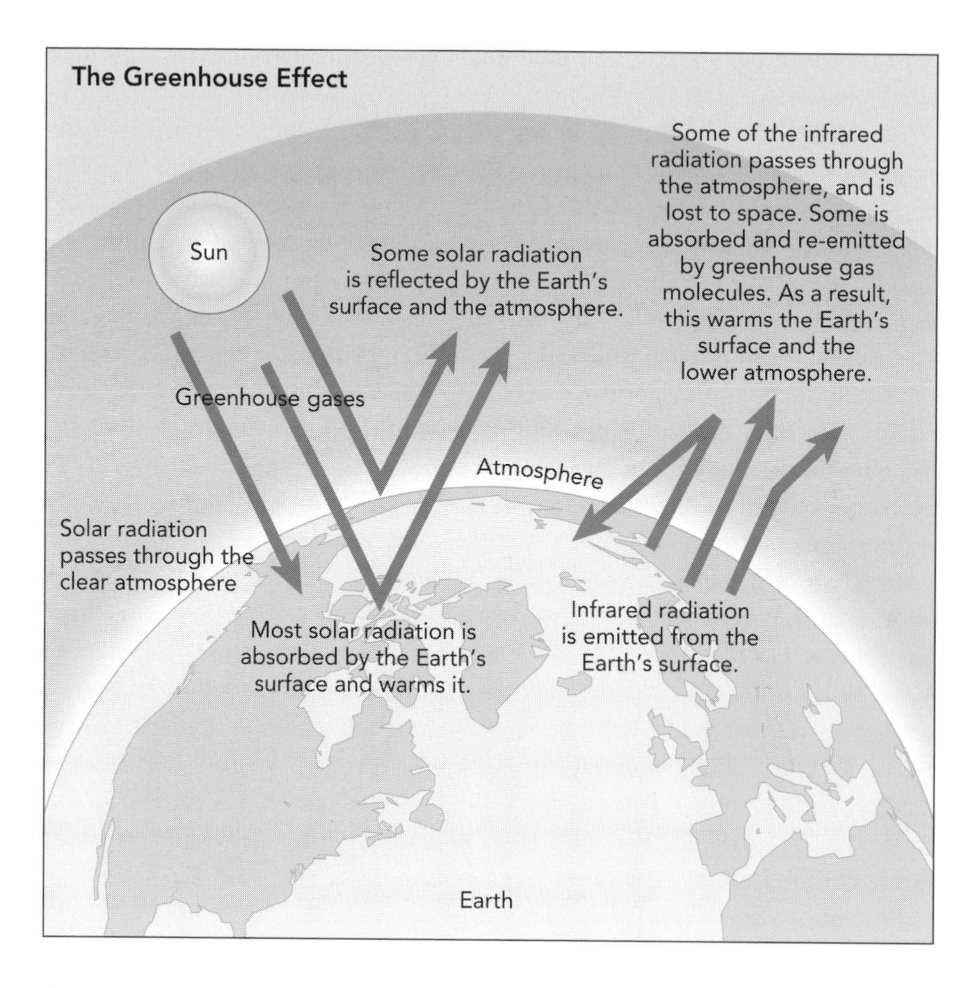

Figure 6.4 The greenhouse effect: trapping of heat by greenhouse gas molecules, such as CO_2 and CH_4, in the Earth's atmosphere. (Adapted from M. E. Mann, 2012, *The Hockey Stick and the Climate Wars*. Copyright 2012 Columbia University Press.)

with the paleoclimate record reveals that today's climate changes are starting to exceed the natural variability inherent in the system. For one thing, the present atmospheric concentrations of carbon dioxide, methane, and nitrous oxide now surpass levels found in the Antarctic ice cores over the last 800,000 years, spanning at least eight glacial-interglacial cycles.[20] Furthermore, the trapped gas bubbles in the ice cores tell us that since the Industrial Revolution began, these greenhouse gases have increased at the fastest average rate of the past 16,000 years!

Another clue that humans are now acting as agents of climate change comes from the dramatic warm-up of the last few decades. The decade of the 2000s was the warmest recorded instrumentally, with 2010 tying 2005 as the warmest ever.[21] Concurrent with the air warming is the rise of ocean temperatures, which contributes to the steadily mounting sea level. Other manifestations of the warming trend have been described above.

The climate of the last 1,000 years has been reconstructed from a broad range of proxies, including ice cores, lake sediments, tree rings, and corals. In spite of differences among records, the early Middle Ages (the Medieval Warm Period) in the Northern Hemisphere between a.d. 900 and 1200 was generally slightly warmer than the 1,000-year average, whereas the period from about 1400 to 1850 (the Little Ice Age) was 0.3°C–0.4°C lower than the long-term average. Since the mid-19th century, when instrumental records became available, the temperature has climbed steadily upward except for a slowdown during the 1940s and 1950s. Since the 1970s, temperatures have soared and now stand well above the natural variability of the last 1,000 years.[22] Global climate models have simulated recent temperature fluctuations, incorporating a number of natural factors that are known to affect climate (such as volcanic eruptions and solar variability).[23] They also include the effects of anthropogenic greenhouse gases, sulfur dioxide, and carbon aerosols (e.g., soot). The computer simulations show that natural driving forces can explain most observed climate changes until around 1960, but since then the anthropogenic factors have dominated.[24]

THE ENCROACHING SEAS

Although the current rate of global sea level rise is still rather small—on the order of several millimeters per year, or just under a foot per century, the specter of the ocean lapping farther and farther inland in the future hangs over the tens of millions of people living along the coasts of the world and on small, low-lying islands. Some regions are already losing ground to the rising seas because they are subsiding due to natural geological processes or excessive pumping of groundwater. Examples include the Mississippi Delta, Chesapeake Bay, and the Nile Delta, as well as cities like Houston, Jakarta, and Bangkok. These places may be like the proverbial canaries in the coal mine—an early warning system that previews what may be in store for many of the world's shorelines in coming years. Before anticipating the future, we will compare today's sea level measurements with the recent past.

Mariners have relied on tide gauges for centuries to track the daily ebb and flow of the tides. In order to derive a more meaningful long-term sea level trend, such instruments should provide data spanning periods over 50 years. However, most of the nearly 2,000 worldwide tidal records are too short or too full of gaps to be useful. Very few records predate 1880. Most of the longer ones are located in northwestern Europe, with some scattered in the United States (New York City, San Francisco, Key West) and elsewhere.

A global compilation of tide gauge measurements shows an upward sea level trend of 1.6 millimeters (0.06 inches) per year from 1880 to 2009, increasing slightly to 1.7 millimeters (0.07 inches) per year from 1900 to 2009 (fig. 6.5).[25] The rise is most likely related to the global warming described above. However, the time series displays considerable decadal variability, with occasional periods of slower growth, for example in the 1960s.

Satellite altimeters (e.g., TOPEX/POSEIDON and Jason) in operation since 1993 measure an even higher sea level trend of ~3.2–3.3 millimeters (0.01 inches) per year between 1993 and 2009—a period, however, that includes several El Niño–Southern Oscillation (ENSO) events characterized by above-average sea surface temperatures (and thermal expansion) over a wide

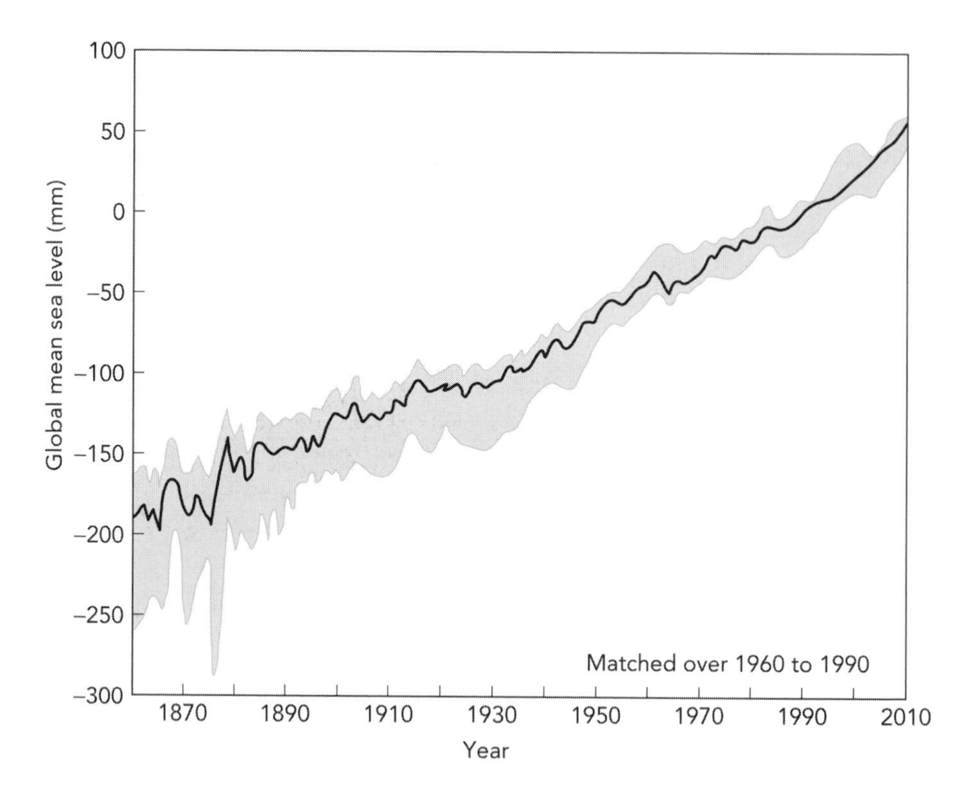

Figure 6.5 Global mean sea level rise from 1860 to 2009 from tide gauge observations and satellite altimeter measurements since 1993 (heavy black line), compared with estimates from other recent studies (shaded area.) A slight acceleration in the trend appears after the early 1990s. (Modified from Church and White, 2011, fig. 6.)

ocean area.[26] Since the satellite record is still fairly short (almost 19 years since 1993), a question remains whether this represents a true acceleration or just normal multi-decadal ocean variability. For example, the rate of sea level rise slowed to 2.5 millimeters per year between 2003 and 2008.[27] Nevertheless, a small but statistically significant acceleration of 0.009 ± 0.004 millimeters per year occurred between 1900 and 2009.[28]

Sources of Recent Sea Level Changes

What is stimulating the recent sea level rise? How is it linked to the mounting global temperature? As discussed in chapter 2, the two dominant processes that can alter ocean height on a centennial timescale include thermal expansion as ocean water warms and addition of meltwater from dwindling mountain glaciers and polar ice sheets with rising global temperatures.

Mountain Glaciers

The introduction to this chapter vividly describes the rapidly vanishing glaciers of the Swiss Alps and the Canadian Rockies. Glaciers throughout the Alps shrank 35 percent in area from 1850 to 1975 and almost 50 percent by 2000.[29] More than a third of the loss has occurred within the past two decades, fueled by global warming and reinforced by natural climate cycles.[30] Approximately 8 percent of the extant ice volume disappeared just in 2003, a year that experienced an extremely hot and dry summer. After a mere five recurrences of such conditions, the European Alps could likely become extensively deglaciated. Inasmuch as 90 percent of the Alpine glaciers are less than 1 square kilometer in area, the likelihood of seeing a nearly ice-free Alpine mountain range within several decades is not unrealistic.

Mountain glaciers and small ice caps worldwide are quickly wasting away. They delivered an estimated 0.50 millimeters (0.02 inches) per year to the ocean between 1961 and 2003 (table 6.1).[31] But this may be an underestimate. A more comprehensive compilation suggests instead that they furnished 0.79 millimeters per year (0.03 inches per year) to the ocean during the same period.[32] This has now increased more recently, between 2001 and 2006, to 1.1–1.4 millimeters (0.04–0.06 inches) per year.[33]

Ice Sheets

Recent signs of ice loss in Greenland and on the West Antarctic Ice Sheet raise growing concern. While meltwater influxes from both Greenland and

Table 6.1 Current rates of ice sheet melting

Greenland Ice Sheet
0.23 mm/yr (1996) to 0.57 mm/yr (2005) (Rignot and Kanagaratnam, 2006)
0.31 mm/yr (1960s) to 0.74 mm/yr (2007) (Rignot et al., 2008)
0.46 mm/yr (2000–2008); 0.75 mm/yr (since 2002) (van de Broeke et al., 2009)
West Antarctic Ice Sheet
0.14 mm/yr (1996–2006) (Rignot et al., 2008)
0.37 mm/yr (2006 only) (Rignot et al., 2008)
0.4 mm/yr (2002–2005) (Velicogna and Wahr, 2006)
0.37 mm/yr (2002–2008) (Chen et al., 2009)
Antarctic Ice Sheet
0.23 mm/yr (1996–2006) (Rignot et al., 2008)
0.54 mm/yr (2006 only) (Rignot et al., 2008)
0.53 mm/yr (2002–2008) (Chen et al., 2009)
Combined Greenland and Antarctic Ice Sheets
0.7 mm/yr (1993–2007) (Cazenave and Llovel, 2010)
1.3 mm/yr (2006) (Rignot et al., 2011)

Antarctica still represent a fraction of the total rise in sea level, they have increased rapidly in recent years.[34] Melting ice in both Greenland and Antarctica added some 0.35 millimeters (0.01 inches) per year to the world's oceans from 1992 to 2006 (fig. 6.6).[35] In 2006 alone, the combined ice loss of Greenland and Antarctica yielded the equivalent of 1.3 millimeters (0.05 inches) per year sea level rise (table 6.1).[36]

Satellites detect a thinning of the Greenland Ice Sheet at lower elevations and ice entering the ocean more rapidly, raising sea level by 0.23 to 0.57 millimeters (0.09–0.2 inches) per year between 1996 and 2005.[37] Newer measurements over Greenland by the gravity satellite GRACE find 0.46 millimeters (0.02 inches) per year between 2000 and 2008, increasing to 0.75 millimeters (0.03 inches) per year since 2006 (table 6.1).[38] Summer surface ice melting from the Greenland Ice Sheet reached record levels in 2010, lasting up to 50 days longer than the average 48-day annual melting between 1979 and 2009.[39] The 2007 and 2008 summer melt seasons did not lag far behind.[40] The

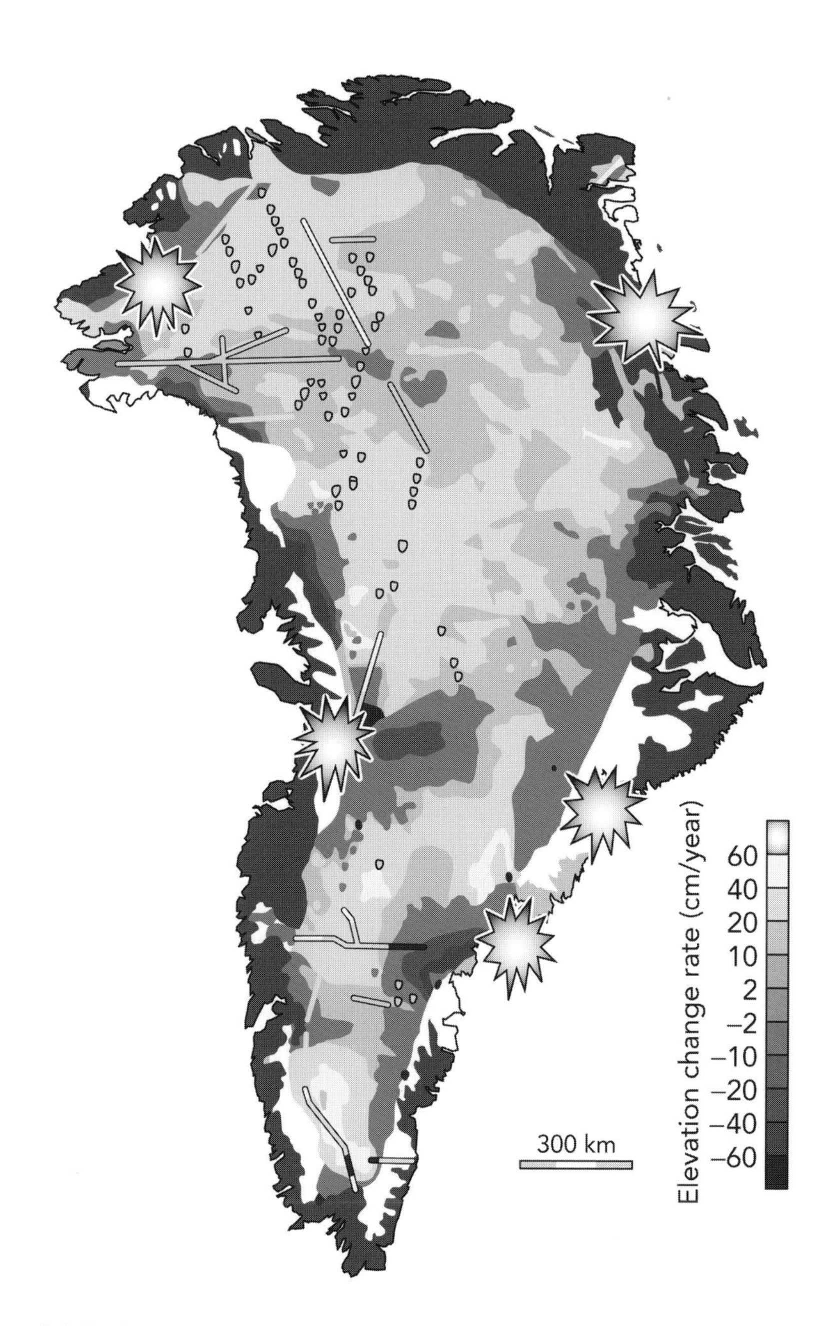

Figure 6.6 Rapid ice thinning and glacier discharge in Greenland. "Explosions" mark accelerating outlet glaciers, buttressed by thinning grounded ice shelves. (Adapted from Robert Bindschadler, NASA GSFC.)

ice losses, hitherto concentrated mainly in the southeast and southwest, may have been spreading northward into northwestern Greenland since 2005.[41]

The reason for the ice loss is no big mystery: summer temperatures in southern Greenland are several degrees higher than in the early 1990s and the melt season has correspondingly lengthened. As a result, water pools on the surface of the ice sheet, enters crevasses and open passageways (**moulins**) in the ice, and cascades down to the bottom of the glacier. There, the accumulating layer of water reduces friction, enabling the glacier to accelerate toward the sea, where large chunks of ice calve off to form icebergs (fig. 6.7).

Calving icebergs trigger glacial earthquakes, increasing glacial velocity, at least temporarily. These events peak in the summer months, falling off in winter and early spring. The number of glacial earthquakes in Greenland increased between 1993 and 2005, but has decreased since then.[42] Antarctic glacial earthquakes have been less well studied, but behave in a similar manner to those in Greenland.

The flow of glaciers is also hastened by warming of the surrounding ocean water. Many glaciers in Greenland and Alaska that reach the sea are held in check by floating ice shelves or ice tongues, which act like buttresses supporting a decrepit, shaky old building. Removing this support is like releasing water from a dam. A warming ocean melts the ice tongue from beneath, undermining it; the glacier then surges forward dramatically, discharging more icebergs (fig. 6.8). For example, the warmer ocean water brought in by changing wind patterns triggered a chain of events that culminated in the acceleration of the Jakobshavn Isbrae glacier in Greenland.[43] Water temperatures just a few degrees above freezing quicken submarine ice melting and iceberg discharge. Observed as well as calculated patterns of glacial retreat in Greenland imply that presently dynamic changes at the glacier's seaward edge may be more important than basal lubrication.[44] To the extent that higher summer air and water temperatures have induced the ice losses already under way (although highly variable on an annual basis), a sustained warming over much longer periods could lead to a significant future drawdown of the Greenland Ice Sheet.

Antarctica, straddling the South Pole, is much colder than Greenland, yet nevertheless it has warmed over the last half century at an average rate of 0.12°C per decade.[45] It contributed the equivalent of approximately 0.23 millimeters per year of sea level rise between 1996 and 2006, and 0.53 millimeters per year between 2002 and 2008.[46]

The Antarctic Peninsula warmed an average of 0.11°C (0.2°F) per decade between 1957 and 2006, while the West Antarctic Ice Sheet (WAIS) has heated up by 0.17°C (0.3°F) per decade.[47] The dramatic breakup of at least 10 fringing ice shelves on the Antarctic Peninsula since the 1990s has been

Figure 6.7 Surface summer ice melting on the Greenland Ice Sheet. Water cascades down a moulin (open passageway) and pools at the bottom of the glacier. The wet bottom ice slides faster to the sea, discharging more icebergs. (Photo: Marco Tedesco, City College of New York CUNY.)

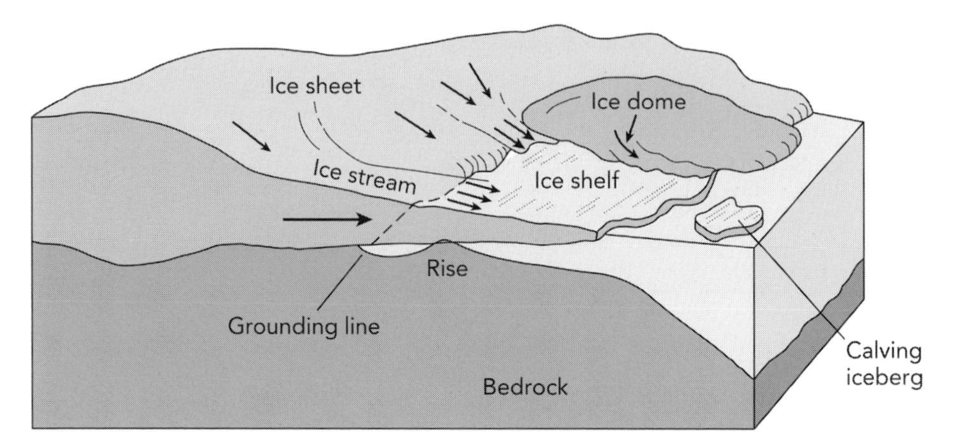

Figure 6.8 Sketch of an ice sheet, showing the grounding line—the dividing line between the portion of the ice sheet resting on solid bedrock and its floating extension on the ice shelf. Icebergs are formed as the ice breaks off the shelf or tongue.

noted above. Following the breakup of the Larsen Ice Shelf, its tributary glaciers began to surge forward.[48] Over the past 60 years, 87 percent of the glaciers on the Antarctic Peninsula have shrunk and the boundary of retreating glaciers has progressively shifted southward (i.e., toward the central landmass of Antarctica) in a manner consistent with the warming trend.[49] At present, the peninsula is adding approximately 0.1 millimeter (0.004 inches) per year to the sea.[50]

The fairly small glaciers on the Antarctic Peninsula would at most add only half a meter (20 inches) to sea level if all of them melted. However, their surging behavior after the demise of the fringing shelves may serve as a model for the future of the much larger West Antarctic Ice Sheet (WAIS). WAIS is potentially unstable because large portions are "grounded" below sea level; in other words, the ice rests on bedrock that lies below sea level. The ice sheet extends out to sea and floats as an ice tongue or shelf. The dividing line between the part of the ice sheet grounded on solid bedrock and its floating extension is called the **grounding line** (fig. 6.8). It is especially sensitive to increasing sea level, which lifts the floating ice and forces the grounding line to retreat inland. The pullback quickens if the bedrock topography slopes inland, as is the case for the WAIS. Once the process gets under way, the ice sheet can surge forward and thin, eventually breaking up. How much and how long it could take to disintegrate under plausible future scenarios of climate warming will be the focus of chapter 7.

The WAIS contains enough ice to raise sea level by 3.3 meters (10 feet) if it were fully melted.[51] Portions of the WAIS are already thinning. Satellite laser altimeter and radar measurements of elevation and ice thickness changes show that the WAIS lost ice equivalent to ~0.14 millimeters (0.01 inches) per year of sea level rise between 1996 and 2006, and 0.37 millimeters (0.015 inches) per year between 2002 and 2008.[52] The Pine Island and Thwaites Glaciers, once described as the "weak underbelly of West Antarctica," on a sector of the WAIS abutting the Amundsen Sea, have been thinning and retreating in recent years. These two glaciers contain the equivalent of 1.5 meters (5 feet) of sea level rise and may be the major source of the current WAIS mass loss. Although this region has warmed slightly less than the Antarctic Peninsula, upwelling of warmer, deeper water by shifting ocean currents may be eroding the undersides of the ice shelves, reducing their buttressing support and enabling the glaciers to surge forward. Unlike the more stable East Antarctic Ice Sheet, the WAIS has waxed and waned in the geologic past. Drill cores data along with modeling experiments suggest an elimination of surface ice if summer temperatures stayed above freezing for much of the season.[53] Furthermore, ocean temperatures above 5°C (9°F) could trigger a collapse of the WAIS, by undermining the fringing ice shelves. Atmospheric CO_2 levels exceeding 400 parts per million would also increase the vulnerability of the WAIS. Under some climate change scenarios, these thresholds could be exceeded by the end of the century, although no one knows for certain how long the WAIS would take to disintegrate—whether a century or two, or a millennium.

Ocean Thermal Expansion

As ocean water heats up, it expands in volume and its density decreases, raising sea level. Between 1961 and 2003, most of the world's upper ocean surface warmed down to depths of 700 meters (2,300 feet). Sea level correspondingly climbed by ~0.3–0.5 millimeters per year during this period.[54] From 1993 to 2007, this increased to ~1.0 millimeter per year, representing roughly a third of the observed rise.[55] But the importance of thermal expansion is likely to decrease if substantial amounts of the Greenland and WAIS Ice Sheets melt in coming centuries.

Ocean Circulation

Ocean height is also influenced by currents and circulation patterns. Unlike water in a bathtub, the dynamic ocean pushes water preferentially in certain directions, elevating sea level in some regions more than in others. Climate warming could influence regional sea level indirectly, for example, by slowing down the ocean conveyor belt system. Sudden changes in thermohaline circulation have been blamed for past abrupt climate shifts (e.g., the Heinrich and Dansgaard-Oeschger events of the last ice age, or the Younger Dryas cold reversal and subsequent warming since the last deglaciation; see chapters 4 and 5). If more ice melted, northern seawater would freshen, weakening North Atlantic Deep Water (NADW) formation due to the lower salinity gradient. This in turn would reduce the delivery of warmer waters to the north, at least temporarily, initiating a regional cooling counter to the worldwide warm-up. Aside from its potential climatic consequences, a weaker NADW would also alter regional sea level, more noticeably along the northeast coast of North America.[56]

Sea level along the U.S. East Coast is generally lower than in the mid- or eastern Atlantic, because a steep sea-surface slope exists between the fast-moving, strong, and narrow Gulf Stream–North Atlantic Current, on the one hand, and more elevated water toward the center of the subtropical North Atlantic ocean gyre on the other (see chapter 1). Thermohaline circulation would slow because of decreasing salinity of sub-polar waters from ice melt and from river runoff due to heavier rainfall. The main locus of North Atlantic Deep Water sinking would shift southward. As a result, water along the western Atlantic continental margin would warm, the sea surface gradient would diminish, and sea level along northeastern North America would rise.[57] However, regional tide gauges along the eastern seaboard have not yet detected an upswing in sea level (beyond the small historic trend). While the North Atlantic may have freshened recently, signaling the onset of the predicted Atlantic heat conveyor slowdown, the record is still too short to establish a true trend, given the high degree of multi-annual to multi-decadal ocean variability.[58]

Gravity

Gravitational effects due to dwindling ice masses can also affect sea level. Gravitational attraction changes as ice sheets and glaciers lose mass to the oceans. A massive ice sheet draws ocean water toward it, elevating nearby sea level. As the ice sheet thins or shrinks, the gravitational pull lessens and nearby sea level drops, in spite of the added meltwater. (However, it rises much farther away from the ice mass).

Land Water Storage

A net transfer of water from continental reservoirs to the ocean, or vice versa, will affect the ocean volume and, hence, sea level. Continental water storage

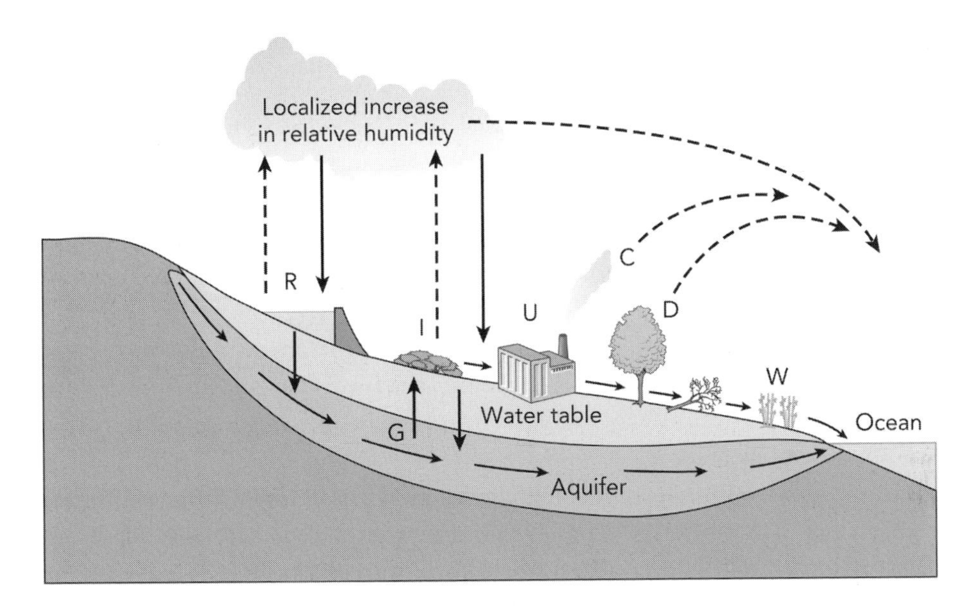

Figure 6.9 Human modifications of the land hydrological cycle. R, reservoir impoundment; G, groundwater mining; I, irrigation; U, runoff from urbanization; C, localized increase in relative humidity, D, runoff due to deforestation; and W, wetland drainage. (Modified from Gornitz, 2001.)

can vary for a number of reasons, some of which are climatic—i.e., related to variations in rainfall, but increasingly since the mid-20th century due to human transformations of the hydrological cycle (fig. 6.9).[59] Groundwater mining (water withdrawal in excess of natural recharge) may add ~0.25 millimeters (0.01 inches) per year to the sea. Since vegetated surfaces absorb more water than bare soil, deforestation and draining of wetlands tend to increase river runoff (at least initially, until plants grow back). Urbanization also increases runoff because of the impermeable surfaces that impede groundwater recharge. These processes transfer continental water to the sea. On the other hand, construction of dams, particularly since the 1950s, has retained the equivalent of 0.5 millimeters per year.[60] Water is also stored on land through seepage into porous rocks beneath reservoirs, particularly in arid lands. Similarly, irrigation for agriculture in arid climates increases water storage due to infiltration into the water table. Except for reservoirs, quantitative figures for changes in continental water storage are highly uncertain or unavailable. Since these counterbalancing processes largely cancel each other out, their effects on sea level have not been included in table 6.2.

Crustal Land Motions

Vertical land movements near the coast with respect to the ocean produce apparent changes in sea level even for a static ocean height. Multiple processes can induce these land motions—fault displacements, glacial isostasy, sediment loading (e.g., subsidence of the Mississippi or Ganges-Brahmaputra Deltas), or excessive pumping of subsurface water or hydrocarbon fluids.

Regional to local sea level changes can deviate widely from the global average, since the above-mentioned processes vary in magnitude from place to place and over time. Therefore, "global" sea level rise is far from uniform and will vary geographically. Figure 6.10 schematically summarizes the main causes of sea level change.

The 2007 UN-sponsored **Intergovernmental Panel on Climate Change (IPCC)** has compiled the major contributions to global sea level rise from 1961 to 2003 (table 6.2; see also fig. 6.10). All sources have increased since the 1990s. The IPCC attributes 57 percent of the total calculated rise in global sea level since 1993 to ocean thermal expansion and 43 percent to ice melt. However, recent studies suggest that thermal expansion accounts for only 30 per-

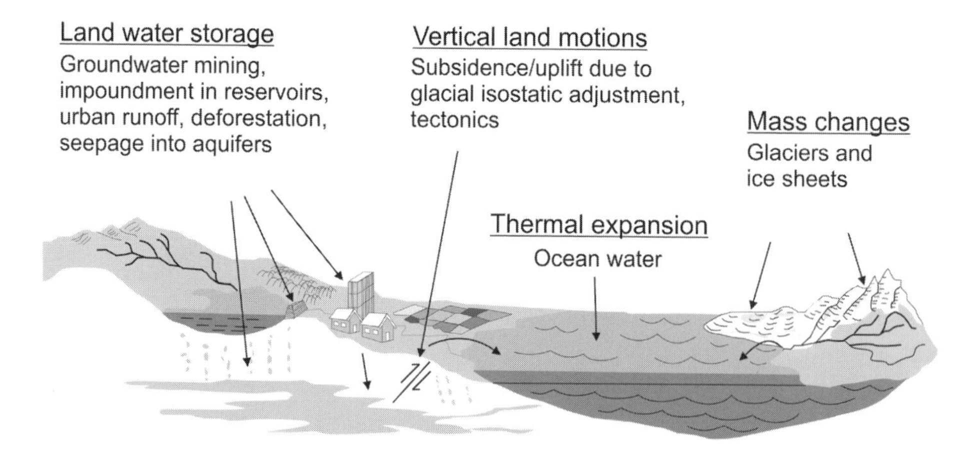

Figure 6.10 Major causes of sea level rise. (1) Climatic: (a) *changes in ice mass*—melting of mountain glaciers and ice sheets; (b) *changes in ocean volume*—thermal expansion. (2) Non-climatic: *vertical land motions*—glacial isostatic adjustments, tectonics; continental water storage (dams, groundwater mining, seepage, urban runoff, irrigation, deforestation, subsurface fluid withdrawal). Not shown: changes in ocean circulation, gravity. (V. Gornitz.)

Table 6.2 Rates of sea level rise, late 20th century

Contributions to SLR	SLR 1961–2003, mm/yr	SLR 1993–2003, mm/yr
Thermal expansion	0.42 ± 0.12	1.6 ± 0.5
Glaciers and ice caps	0.50 ± 0.18	0.77 ± 0.22
Greenland Ice Sheet	0.05 ± 0.12	0.21 ± 0.07
Antarctic Ice Sheet	0.14 ± 0.41	0.21 ± 0.35
Total	1.1 ± 0.5	2.8 ± 0.7
Observed	1.8 ± 0.5	3.1 ± 0.7
Difference (Obs. - total)	0.7 ± 0.7	0.3 ± 1.0

Source: IPCC (2007).

cent of the observed rise as compared with ~55 percent for ice melt, which may have increased even further proportionally within the last few years.[61]

The IPCC found an unexplained gap of 0.7 millimeters per year in observed versus calculated sea level rise between 1961 and 2003 (i.e., 1.8 vs. 1.1 millimeters per year), sometimes referred to as a "sea level enigma." This gap narrowed considerably between 1993 and 2003 (table 6.2).[62] The IPCC probably significantly underestimated the growing contribution of ice melt and overestimated that of thermal expansion.

The foregoing review demonstrates the close relationship between the recent sea level speedup and the worldwide warming trend that began sometime in the mid- to late 19th century but has accelerated in recent decades, largely associated with growing anthropogenic inputs of greenhouse gases into the atmosphere. Ocean water has warmed and expanded, raising the water height. But even more dramatic are the changes in the **cryosphere**—the frozen portions of the Earth's surface. Glaciers, ice sheets, sea ice, snow cover, and permafrost have all receded markedly. Of these, only glaciers and ice sheets contribute directly to sea level rise, and as mentioned above, they now constitute the dominant source.

THE BEGINNINGS OF THE MODERN SEA LEVEL SPEEDUP

After the large ice sheets that blanketed much of North America and northern Eurasia during the last ice age vanished after ~7,000–5,000 years ago, eustatic sea level rise slowed to a tiny trickle. The IPCC report, for example, notes that sea level changes over the last 2,000 years were not more than

0 to 0.2 millimeters per year.[63] However, local (relative) sea level still varied because of ongoing, although diminishing, glacial and hydro-isostatic (and other vertical crustal) movements. For the 20th century, tide gauges register a global average sea level rise of 1.7 to 1.8 millimeters per year, accelerating further to more than 3 millimeters per year since 1993. When did modern sea level increase begin to pick up speed? How does the current trend compare with that of the period before tide gauges?

Tide gauge data over the past century (and longer, in a few cases) have been matched to dated geological materials from the same locality.[64] The 20th-century sea level trend generally exceeds that of the last few thousand years by 1–2 millimeters per year. In southern New England, for instance, the average rate of sea level rise was 1 millimeter per year from around 1300 to 1850, as compared to 2.2–2.8 millimeters per year from nearby tide gauges with records dating back to 1856.[65] In Nova Scotia, sea level rose by 1.6–1.7 millimeters per year prior to 1900; thereafter the rate nearly doubled, to 3.2 millimeters per year.[66] From 5,000 to 500 years ago, the New Jersey sea level trend held steady at 1.8 millimeters per year,[67] in stark contrast to 20th-century rates of around 4 millimeters per year, based on local tide gauges.[68] The recent spurt in sea level rise is not confined just to North America. In New Zealand, sea level rose at a leisurely pace of 0.3 millimeters per year from ~ a.d. 1500 to 1900 but accelerated to 2.8 millimeters per year during the 20th century.[69] Some older studies covering northwestern Europe, Australia, eastern North America, and New Zealand also detect increases of ~1–2 millimeters per year in modern over late Holocene rates of sea level rise.[70]

The precise date of onset for the current accelerated phase is uncertain because of limited instrumental data before the 1850s and lack of high-resolution proxy sea level data for the last few thousand years. Most of the above studies suggest that the modern acceleration may have started between 1850 and 1900—a period also corresponding to the beginning of the rise in greenhouse gas emissions, the warming trend, and the retreat of mountain glaciers. Sea level may be accelerating even more since the 1990s, when satellite altimeters began to pick up a much higher sea level trend of more than 3 millimeters per year. While the satellite record is still too short to rule out possible inter-annual variability, the longer tide gauge record from 1870 to 2004 detects a small acceleration during this period.[71]

A LOOK AHEAD

If global warming continues and intensifies, as most computer climate models predict, ice melt will eventually dominate future sea level rise, overshadowing the role of thermal expansion. The ice locked up in mountain glaciers can add no more than 0.3 to 0.5 meters to the sea.[72] Even so, a small residual fraction will probably survive at the highest elevations and in colder, high-latitude regions. Accelerated melting of the Greenland Ice Sheet due to attrition and thinning at its terminus may eventually slow down and end, since surviving ice lies at much higher altitudes. While this particular retreat mechanism may diminish in importance, surface melting will likely increase if the recent warming trends persist and intensify.

The past history of the ice sheets poses sober lessons for the future. The last interglacial, 125,000 years ago, when sea level stood 4–6 meters higher than at present, experienced climate conditions similar to those expected in the next century or two. Will the Earth's climate heat up enough to push the ice sheets into an irreversible downward spiral? Even if greenhouse gas emissions are curbed or halted within the next few decades, thermal expansion will continue to elevate sea level as heat slowly penetrates deeper and deeper into the oceans. These effects may last for many centuries. Just how much more sea level rise can we expect? Will this planet become a "water world"? Will most coastal cities resemble Venice, where most traffic moves by boats on canals, high waters now flood historic piazzas nearly every winter, and people wear knee-high boots and walk around the piazzas on wooden pedestrian walkways? Chapter 7 explores plausible scenarios of sea level rise as the world warms. Later chapters will examine ways of coping with the rising waters and offer hints on means of slowing down or stopping the impending deluge.

We can't rely on future volcanic eruptions slowing ocean warming and sea level rises.
—PETER GLECKLER, LAWRENCE LIVERMORE NATIONAL LABORATORY

7 Sea Level Rise on a Warming Planet

A HARBINGER OF THINGS TO COME?

In late April 2009, huge chunks of ice broke off a shelf on the western Antarctic Peninsula in Antarctica. Within days some 700 square kilometers (270 square miles) of ice had dropped off the Wilkins Ice Shelf, adding to the 330 square kilometers (127 square miles) of ice that crumbled within a week earlier that month, and 400 square kilometers (160 square miles) from the previous year (see fig. 6.2). Since the 1990s, the Wilkins Ice Shelf has lost 40 percent of its ice in a series of dramatic disintegrations, joining a growing list of ice shelves on the Antarctic Peninsula that have shattered within the past 20 years. On the east side of the Antarctic Peninsula, the Larsen A Ice Shelf broke apart in 1995, followed in 2002 by the Larsen B Ice Shelf, which once covered an area the size of Rhode Island. At least 10 ice shelves fringing the Antarctic Peninsula have retreated in recent decades (table 7.1).[1] Jutting out farther north than the rest of Antarctica, this region has warmed by 1–2°C (1.8–3.6°F) within the past half century.[2]

Ice shelves fringe large segments of the Antarctic continent and some fjords and bays of Greenland and Ellesmere Island, Arctic Canada.[3] Episodic calving of icebergs is normal. But the disintegration of multiple ice shelves in the Antarctic Peninsula (and also in Greenland, although on a much smaller

Table 7.1 Retreat of Antarctic Ice Shelves

Ice Shelf	*Period*	*Cumulative Loss (square kilometers)*
Jones	1947–2003	25
Larsen Inlet	1986–1989	407
Larsen A	1986–1995	2,168
Larsen B	1986–2008	17,411
Larsen C	1976–2003	10,600
Müller	1956–1993	31
Northern George VI	1974–1995	993
Prince Gustav	1945–1995	2,000
Wilkins	1990–2009	5,141
Wordie	1966–1989	1,300

Based on data from the National Snow and Ice Data Center (http://nsidc.org/sotc/iceshelves.html) and the European Space Agency (http://www.esa.int/esaCP/SEMPAVANJTF_index_2.html) as of December 2009 (ice loss values are approximate).

scale) over the last few decades is a more recent phenomenon, closely associated with regional and global warming. The breakup and melting of the floating ice by itself does not raise sea level, inasmuch as it has already displaced its mass in water. However, the attached shelf acts as a barrier, retarding even further the proverbially slow forward motion of the glacier or ice stream (see fig. 6.8). After removal of such a barricade, the glacier accelerates forward, as has happened to a number of glaciers on the Antarctic Peninsula over the past half century, once their fringing shelves disintegrated.[4] Icebergs shed by surging glaciers add to sea level rise, because that ice was originally land-based.

Although the small glaciers on the Antarctic Peninsula would contribute a mere half meter (20 inches) to sea level if all of them melted, their speedup after removal of the fringing shelves may foreshadow the future behavior of the much larger West Antarctic Ice Sheet (WAIS). The WAIS holds enough ice to raise sea level by more than 3 meters (>10 feet)[5] and is potentially unstable, because large portions are grounded below sea level.[6] Melting ice would flood much of the area. To make matters worse, rising sea level would also lift the floating ice shelves, push the grounding line farther inland, and speed up the ice stream or glacier flow. Warming ocean waters may further melt and undermine the bases of the ice shelves.

Just as rivers that drain continents ultimately reach the sea, most of the ice in Antarctica eventually enters the ocean, traveling along faster-moving ice streams that usually terminate in broad, floating ice shelves. One such frozen stream is the Pine Island Glacier, which feeds the Amundsen Sea sector of the West Antarctic Ice Sheet (WAIS). It has been thinning and speeding up since the mid-1990s.[7] While the cause is uncertain, a leading candidate is ocean warming beneath the fringing ice shelf, which may be undermining it, thereby enabling the Pine Island Glacier to quicken its downstream pace. In the Northern Hemisphere, the Jakobshavn Isbrae glacier on the west coast of Greenland has also thinned and doubled its velocity within the last decade.[8] These recent observations may be part of a more widespread phenomenon.[9] The ICESat (Ice, Cloud and land Elevation Satellite) laser altimeter has detected extensive thinning on the edges of both the Greenland and the Antarctic Ice Sheets. Southeast and northwest Greenland ice margins also show marked thinning.[10] In Antarctica, the most pronounced effects appear along the Pine Island, Smith, and Thwaites Glaciers near the Amundsen Sea Embayment, and elsewhere along the margins of the WAIS, the Antarctic Peninsula, and to a lesser extent even in spots in East Antarctica. Are these the first manifestations of a recent meltdown induced by global warming?

As mentioned in chapter 6, atmospheric greenhouse gas levels now surpass those of the last 800,000 years, the upward trend in CO_2, first recorded at Mauna Loa in 1959, continues its ascent unabated, temperatures have begun to exceed those of the last millennium, and the average 20th-century rate of sea level rise has jumped from 1.7–1.8 millimeters (0.07 inches) per year to more than 3 millimeters (0.01 inches) per year since 1993. While ocean thermal expansion remains an important source of sea level rise, the fraction due to ice melt (from all sources) is increasing. What does this portend for the future?

As atmospheric greenhouse gases continue to accumulate, will soaring temperatures provoke another major ice sheet meltdown, with sea level rising several meters, drowning low-lying Pacific islands, major deltas, and sections of most of the world's large coastal cities? Or will the increase in sea level stay below a meter, producing consequences that, although troubling, are more manageable?

Climate scientists look ahead by using computer-generated global climate models (GCMs) that attempt to replicate mathematically, the major natural atmospheric and oceanic processes, based on the known laws of physics that govern our climate. In 2007, the Intergovernmental Panel on Climate Change (IPCC) compiled the most comprehensive synthesis of recent computer model studies, reflecting the consensus view of several hundred of the world's leading experts in diverse fields in the natural and social sciences.[11]

These models incorporate current trends in various components of the climate system, and also provide a suite of possible future outcomes, based on admittedly simplified scenarios of economic development, fossil fuel consumption, and population growth, in addition to geophysical processes. The following sections summarize the main IPCC sea level rise projections and discuss why they may paint an overly conservative picture of the future.

ANTICIPATING THE FUTURE

The IPCC sea level rise projections employ a suite of 17 atmospheric and oceanographic global climate models (AOGCMs) from leading academic and government research institutes worldwide, encompassing multiple greenhouse gas emissions scenarios resulting from a broad range of distinct pathways of economic, demographic, and technological change. The scenarios span the current range of uncertainties in anticipating future greenhouse gas emissions. The resulting set of scenarios can be grouped into four main "storylines" according to different views of regional versus globalized economic development, rates of growth, and degrees of environmental protection. Each individual scenario represents a specific outcome of one of the four storylines. Because most AOGCMs calculate only the thermal expansion component of sea level rise, the meltwater components (i.e., from mountain glaciers, Greenland, and Antarctica) must be estimated separately. On the basis of past records, these ice masses are assumed to melt and add ice to the oceans in proportion to the increase in global average temperature. Adjustments are made for reduction in glacier or ice sheet area with rising temperature. Some allowance is also made for even more rapid ice discharge, by adding an extra 0.1–0.2 meters (0.3–0.7 feet) to the sea level rise.[12]

Table 7.2 shows the increase in sea level for six representative emission scenarios for the late 21st century (2090–2099) over that of the late 20th century (1980–1999). By the end of this century, the total rise in global mean sea level for these six scenarios lies between 18 and 59 centimeters (7–23 inches) (table 7.2). Although thermal expansion constitutes nearly a quarter of the total observed sea level rise between 1961 and 2003 (table 6.1), its share could represent 70–75 percent of the total by 2100, according to the IPCC. Meltwater from both mountain glaciers and Greenland would furnish most of the balance. However, the climate models foresee that Antarctica will add nothing to the rising seas, or may even remove water, because global warming will augment precipitation over that continent and the increased snowfall will more than outweigh any expected ice melting. On the other hand, as emphasized in chapter 6 and below, the ice melt components already

Table 7.2 IPCC projection of sea level rise between 1980–1999 and 2090–2099 (in meters). The numbers represent the 5% to 95% confidence intervals.

Contributors to SLR	B1	B2	A1B	A1T	A2	A1F1
Thermal expansion	0.10 to 0.24	0.12 to 0.28	0.13 to 0.32	0.12 to 0.30	0.14 to 0.35	0.17 to 0.41
Mountain glaciers	0.07 to 0.14	0.07 to 0.15	0.08 to 0.15	0.08 to 0.15	0.08 to 0.16	0.08 to 0.17
Greenland	0.01 to 0.05	0.01 to 0.06	0.01 to 0.08	0.01 to 0.07	0.01 to 0.08	0.02 to 0.12
Antarctica	−0.10 to −0.02	−0.11 to −0.02	−0.12 to −0.02	−0.12 to −0.02	−0.12 to −0.03	−0.14 to −0.03
Land ice sum	0.04 to 0.18	0.04 to 0.19	0.04 to 0.20	0.04 to 0.20	0.04 to 0.20	0.04 to 0.23
Total SLR	**0.18** to 0.38	0.20 to 0.43	0.21 to 0.48	0.20 to 0.45	0.23 to 0.51	0.26 to **0.59**

dominate recent sea level rise, and their share is only likely to increase in the future.

The IPCC sea level rise predictions for 2100 already appear too conservative. The global climate models may be underestimating today's sea level rise, which already lies toward the upper end of this decade's model forecasts, probably because they are unable to adequately represent all of the physical processes involved in ice melting (fig. 7.1).[13] Therefore they are missing the recent speedup of melting land ice and will probably underestimate future ice sheet inputs. Nor can the models yet predict the likelihood or timing of any future ice sheet breakup.

As indicated previously, recent trends from Greenland and the West Antarctic Ice Sheet (WAIS) raise growing concern.[14] Melting ice on Greenland and Antarctica have added ~0.7 millimeters (0.03 inches) per year to the sea since the 1990s (table 6.1).[15] Global warming would thin these ice sheets even further. Together, Greenland and WAIS ice, if fully melted, could raise sea level by ~10 meters (~33 feet). A few degrees Celsius of global warming

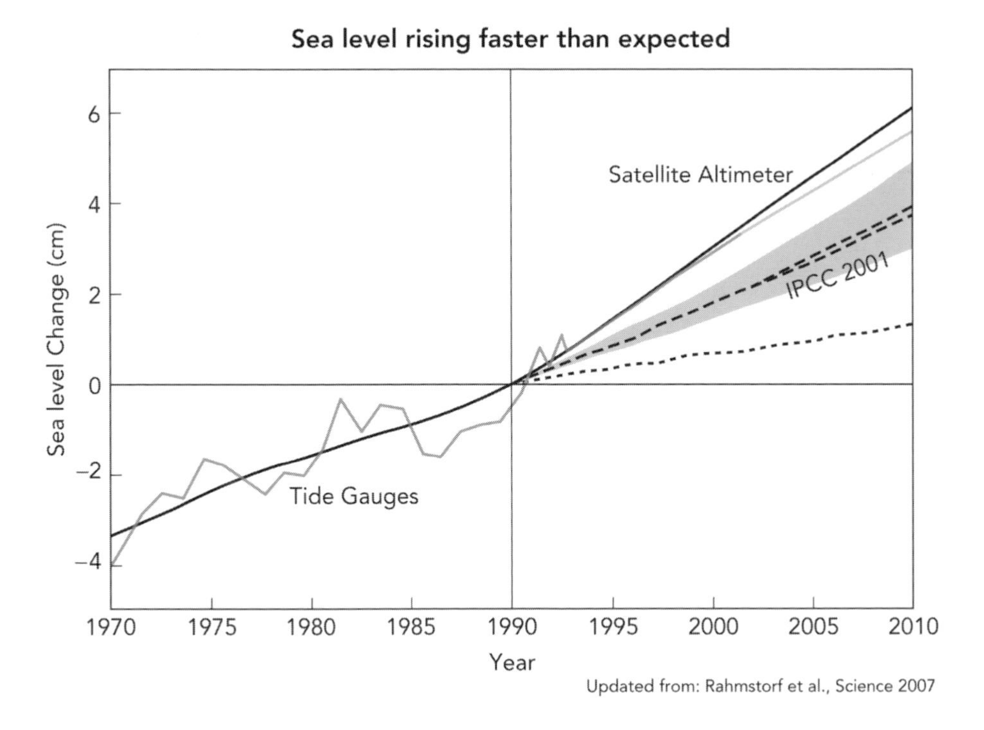

Figure 7.1 Observed versus modeled sea level rise. (updated from Rahmstorf et al., 2007). The top curve shows sea level data from tide gauges (red) and altimeters (blue); the gray area and dotted lines are IPCC-projected sea levels, starting in 1990.

could initiate an irreversible deglaciation of Greenland. While several global climate models predict comparable temperature increases by the end of this century, how fast the meltdown would unfold remains a major unknown.

The IPCC estimates the response of glaciers or ice sheets to rising air temperature based on past observations of changes in global mean air temperature and ice melting. Since mountain glaciers shrink in area and volume as they melt, a scaling relationship is applied to adjust for such changes with rising temperature. This relationship is assumed to hold into the future as well. Accelerated ice melting resulting from global warming is simulated by adding 10–20 centimeters (4–8 inches) to the upper bound of the projections. However, this assumption, linked simply to temperature rise, does not realistically portray dynamic ice behavior such as increased fracturing of crevasses, downward percolation of surface meltwater to the base of the glacier, and lubrication by the water layer, which accelerates glacier flow. Nor does the IPCC include effects of thinning of ice shelves, particularly in West Antarctica, as the ocean warms. If these ice shelves thin sufficiently and begin to disintegrate, as already observed on the warmer Antarctic Peninsula (table 7.1), glaciers with ocean outlets that had been held back by ice shelves may begin to surge forward (e.g., the Jakobshavn Isbrae glacier, Greenland, and some glaciers feeding the Amundsen Sea Embayment, West Antarctica.[16]

Taking a different approach, semi-empirical calculations extrapolate observed relationships between recent global sea level and temperature to obtain future ocean heights.[17] Sea level escalates in direct proportion to projected temperatures, at least for the next century or two. By 2100, sea level could climb between 0.75 and 1.9 meters (2.5–6.2 feet) above the 1990 level (fig. 7.2).[18] The high-end rate exceeds the average rate of the last deglaciation. While semi-empirical calculations yield higher sea levels than those of the IPCC, they too have not taken future dynamic ice sheet behavior fully into account. Meanwhile, the cryosphere—the frozen world of glaciers, ice sheets, permafrost, and sea ice—continues to thaw at an accelerating rate.

The foregoing discussion illustrates some shortcomings of the current generation of computer models. The models fail to capture fully the complexities of ice flow behavior; for example, what makes the ice stream surge forward? Is it melting on the surface of the glacier or ice sheet (which implies increasing air temperatures), with water trickling down to the base and lubricating the glacier, or melting from beneath the ice shelves, leading to their disintegration (which implies warmer ocean water)? Recent observations lean toward the latter hypothesis in a number of cases, but the balance may shift as air temperatures continue to rise.

Another major uncertainty lies in the role of clouds in climate change. In general, more water evaporates as temperatures increase. Since water vapor

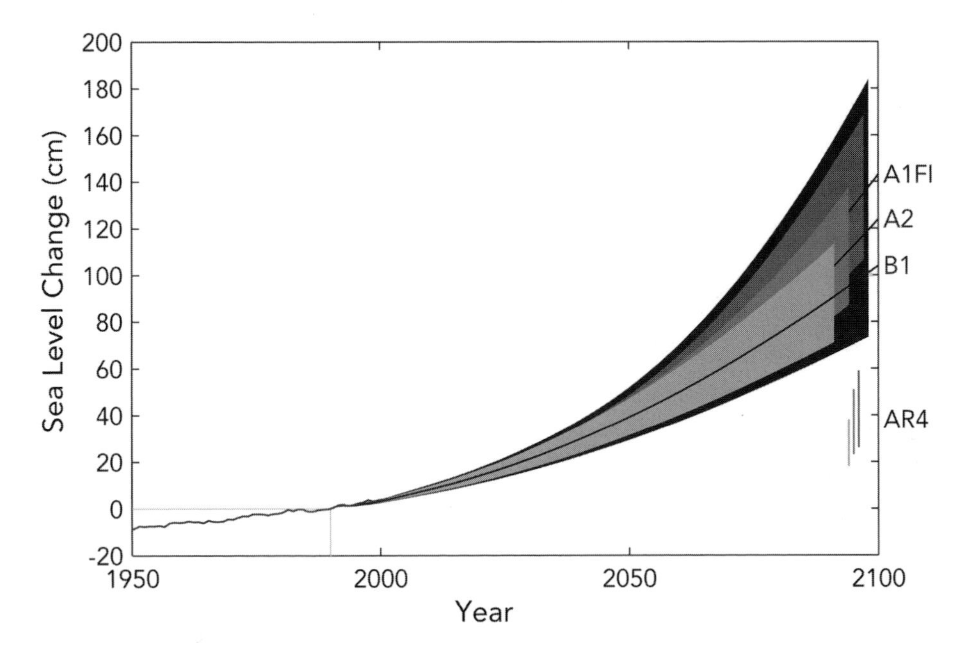

Figure 7.2 Projected sea level rise from 1990 to 2100, using a modified semi-empirical approach, for three different IPCC greenhouse gas emissions scenarios. The observed sea level for 1950 to 2000 is also shown. The vertical bars represent the IPCC AR4 2007 sea level rise projections for these scenarios. (M. Vermeer and S. Rahmstorf, 2009, "Global Sea Level Linked to Global Temperature," *Proceedings of the National Academy of Sciences* 106 (51):21527-21532, fig. 6.)

is a powerful greenhouse gas, an increase in cloudiness would presumably cause more warming—a positive feedback. On the other hand, since clouds are much brighter than most land surfaces or open ocean, they would reflect much of the incoming sunlight back to space—a negative feedback that could "put the brakes" on global warming. At this point we are still unsure which of these two feedbacks would predominate. Another wrinkle involves which type of cloud will increase. Thin and wispy high-altitude cirrus ice clouds admit the rays of visible light, but trap outgoing infrared radiation (fig. 7.3a). Thus, an increase in high-altitude clouds could heat up the atmosphere. Conversely, if thick low-level cumulus or stratus water clouds increase instead, they reflect more sunlight than they absorb, leading to more cooling (fig. 7.3b). Thus, the net effect of clouds ultimately depends on which type prevails and where in the atmosphere they form. The processes occurring within a cloud are complex and difficult to model, inasmuch as the basic physics is not always well developed.

Figure 7.3a High-altitude cirrus ice clouds. (NASA JPL.)

Figure 7.3b Low-altitude cumulus water clouds. (NASA JPL.)

Other major unknowns include future trends in population growth, economic development, and technological change. Although a detailed consideration of these processes lies beyond scope of this volume, and is not further considered here, it suffices to point out that these trends will ultimately affect sea level change indirectly, through changes in energy demand, utilization of fossil fuels, and the ensuing greenhouse gas emissions, which lie at the root of the current climate change debate.

The sea level rise projections summarized here undoubtedly will change as our databases expand, our understanding of the underlying processes grows, and our computer models become more sophisticated. The incomplete representation of dynamic ice processes in current climate models introduces a layer of uncertainty in future sea level rise projections. Thus analogies are frequently drawn from paleoclimate records. Geologic history teaches us that climate and sea level change are closely intertwined and that former warm periods generally correspond to higher water levels, at times much higher than present. Therefore, we recap several well-documented episodes of rapid sea level rise, from the last interglacial (Eemian, Sangamonian) circa 125,000 years ago, and from the great meltdown at the end of the last ice age, described in chapters 4 and 5, in order to anticipate potential future outcomes.

LESSONS FROM PAST EPISODES OF RAPID SEA LEVEL RISE

Though past analogs are not always strictly equivalent, because major continental ice sheets are absent and climate-driving forces differ, the deglaciations and meltwater pulses can nonetheless set plausible upper bounds on what we might expect in the foreseeable future.

The Last Interglacial (Eemian)

During a relatively balmy period, between 130,000 and 116,000 years ago, high-latitude temperatures averaged some 3°C–5°C above present levels—within the range of some IPCC projections.[19] Sea level stood at ~6 meters (13–20 feet), possibly as high as 6.6–9.4 meters (21.7–30 feet) higher than today.[20]

Over a 3,000-year period in a computer simulation, the thinning Greenland Ice Sheet caused sea level to rise by 3.4 meters (11.2 feet), corresponding to an average rate of increase of 11.3 centimeters (4.4 inches) per century.[21] Antarctica probably supplied the balance of the higher sea level.[22] Thus, the combined melting of both ice sheets over 3,000 years raised sea level ~1.33–2 millimeters per year—comparable to early-20th-century rates.[23]

However, oxygen isotope ratios in Red Sea foraminifera imply significantly higher rates of sea level rise during the last interglacial, averaging ~1.6 meters (5.2 feet) per century—fast enough to wipe out a Greenland-sized ice sheet within four centuries.[24] But even then, not all of Greenland melted—its core likely remained intact.[25]

The Last Deglaciation

The last glacial meltdown spanned many millennia, from around 20,000 years ago to approximately 7,000 years ago (chapter 5). Over this extended period, rates of sea level rise averaged ~10 millimeters per year, punctuated by a number of rapid outbreaks that lasted a few centuries. Maximum rates of sea level rise reached 40–65 millimeters per year during the largest meltwater pulse (1A) around 14,000 years ago, lasting several hundred years (table 5.1). This was later followed by several smaller meltwater pulses.

However, meltwater pulses succeeded long periods of warming. Steady ice deterioration over centuries eventually primed the ice sheets for rapid bursts of disintegration. For example, meltwater pulse 1A (beginning 14,300 years ago) lagged several centuries behind a warm interlude that started 14,640 years ago.[26] Therefore, even if Greenland temperatures were to exceed a critical threshold by 2100, higher temperatures would probably need to persist a few centuries before major ice breakup would occur, with rapid disintegration more likely toward the middle or final stages of the process than at the beginning.

Implications for the Future

Uncertainty still surrounds the fate of the West Antarctic Ice Sheet. A core drilled beneath the Ross Ice Shelf in western Antarctica reveals recurrent expansion and contraction of the WAIS at ~40,000-year intervals during the early Pliocene, ~5–3 million years ago.[27] During the warmer phases, sea surface and air temperatures remained above freezing for much of the Antarctic summer. Such conditions may prevail once more by the end of the century if atmospheric CO_2 levels continue to rise unabated.

The potential fragility of the WAIS is underscored by the fate of its fringing ice shelves (figs. 6.2, 6.8). The WAIS could disappear within centuries if the surrounding ocean warms by 1°C and the ice shelves buttressing the Antarctic ice sheets start to melt from beneath at rates exceeding 0.5–10 meters per year per degree Celsius.[28] Recent ocean temperatures around Antarctica have already increased by ~0.2°C (0.4°F); further increases would enable the WAIS to play an important role in future sea level rise.[29]

Mountain glaciers, being relatively smaller ice masses, respond much faster to changes in climate. For example, most glaciers in the Alps and Scandinavia advanced during the Little Ice Age, a cool period lasting approximately from 1400 to 1850.[30] Glaciers have been retreating worldwide, as the Earth warms, starting toward the end of the 19th century, and especially since the mid-20th century (see chapter 6).

What fate awaits the mountain glaciers if our planet continues to warm up? Are they likely to disappear by the end of this century? How much more would this add to sea level rise? The mass balance of a glacier (i.e., its gain in mass due to snow accumulation minus losses due to melting or other processes) is largely determined by climate. In general, climate warming would lengthen the summer melting season and shorten the winter accumulation period, leading to a negative mass balance. Large, cold glaciers would thin vertically before they retreat horizontally, taking several centuries. But glaciers that feed directly into the sea (as in Alaska) are especially susceptible to out-of-equilibrium instabilities at their edges, promoting increased surging and calving. They are retreating and thinning rapidly, and represent a major source of recent ice loss.

In a 2007 study, Colorado glaciologist Mark Meier and his colleagues firstly assumed that glaciers will continue to lose ice mass at a constant present-day rate, then secondly, that the present acceleration of mass loss continues.[31] The first assumption generated a rise in sea level of 0.1 meter (4 inches) by 2100; the second yielded another 0.24 meters (9 inches). This compares with the IPCC projections of 0.07–0.17 meters (3–7 inches) from glaciers alone and 0.18–0.59 meters (7–23 inches) from all sources. Meier et al. foresee that up to 35 percent of the total glacier volume would be lost by 2100. They further point out that while many of the smaller glaciers are likely to vanish by the end of the century, most of the ice remains in large glaciers or polar glaciers that are not likely to shrink significantly in area, but instead would become much thinner. More than half of the ice volume resides in relatively thick, large masses.

Drawing upon former climate and sea level analogs, while illuminating, will not exactly duplicate future conditions. Although past changes in climate may resemble the computer model simulations, the basic underlying causes may differ. In contrast to the Eemian and the Last Glacial Termination, which were triggered by high northern latitude summer insolation, according to Milankovitch (e.g., figs. 4.6, 4.7), the present thaw derives mainly from increasing anthropogenic atmospheric greenhouse gas emissions. As a result, the hemispheric and seasonal distributions of warming will likely differ. During the lengthy deglaciations, prolonged periods of prior warming weakened the ice sheets, leading to subsequent rapid breakdown. Furthermore, since the massive continental ice sheets (the Laurentide, Fennoscandian, etc.) that formerly covered much of North America and northern Europe are gone, the overall ice supply is now much less (fig. 4.12). Therefore, sudden ice sheet breakup that once created the major meltwater pulses will play a reduced role in the future. Even if a substantial fraction of Greenland ice were to melt rapidly within the coming century, the total volume of available ice (corresponding to a sea level rise of 7 meters (23 feet) constitutes just a fraction of the ice derived from the ice sheets during meltwater pulse 1A (i.e., 16–24 meters, or 52–79 feet), placing an extreme upper limit on how much sea level can rise within this century.

However, global sea level is highly unlikely to exceed 2 meters (6.6 feet) by 2100, since modern glaciers cannot discharge enough ice into the ocean fast enough to accommodate higher increases.[32] Even although sea level may have climbed an average of ~1.6 meters (5.2 feet) per century at the peak of the last interglacial, constraints on Greenland's current ability to discharge ice fast enough today limit this particular source to slightly over half a meter within a century.[33] Given the overall lower ice supply (as compared to the last glaciation), therefore, very high rates of ice discharge are unlikely to be sustained for long. If global warming continues unabated, most of the sea level speedup will probably occur late in this century and beyond, inducing a current sense of complacency, as present-day rates still remain quite low. A plausible high value lies near 1 meter of ice melt by 2100, assuming an average rate near 10 millimeters per year like that of the Last Glacial Termination (excluding the meltwater pulses), or roughly 3 times the current total rate (see chapter 5).[34]

ICE SHEETS AT RISK—POTENTIAL INSTABILITIES

James Hansen, director of the NASA Goddard Institute for Space Studies in New York City, has argued, using the last interglacial (and other episodes of rapid sea level change) as analogies, that continuing our upward trajectory of greenhouse gas emissions unabated on a "business as usual" course would melt down the ice sheets enough to raise sea level by several meters by the end of this century![35] But how realistic is such a catastrophic scenario? The paleoclimate records show that outbursts of accelerated sea level rise—the meltwater pulses—were generally preceded by centuries of prior sustained warming, which weakened the ice sheets, rendering them more susceptible to rapid disintegration once a critical threshold was passed. Therefore, any future drastic ice sheet breakup that could raise sea level by several meters would probably require more than one century.

Can we, nevertheless, expect any unforeseen climate "surprises" in our future, given the history of past abrupt climate change and rapid meltwater pulses? Could the ice sheets melt much faster than anticipated and raise sea level by more than a meter by 2100, as Hansen has claimed? While the total or even partial meltdown of Greenland and the West Antarctic Ice Sheet remains a long shot, the consequences of such a calamity would be immense. How likely is such a meltdown and how fast could it take place?

What is happening to the Greenland Ice Sheet at present? Greenland could be irreversibly destabilized by a regional temperature rise comparable to that of the last interglacial and also of several IPCC climate simulations for 2100 (see below). Therefore, if the Earth continues to heat up, it is not unreasonable to expect the Greenland Ice Sheet to shrink considerably within the next few centuries. As mentioned in chapter 6, the Arctic warming appears to have already begun. An increase in snowfall at higher elevations in Greenland only partially compensates for the accelerated melting of glaciers reaching the sea during summer, when expanding pools of water collect on the surface and water cascades down crevasses and moulins to the base (fig. 6.8). However basal lubrication may not be the main cause of present glacial retreat in Greenland.

Computer modeling of ice behavior shows that as a glacier thins upstream—for instance as the result of increased surface melting and downward draining of meltwater—it exerts less of a forward push downstream, which will slow it down. On the other hand, the thinning of ocean-reaching ice at its terminus reduces resistance to the glacier's forward motion, allowing it to accelerate forward and become thinner. These changes propagate upstream, leading to further retreat (table 6.1).[36] The high degree of sensitivity of a number of Greenland glaciers to terminus conditions (e.g., the Jakobshavn Isbrae glacier) may account for the recent observations.[37] With a prolonged bout of climate warming, however, surface melting and basal lubrication may ultimately predominate over attrition of the glacier's edge.

Satellite radar and laser altimeter observations have led to important insights into the movements of Antarctic ice streams—the faster-moving ribbons within the larger ice mass that convey 90 percent of the ice seaward. Radar offers a subsurface view of the terrain; the laser altimeter detects changes in surface topography. These observations show that in faster-flowing regions, the river of ice is sliding over either a shallow pool or lake or a layer of easily deformed, water-saturated sediment, such as glacial till—rocks and

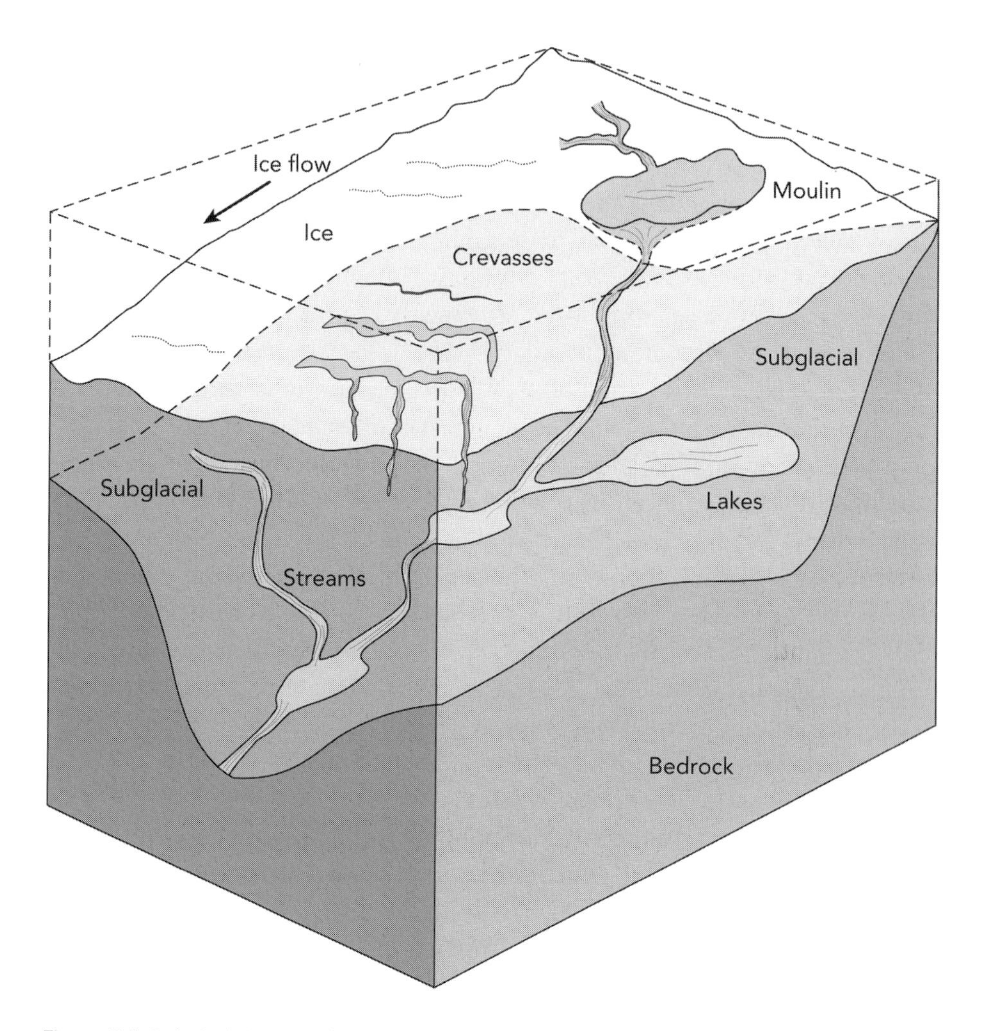

Figure 7.4 Subglacial lakes and rivers. Surface meltwater flows downward through crevasses or moulins, connecting to a network of subglacial lakes and streams. The layer of water at the base facilitates the forward motion of the ice sheet.

sediments dragged along by a glacier at its base as it inches forward. The glacier glides forward smoothly over water with less resistance or friction, like slipping on wet ice. Entire networks of subglacial streams and more than 160 subglacial lakes have now been discovered in Antarctica (fig. 7.4).[38] As temperatures climb and more meltwater reaches the base of the ice sheet, ice streams could pick up speed and deliver more icebergs to the ocean. On the other hand, loose rock and sediment may accumulate into a wedge near the grounding line, acting to stem the rapid advance of the ice stream if sea level rise does not exceed several meters.[39] Yet even a slight increase in nearby sea surface temperature could in turn counteract the stabilizing effect of the sedimentary wedge by undermining the floating ice shelf buttress. As mentioned earlier, an upswing in Antarctic land and ocean temperatures is already thinning the margins of the ice sheets in many places.[40]

A key remaining question lies with the stability of the WAIS and what it would take to break it apart. The ICESat laser altimeter detects pervasive ice thinning over much of West Antarctica, particularly over the glaciers that feed into the Amundsen Sea. The thinning extends inland in places to within 100 kilometers of the ice divides.[41] In 2006, the sea level contribution of the WAIS was around 0.4 millimeters per year and that of Antarctica was 0.5 millimeters per year (table 6.1).[42] Nearby ocean temperatures have already increased by ~0.2°C (0.4°F).[43] Even this slight increase may have been enough to initiate the recent acceleration of the Pine Island and neighboring Thwaites Glaciers that feed the Amundsen Sea Embayment, by thinning the undersides of their ice shelves. By 2006–2007, Pine Island Glacier alone accounted for ~2 percent of present sea level rise. While this amount may seem inconsequential at first glance, the glacier's annual volume loss quadrupled between 1995 and 2006.[44] A doubling of glacier speeds in this region would add 5 centimeters (2 inches) per century to the world's oceans, but if they sped up as much as Jakobshavn Isbrae (Greenland), they could contribute as much as 30 centimeters (1 foot) per century.[45] This area—the "weak underbelly of the West Antarctic Ice Sheet"—contains enough ice to elevate sea level by ~1.5 meters (5 feet). Furthermore, recent changes underline the potential ability to respond more rapidly to future climate change, which could therefore significantly affect sea level within the next few centuries.

OCEAN SURPRISES—SUDDEN SHIFTS OF THE GREAT CONVEYOR BELT

Abrupt climate changes in the geologic past have been linked to major shifts in ocean circulation (see chapters 1 and 4). For example, during the Heinrich events of the last Ice Age, repeated massive iceberg incursions within the last 70,000 years freshened and cooled the North Atlantic, causing a breakdown of North Atlantic Deep Water (NADW) formation. This in turn disrupted the normal flow of the oceanic conveyor belt, hindering northward transport of heat and moisture. As a result, the region plunged into a deep freeze, until the anomalous currents reversed and normal circulation was restored. Could future climate change provoke a similar sequence of events? On a hotter Earth, the higher sea surface temperatures and lower salinity due to increased freshwater from rivers, glaciers, and the Greenland Ice Sheet could block the sinking of northern deep water. Some signs of ocean freshening and slowdown of the circulation may have already appeared.[46] However, given the variable mood swings of the ever-restless ocean, it may be too soon to discern whether these preliminary observations represent a true trend, rather than natural year-to-year or longer fluctuations. It could take at least a decade or more of continuous observations to sort out the effects of the large inter-annual variability from a genuine trend or sudden shift in circulation.

Meanwhile, several recent computer model studies, in common with earlier ones, have found that higher ocean temperatures and increased ice melting (particularly from the Greenland Ice Sheet) would indeed freshen northern waters, weaken the NADW, and shift the sites of its formation farther to the south (fig. 7.5).[47] Sea level would rise faster in the northwestern Atlantic Ocean than elsewhere because a weakening of deepwater formation would produce additional warming along the Gulf Stream and drive ocean water toward the coast. Thermal expansion of the heated water would elevate the nearshore ocean surface. (A simple way of thinking of it is to visualize a piling up of warm water that otherwise would have flowed farther north). Therefore, cities such as Boston, New York City, and Washington, D.C., could eventually experience a sea level rise of 20 to 50 centimeters (8 to 20 inches) on top of that coming from thermal expansion and ice melt. These studies single out the northeastern United States as a region that can expect the greatest additional sea level rise from such shifts in ocean circulation. While these model experiments suggest that the projected Greenland ice melt may not have a substantial effect on global temperatures, it could cut down North Atlantic regional warming by several degrees.[48] While Europe and northeastern North America may not necessarily grow cooler, they could instead warm less or more slowly than other regions.

These model projections underline the particular vulnerability of the northeast coast of the United States to future sea level rise. The higher anticipated ocean levels place the large northeastern cities at greater risk during coastal hazards such as the storm surges from major hurricanes and nor'easters. Other vulnerable regions will be examined in chapter 8.

A LONG-TERM "CLIMATE COMMITMENT"

Looking ahead just one century does not end the story, however. Even if we manage to curb further greenhouse gas emissions in the near future, we have already "committed" ourselves to further warming and sea level rise centuries and perhaps millennia hence. Because of the high **thermal inertia**[49] of ocean water, it may take several centuries for the heat that has already been absorbed near the surface of the ocean to penetrate all the way to the

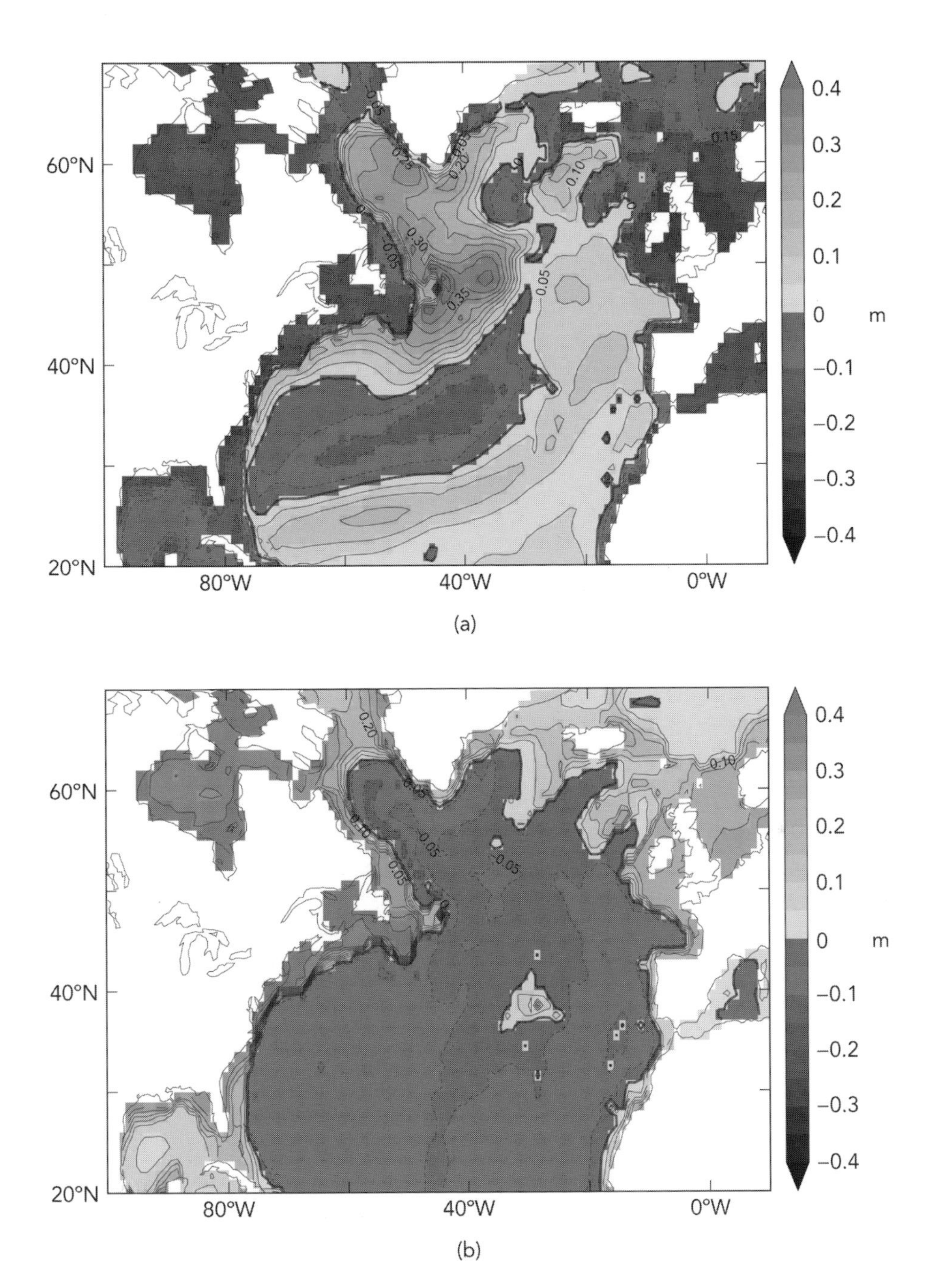

Figure 7.5 Sea level rise along the East Coast of North America resulting from Greenland icemelt and slowdown of the NADW. Top: The greatest steric effect occurs just beyond the continental shelf, where the water column deepens. Bottom: Water flows toward the coast, causing sea level to rise. (Reprinted by permission from Macmillan Publishers Ltd: NATURE GEOSCIENCE. J. Yin, M. E. Schlesinger, and R. J. Stouffer, 2009, "Model Projections of Rapid Sea-Level Rise on the Northeast Coast of the United States," *Nature Geoscience* 2:262–266. Copyright 2009.)

bottom. Therefore, even with a stabilization of greenhouse gases at year 2000 levels, temperatures could rise an additional 0.6°C (1.1°F) by the 2090s over those of the late 20th century.[50] The ocean warming continues for many centuries, although at diminishing rates.

The climate commitment for sea level stems from two sources: the slow journey of warm surface water to the ocean depths, and once under way, the extent of the ice sheet meltdown beyond 2100. Neither of these processes can be easily reversed even after global warming has been curbed or halted.

Thermal Expansion

The IPCC report foresees an additional sea level rise of 0.3–0.8 meters (1–2.6 feet) in 2300 due to thermal expansion, even if greenhouse gases were stabilized at levels corresponding to those of the A1B emissions scenario in 2100. The ocean surface would continue to rise for several centuries or longer from ongoing thermal expansion as surface heat slowly diffuses downward, although at gradually declining rates. When deepwater formation finally reaches a steady state, the oceans would warm up fairly uniformly, resulting in a 0.5-meter (1.6-foot) rise in sea level for each degree Celsius of warming, as compared with 2000. If for some reason deepwater formation were sharply curtailed (say by tripling CO_2 levels), the deep ocean could warm up even more and by the time a steady-state thermal expansion was reached, sea level could rise by as much as 4.5 meters (14.8 feet), not counting the ice sheet contributions![51] These changes would develop over many centuries to a millennium.

Mountain Glaciers

It is anticipated that most mountain glaciers will vanish within the next 1–2 centuries. A 5°C (9°F) increase in summer air temperatures would reduce the Alpine glacier area by more than 90 percent relative to 1971–1990.[52] Since 90 percent of all Alpine glaciers are already smaller than 1 square kilometer, it is not inconceivable that most could disappear within the next few decades, especially if Europe experiences more hot and dry summers like that of 2003. Yet even in a warmer world, some glaciers will survive at the highest altitudes. The IPCC estimates that if global temperatures 4°C (7.2°F) above 1900 levels persist beyond 2100, around 60 percent of mountain glacier ice would disappear by 2200 and nearly all by 3000.[53] However, the glacier commitment to sea level rise is relatively small as compared to the ice sheets; at most they would add 0.3–0.5 meters (1–1.6 feet) to the world's oceans.

Greenland

The Greenland Ice Sheet contains the equivalent of 7 meters (23 feet) of sea level rise. Annual average regional temperatures of 3.2°C–6.2°C (5.8°F–11.1°F) and global average temperatures of 1.9°C–4.6°C (3.4°F–8.3°F) above pre-industrial temperatures could initiate a major meltdown of Greenland.[54] These projected temperature rises are comparable to those of the last interglacial and several IPCC model projections for 2100. However, multiple centuries of warming over Greenland could raise sea level significantly. Even a rise of 1 or 2 meters (3.3–6.6 feet) could endanger many low-lying coastal regions.

Further ahead into the future, in a joint ice-atmosphere-ocean computer model simulation in which atmospheric CO_2 concentrations have quadrupled, sea level rises by 5.5 millimeters (0.2 inches) per year over the first 300 years, due to Greenland Ice Sheet attrition, but the rate diminishes thereafter as the ice sheet continues to contract. According to the model, Greenland's ice volume shrinks to 80 percent within 270 years. Only 60 percent remains after 710 years and just 40 percent after 1,130 years (fig. 7.6).[55]

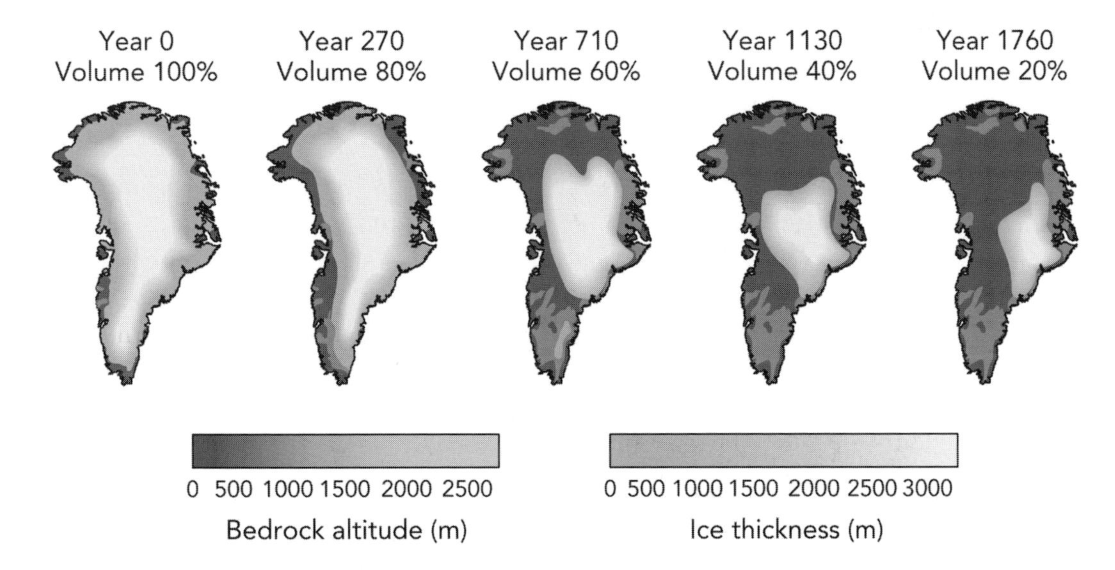

Figure 7.6 The dwindling Greenland Ice Sheet at various future times, according to a combined ice sheet and climate model, assuming a climate with a constant quadrupled pre-industrial atmospheric CO_2. (Modified from IPCC, 2007a, Climate Change 2007: The Physical Science Basis. Working Group I Contribution to the Fourth Assessment Report of the Intergovernmental Panel on Climate Change, Figure 10.38. Cambridge University Press.)

Antarctica

Most climate models project greater accumulation of snow and ice in Antarctica, because of increased precipitation there. Since the enhanced snow/ice accumulation exceeds future melting and other ice losses (in the model), the net sea level contribution is negative; for example, if greenhouse gas concentrations remain constant at 2100 levels in the IPCC A1B scenario, sea level falls by 0.4–2.0 millimeters (0.02–0.01 inches) per year.[56] However, increased ice discharge, due to weakened buttressing by the major West Antarctic ice shelves and greater surface or basal ice melting, could override the higher accumulation rates. Thinning rates of 0.5–10 meters per year per degree Celsius would suffice to destabilize the West Antarctic Ice Sheet. The WAIS could begin to collapse if surrounding ocean water temperatures increase by 5°C (9°F).[57] We may already be setting the stage for the future demise of the WAIS, since regional sea surface temperatures have already warmed slightly and several glaciers feeding the Amundsen Sea Embayment have begun to surge forward.

THE COMING WATERWORLD?

Our planet has been warming since the late 19th century—a trend that is largely the consequence of human activities, especially within the past 50 years. Twentieth-century sea level rise exceeds that of the preceding millennium by at least 1 millimeter per year. The IPCC anticipates that if global warming continues unabated, sea level will rise by 18–59 centimeters in 2100. However, a comparison with recent sea level trends suggests that the IPCC projections may already be too conservative. Satellite and field observations clearly demonstrate that much of the cryosphere—mountain glaciers, Arctic sea ice, permafrost, even major ice sheets—is thawing to some extent. Among these frozen realms, only mountain glaciers, ice caps, and ice sheets will affect sea level directly; the other components will exert their influence more indirectly through various climate feedbacks (for example, the ice-albedo feedback and ocean freshening). To supplement the computer model outputs, with their inherent uncertainties, we also draw upon past periods of faster sea level rise, keeping in mind that present or future circumstances may differ in important ways from those of the past.

The primary unknown in assessing future ocean elevations stems from anticipating the upcoming course of the large ice sheets—the length of time and degree of warming needed before ice streams speed up significantly and widespread ice shelf breakup occurs. It is still too early to know whether the recent acceleration simply reflects a short-lived climate anomaly or indeed marks the onset of a long-term warming trend. Even with expanding ice melt, it is unlikely that global sea level rise will exceed 1 meter by much and even less probable that it will top 2 meters (6.6 feet) by the end of the century, especially since glaciers cannot discharge their ice fast enough into the ocean to generate greater increases.[58] A comparison with the average rise in sea level over the last deglaciation and semi-empirical calculations sets a plausible upper bound on global sea level rise on the order of 1 meter or somewhat higher by 2100. (Local sea level could climb even further, due to land subsidence and other factors.)

However, as mentioned above, the story does not end in 2100, and if global warming persists into the future, a significantly higher sea level cannot be completely ruled out. Clearly, some of the current sea level projections are likely to change as more data become available and as we refine our computer models on the basis of a more complete mathematical portrayal of the physical processes underlying glacier movement and cloud formation, and other elements of the climate system that are currently less well understood.

While it is difficult to quantify exactly the timetable of the mounting ocean trajectory, we can be reasonably certain that increasing atmospheric greenhouse gases and soaring temperatures can only push the upward-creeping ocean to much greater heights. Even at present, living along the shore presents a host of dangers that are likely to be exacerbated by rising sea level. The most prevalent of these hazards today include beach erosion and damages due to severe storm surge. Chapter 8 examines the factors that render a shoreline more vulnerable and how these risks may intensify with the rising seas.

Even castles made of sand fall into the sea, eventually.
—JIMI HENDRIX

8 Shorelines at Risk

ENDANGERED COASTS

Vanished Islands of Chesapeake Bay

Chesapeake Bay, the largest estuary in the United States, harbors extensive salt marshes, abundant waterfowl, and bountiful sea life. Its productive waters have furnished livelihoods for many generations of fishermen, and it also offers ample recreational opportunities today. The advancing sea began to inundate the Chesapeake Bay 9,000–8,000 years ago, drowning the ancestral Susquehanna River valley. As the submergence slowed, between 7,000 and 6,000 years ago, the estuary gradually assumed its modern shape. Few people realize that the sea is rising faster in Chesapeake Bay than elsewhere along the U.S. East Coast today. Seawater there is creeping upward at a rate of 3–4 millimeters (0.12–0.16 inches) per year, double the global rate of 1.7–1.8 millimeters (0.07 inches) per year in places. Low-lying mid-Atlantic coasts are affected not only by global sea level rise but also by crustal subsidence south of the former glaciers. Acting like a seesaw, land in Canada, once bowed under the load of the ice sheets, continues to rebound isostatically. In addition, a long-buried geologic sag in the Earth's crust may still be slowly sinking,[1] adding to the above-average sea level rise. A buried impact crater at the mouth of the Chesapeake Bay that is 35–36 million years old has created a hole into which more recent sediments have piled. Compaction and subsidence of these sediments over time have shaped the development of local landforms, faults, groundwater flow, and even local sea level.[2] Finally, the Virginia coastal towns of Norfolk and Hampton Roads cause additional ground subsidence by extracting water faster than nature can replenish local aquifers. Nonetheless, the pace of sea level rise in Chesapeake Bay, as elsewhere, has speeded up since the beginning of the 20th century as the Earth's temperature has soared.

A number of small islands, some occupied since European settlement, have shrunk considerably in area or have almost totally disappeared. Sharps Island was one such island. Old deeds show that the island, once located at the mouth of the Choptank River, Maryland, covered 360 hectares (889 acres) in 1660.[3] Even during the 19th century, the upward-inching sea and waves had slowly nibbled away at the edges of the island, and by 1848, only 175 hectares (432 acres) remained. Although some optimists constructed a hotel in 1895, by 1900 erosion had reduced the island to 36 hectares (89 acres). The hotel was finally abandoned in 1910, when a mere 21 hectares (52 acres) of land were still left. The only trace of the island today is a partially submerged, ruined lighthouse (fig. 8.1).

Another abandoned island is Holland Island, Maryland. Covering 121 hectares in 1668, the area had dwindled to 80 hectares (197.5 acres) by 1901, and in 1989 only 43 hectares (106 acres) remained. A thriving community of 253 people lived there in 1900, mostly fishermen.[4] After 1900, population declined over the years because of increasing erosion problems, and by 1920 the island was abandoned. The final decision to abandon it completely was determined not just by erosion and land loss, but by individuals' gradually moving away, which eventually led to a sense of lost community.[5]

Other Chesapeake Bay islands, some formerly inhabited, whose size has diminished sharply thanks to the lapping waves and rising waters include Barren Island, Poplar Island, James Island, Tilghman Island, Hambleton Island, and Royston Island. The last two have all but vanished. Prior to 1850, the area of these islands held steady or decreased gradually; subsequently, global sea level accelerated by the late 19th century.[6] The still-inhabited Smith and Tangier Islands, home to a unique fishing-based culture, are threatened by the rising seas that continue to encroach on the remaining land (fig. 8.2). Home to 700 people, Tangier Island lies 1–2 meters (3.3–6.6 feet) above sea level. The island has been eroding since the 1850s, especially along its west-

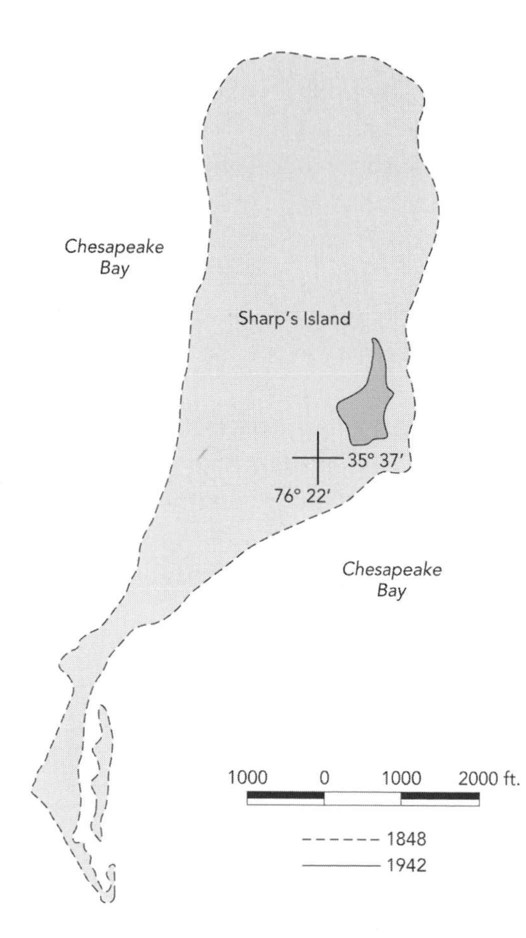

Figure 8.1 Map of Sharps Island, Chesapeake Bay, showing land loss between 1848 and 1942. (Modified from Leatherman, 1992, fig. 2.3. Maryland Geological Survey.)

ern shoreline, which has retreated 500–800 meters (1,600–2,600 feet) during this period.[7]

The rising seas are also attacking cliffs along the adjacent western shore and low-lying coastal wetlands on the eastern side of Chesapeake Bay, especially at Blackwater Wildlife Refuge, near Cambridge, Maryland. Old aerial photographs and more-recent Landsat Thematic Mapper satellite imagery show that nearly 1,700 hectares (4,198 acres) of degraded salt marshes in the Blackwater Wildlife Refuge area became open water between 1938 and 1989.[8] During the same period, the area of healthy salt marshes shrank from ~2,150 hectares (5,309 acres) to only ~950 hectares (2,346 acres). Rising sea level is only partly to blame, however. Upstream dams trap sediments that otherwise would have deposited in the marshes. Furthermore, sand that washes away from islands and along bluffs during storms is not fully replaced by fresh river-borne sediments. This problem is unique neither to Chesapeake Bay nor to its close neighbor, Delaware Bay (which will be discussed below).

Figure 8.2 Tangier Island, Chesapeake Bay. (Photo: V. Gornitz.)

Abandoned Rice Plantations

The islands of Chesapeake Bay are not the only casualties of the encroaching seas along the East Coast. Rice cultivation once thrived in the rich soils of the Low Country of coastal South Carolina and Georgia (recall Carolina long-grained rice). Rice plantations were first established in the late 1600s and prospered for well over 200 years. South Carolina soon secured a place among the wealthiest of the American colonies, and Charleston, its principal port, boasted many fine mansions, some of which are still standing today. Rice growing was initially confined to the coastal wetlands along rivers, such as the Cooper and Ashley Rivers of South Carolina. The early rice plantations employed a unique and ingenious system of cultivation, adapted to the daily rise and fall of the tides.[9] The incoming tide pushed freshwater ahead of salt water, which raised the freshwater level farther upstream. Therefore, specially constructed water gates opened to receive this incoming freshwater that irrigated the growing rice, but closed once the tide ebbed. In other words, the dike and water gate system kept the freshwater on the fields, but prevented the salt water from entering and damaging the crop. This labor-intensive cultivation system relied heavily on slaves already familiar with rice planting and harvesting techniques used in West Africa. They also contributed to the development and maintenance of the extensive waterworks.

Over time, however, erosion removed topsoil, reducing soil fertility and lowering the average field level. Tidal gates and dikes weakened and fields experienced more river flooding. Yet this clever agricultural system worked quite well, on the whole, as long as environmental conditions remained stable. Because of its close adaptation to the tidal cycle, however, it grew increasingly vulnerable to changes in that cycle accompanying rising sea level, and to storm surges from tropical storms and hurricanes. While historians generally attribute the downfall of the coastal rice-growing economy in the Carolinas to factors such as the end of slavery following the Civil War and the mechanization of rice cultivation in Louisiana, a number of destructive hurricanes at harvest time and rising sea level took their toll as well.[10] During the second half of the 19th century, a series of severe storms buffeted the Savannah-Charleston region, particularly in late summer and early fall. Major hurricanes struck the area in 1854, 1885, and 1893, with frequent tropical storms in intervening years. As the ocean level crept upward, so did the groundwater interface between salt water and freshwater, and agricultural soils grew increasingly saline. Over time as the tides pushed their way farther and farther upstream, any given field would need to keep its water gates shut for longer periods to keep out the seawater. River flooding became more commonplace and hurricane damage reduced crop yields to non-economic levels.

Similar problems face the farmers of Bangladesh and other fertile river deltas, particularly in Asia. In Bangladesh in the late 2000s, farmers began to notice white films of salt in their rice paddies.[11] The rising soil salinity, a consequence of a slowly encroaching sea and frequent flooding by severe tropical cyclones, has been cutting down rice yields. Unlike the early rice farmers of the Carolinas, the Bangladeshi growers have nowhere else to go in a country where a quarter of the habitable land lies less than 3 meters above sea level. While they could switch (and are switching) to other types of cultivation (for example, shrimp farming or planting other, more salt-tolerant crops), the people nonetheless depend heavily upon rice, their main food staple.

Moving the Cape Hatteras Lighthouse

The Outer Banks, a group of sandy barrier islands on Cape Hatteras, North Carolina, have been labeled the "graveyard of the Atlantic" because of the numerous ships that sank offshore during powerful ocean storms. To warn passing ships of dangerous shoals and strong currents, a lighthouse was first erected in 1803. The current tower was constructed in 1870 to replace the original structure, which had been damaged during the Civil War. At that time it stood safely 1,500 feet from the ocean. The sea gradually wore away the beach until 1935, when the surf finally reached the lighthouse. The National Park Service acquired the site and by 1950 had restored the lighthouse. By 1970, however, only 120 feet separated the lighthouse from the ocean's edge. The shoreline continued to erode, threatening to topple the landmark into the ocean. After prolonged and acrimonious public debate, Congress finally approved funding in 1998 and the lighthouse and some adjacent buildings were successfully relocated inland 2,900 feet from the 1870 site in the summer of 1999. Safe in its new location, the Cape Hatteras Lighthouse remains operational to the present (fig. 8.3).

The Shrinking Mississippi Delta

The Mississippi Delta is drowning—the result of various factors that have joined forces to lower the ground elevation. Coastal Louisiana has been described as "ground zero for sea level rise in the United States."[12] The Mississippi Delta began to build up once the sea level rise following the last ice age decelerated, 7,000–6,000 years ago. Sea level inched upward at declining rates until the late 19th and early 20th centuries, when the speed picked up once again. Because of the unusually high land subsidence in the delta, local sea level rises up to 10 millimeters per year.[13]

Louisiana's coastal plains have been subsiding for millennia because of natural geological processes. The Mississippi River flows down a natural crustal depression into which thick sediments piled up—initially, debris from melting glaciers far to the north and then later, fine-grained soils and sediments washed down over a vast drainage network covering much of the United States, stretching from the Rockies across the Great Plains and the Allegheny Plateau, and finally moving into the Deep South. The crust has sagged under this heavy load—a sinking reinforced by recent compaction of porous, waterlogged silts, muds, and peats.

However, human activities have exacerbated the natural land subsidence. Since the late 19th century, people have tried to tame the Mississippi River by building dams along its tributaries and levees along its banks in order to contain the floods that threatened growing cities like Baton Rouge and New Orleans. Narrow shipping channels and high levees divert the river water directly out to sea. As a result, the mighty Mississippi River no longer delivers enough sediment to the Gulf of Mexico to make up the losses of sand and silt

Figure 8.3 The historic Cape Hatteras Lighthouse in its new, safe location. (Photo: Captain Albert E. Theberge, NOAA Corps (ret.), National Oceanic and Atmospheric Administration/ Department of Commerce.)

trapped behind upstream dams. Sediments that would otherwise have deposited on the delta wetlands, renourishing and sustaining them, are channeled way out into the Gulf instead, creating a distinctive "bird's foot" delta (fig. 8.4). From space, the delta now looks like a piece of Swiss cheese nibbled by a mouse, full of holes and indentations where once-thriving marshes, home to a rich diversity of wildlife, have disintegrated. Over-pumping of water, oil, and gas exacerbates the normal sediment compaction, adding to the land subsidence.

The result has been rapid erosion of the entire Gulf coast of Louisiana. The rate of erosion has progressively increased since the mid-19th century,

Figure 8.4 The "bird's foot" Mississippi River Delta. The highly indented and crenellated coastline is a sign of land submergence and erosion. (ASTER satellite image, May 24, 2001, NASA GSFC.)

threatening the survival of the coastal wetlands and the livelihoods of the people who depend on the rich natural resources of the Mississippi Delta mangrove swamps and salt marshes. Statewide, the shoreline of Louisiana retreated at an average rate of 9 feet per year between 1855 and 2005. The shoreline retreated 14 feet per year between 1922 and 2005, 27 feet per year between 1996 and 2005, and a whopping 189 feet per year in 2004–2005, thanks to Hurricanes Katrina and Rita.[14] Some of the barrier islands near the Mississippi Delta suffer the highest rates of erosion.

The river has switched course many times, creating a new delta roughly every 1,000 to 1,500 years. Abandoned delta lobes no longer received a regular sediment supply and slowly eroded as the land subsided. Former delta lobes survive as narrow, slightly curved barrier islands, such as the Isles Dernieres (fig. 8.5).

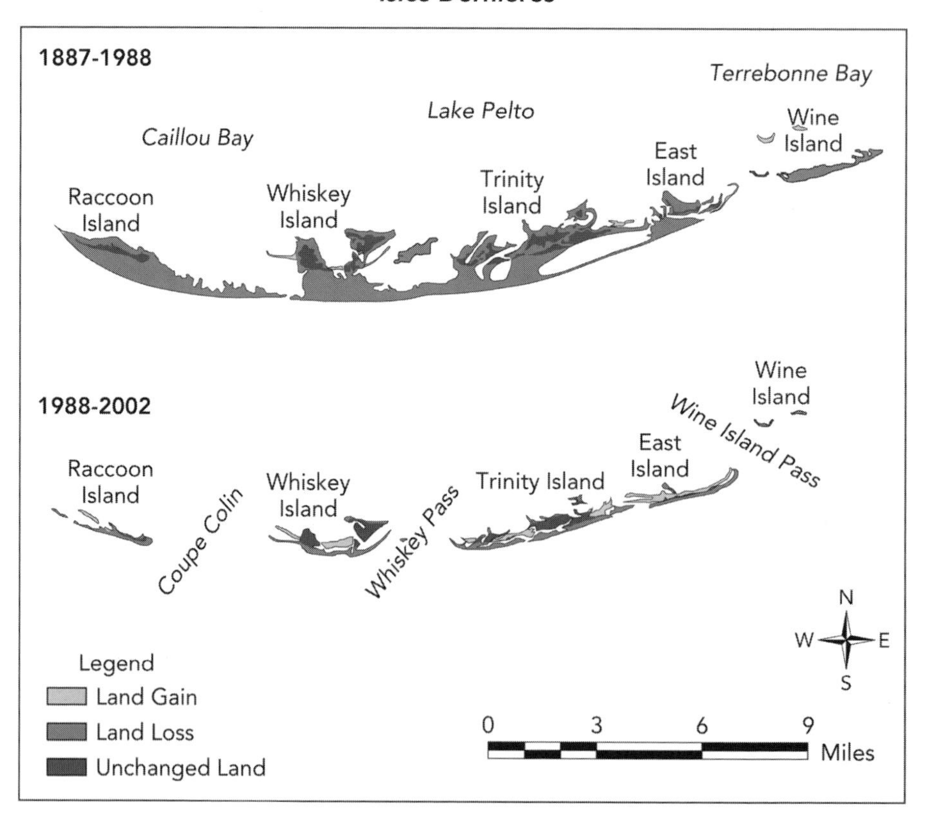

Figure 8.5 Maps of the Isles Dernieres showing landward migration and narrowing of the barrier islands between 1887 and 2002. (Modified from Penland et al., 2005. Copyright 2005, Coastal Education and Research Foundation, Inc. Used with permission.)

In the mid-19th century, the Isles Dernieres, now a chain of four narrow barrier islands, were a single island known as Isle Derniere (the Last Island)—well forested and a popular resort site. The resort was destroyed by a devastating hurricane in 1856 and the island later split in two. In 1887 the islands occupied 8,724 acres. After Hurricane Andrew in 1992, they covered only 1,267 acres.[15] The islands shrank further after Hurricanes Katrina and Rita in 2005, and Hurricane Gustav in 2008, and are now reduced to thin slivers. Soon the last of the Last Islands will become sandbars and mudflats—mere phantoms like Sharps Island, Hambleton Island, and Royston Island in Chesapeake Bay.

The First Wave of Coastal Refugees

Several small tropical island nations share a number of common attributes: land perched on top of narrow atolls and coral reefs, low elevation, and exposure to high tides and storm waves that periodically sweep over much of the islands. The island state of the Maldives in the Indian Ocean, home to about 370,000 people, can claim the dubious honor of being the lowest country in the world, with an average elevation of only 1.5 meters (4.9 feet). Its highest point sits a mere 2.3 meters (7.5 feet) above the ocean. Nearly all of Kiribati, a Pacific island nation of slightly over 100,000 people occupying 810 square kilometers (313 square miles), is lower than 2 meters (6.5 feet). Tuvalu, a tiny nation of 12,400 inhabitants, occupies several islands with an area of 26 square kilometers (10 square miles). Its maximum altitude lies only 4.5 meters (14.8 feet) above sea level.

These three island nations are among the most vulnerable places on Earth to mounting sea level. They are already noticing some effects. Extensive coastal erosion and high waves gnaw away at several Kiribiti islands, along beaches and even farther inland, displacing some people from their traditional homes and fields near the shore (fig. 8.6). But the islands are so narrow that there is little available space inland. Furthermore, populations have increased. The main island, Tarawa, is already overcrowded.[16] Tropical storms, with their high waves and tides, have become more numerous in recent years, threatening shorelines and seawalls. The coastal flooding and accompanying **saltwater intrusion** have contaminated groundwater wells, rendering the water unfit to drink. Similarly, coconut trees and crops planted near the shore are suffering from excess salt. Island leaders are acutely aware of the threat that rising seas pose to their future existence and are already exploring the possible alternative of relocation elsewhere to potentially receptive host countries such as Australia, New Zealand, or Fiji. In March 2012,

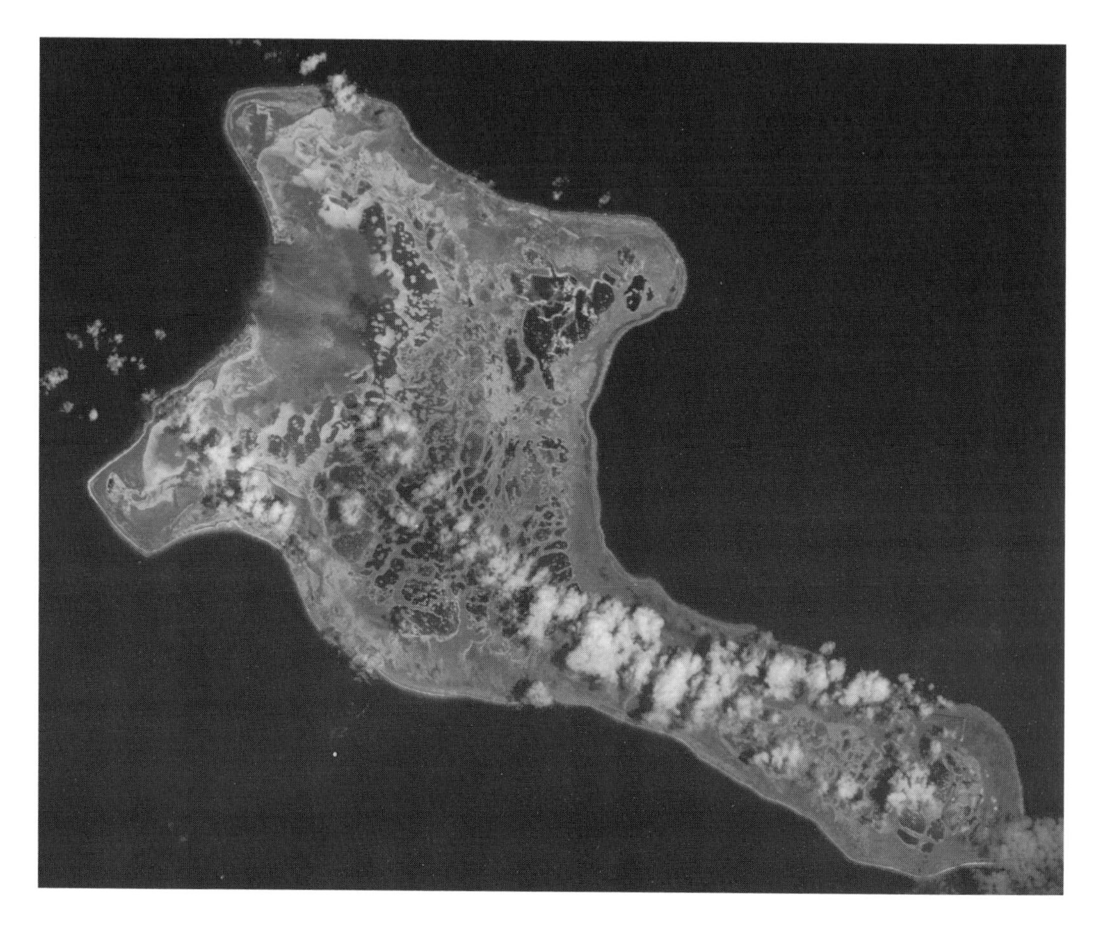

Figure 8.6 Satellite view of Kiribati. (Also known as Christmas Island.) Note the numerous lagoons (dark areas) within the island. (Earth Observatory, NASA Johnson Space Center.)

the government of Kiribati announced plans to buy 6,000 acres of land in Fiji as "climate change insurance" for relocation, as a last resort.

On Tuvalu, the story is similar: increasingly severe cyclones and tropical storms in recent decades, more flooding from higher tides and waves. Worse yet, the water table may be rising too. "In the late 1990s, water started coming out of the ground—first puddles, then a whole sea. That had nothing to do with rain," says Hilia Vavae, Tuvalu's chief meteorologist.[17] The Tuvaluans are also considering future relocation to New Zealand, which already accepts a rather limited number of new immigrants each year.

The Maldives, in spite of extremely low elevation, may have other options, at least in the near future. The island nation is less vulnerable to tropical storms than the small Pacific islands (but not immune to other marine disasters, such as the devastating 2004 tsunami). However, thanks to a thriving tourist-based economy, Maldives is debating whether to set money aside to purchase land elsewhere for relocation, should future sea level changes threaten the very existence of the people.

Such drastic steps raise serious questions. How can national and cultural identity be preserved under forced migration? Even if Maldives buys suitable land elsewhere, say in Australia, could it still claim national sovereignty? What becomes of the nation's legal claims to the submerged reefs and shoals, which could still harbor a wealth of sea life and possibly mineral resources as well?[18] Similar issues confront the other small island nations.

A problem not restricted to tropical islands, retreating shorelines threaten the viability of many coastal communities around the world even today. One striking example is the tiny village of Shishmaref, Alaska, located on a small barrier island, way to the north. This traditional Inupiaq Eskimo village, which subsists on hunting and fishing, may soon wash into the Chukchi Sea, as relentlessly pounding waves are eating away at its shoreline. Beach erosion has always posed a problem because of an exposed coastline and the high tidal range. Soaring temperatures throughout the Arctic (and Alaska) in recent years have significantly reduced summertime **sea ice** (slabs of ice floating on the surface of the ocean).[19] A smaller sea ice cover, which otherwise would buffer storm surges, exposes the shore to more wave damage. Erosion has intensified, in spite of seawalls and other "hard" shoreline defenses intended to keep the sea at bay. The sea has already claimed a number of homes. Warming also thaws the permafrost—permanently frozen ground—further undermining village building foundations. As in Kiribati, Tuvalu, and the Maldives, the 600 villagers of Shishmaref are seriously considering relocating to a safer spot on the nearby mainland, but the move may cost an estimated $200 million. Alternatively, the villagers could build stronger seawalls and strengthen existing ones, or even move houses to other locations on the island. (But since the island is small and elevations are very low, there really is no "safe" area.) Finally, they could move to larger towns such as Nome, Kotzebue, or Anchorage, but then they would likely lose their cultural identity—their unique dialect, hunting skills, traditional songs and dances, and arts (Shishmaref whalebone and walrus ivory carvings are highly collectible). Shishmaref's dilemma is by no means unique; other native Alaskan villages face similar choices. Regardless of which option is finally selected, relocation costs will be high.

The Shishmaref Islanders and citizens of the low-lying Pacific island states are rightfully concerned about future ocean levels and survival of traditional lifestyles, or even national independence. However, the rising ocean may not be the only culprit. Storm- and wave-induced erosion increases as more sea ice melts over longer periods each summer. Warming oceans may also be strengthening Indian Ocean cyclones.[20] ENSO events have become more frequent. Stronger storms cause greater destruction of coral reefs. Another consequence of higher ocean temperatures is an upswing in coral bleaching.[21] During the unusually intense 1998 El Niño, for instance, 16 percent of the world's corals died, primarily in the western Pacific and Indian Oceans. Because of losses due to storms, bleaching, and also mining for buildings and roads, the ability of damaged coral reefs to dampen the force of powerful waves may have been impaired. This could, in part, explain the higher tides and stronger waves reported by the islanders.

THE RESPONSE OF THE COAST TO A RISING SEA

Deserted estuarine islands, abandoned plantations, retreating beaches, once lushly vegetated delta wetlands turning into spongy mudflats, and coastal refugees vividly illustrate some of the initial impacts of sea level rise. But are these uniquely the handiwork of upwardly encroaching marine waters? The shoreline, as active interface between land and ocean, is molded by many natural geological and oceanographic processes that are increasingly influenced by human activities.[22] Until now, these other processes have largely masked the signs of sea level rise, but this situation will undoubtedly reverse as the ocean accelerates upward in the future. Nonetheless, because of the constant action of tides, waves, storms, land movements, and changing sea levels, the shoreline has always been a dynamic and at times hazardous place to live. As ocean waters lap higher and higher upon the beaches, the shoreline will continue to adjust to the shifting conditions. The following sections examine more closely how the rising sea is likely to amplify the present hazards of beachside living.

Inundation—Drowning Lowlands and Shrinking Beaches

The shore responds to the rising sea in two basic ways—by submergence and by erosion. Drowning lowlands (like the bayous of Louisiana) and shrinking beaches (like the Carolina shores) are examples of this process. Inundation means that the land loss is permanent: the land stays underwater indefinitely, not just at high tide or because of a storm flood, but until sea level falls once more at some distant time in the future. The beach also narrows as the water laps farther inland. The waves, breaking higher, wash away more sand, increasing coastal erosion. Meanwhile, salt water penetrates deeper into coastal lagoons, estuaries, up rivers, and into nearshore aquifers, affecting coastal vegetation, fisheries, and freshwater resources. The most direct effect of a rising sea level, however, is land inundation.

As the ocean gradually inches upward, the average land-water boundary slowly edges landward, as do the boundaries of high and low tide. As time progresses, more and more of today's **intertidal zone**—the portion of the shore affected by the tides—will stay submerged over longer and longer periods, until eventually a point is reached where it will become entirely marine, and the new intertidal zone will have shifted farther inland.

While the concept of land inundation is straightforward, deriving an accurate estimate of the area submerged under a given amount of sea level rise is difficult. Sufficiently reliable elevation data to determine the area that is currently above 1 or 2 meters of mean sea level are simply unavailable on a global scale. In the United States, topographic contour intervals for widely accessible elevation maps range between 1.5 and 6 meters (5–20 feet). Although higher-resolution government agency maps exist for some places, they are not readily available to the general public. However, recognizing the need to better assess present and future coastal flooding risks, the U.S. Geological Survey, NASA, and the Coastal Services Center of the National Oceanic and Atmospheric Administration (NOAA) have begun mapping vulnerable shorelines with LIDAR.[23] LIDAR-produced maps can achieve accuracy to a meter or less in the horizontal direction and approximately 15 centimeters in the vertical direction. Elsewhere, commonly used map contour intervals range between 16.4 feet (5 meters, United Kingdom) and 65.6 feet (20 meters, India).[24]

Coastal zones most vulnerable to permanent inundation include those at low elevation, that have a gentle seaward slope, and that are subsiding,[25] among them such coastal landforms as sandy beaches, deltas, mudflats, estuaries, lagoons, and bays (fig. 8.7). In the United States, this description encompasses most of the U.S. Gulf Coast (particularly Louisiana), barrier beaches, and marshes along the eastern U.S. shore, such as Chesapeake and Delaware Bays, and much of the Florida Everglades. Elsewhere, other high-risk areas include the estuaries of the eastern United Kingdom, the European Low Countries,[26] the southern Baltic region, and major river deltas such as those of the Chao Phraya River (Thailand), Irrawaddy River (Myanmar/Burma), Mekong River (Vietnam), Yangtze River, Yellow River (China), Pearl River (China), Niger River (Nigeria), and Nile Delta (Egypt).

The large deltas of southeastern and eastern Asia support dense, predominantly agrarian populations that are already exposed to extensive episodic flooding by frequent severe tropical cyclones. The risk to several other major deltas (such as the Amazon, Orinoco, or Niger Deltas) may be somewhat lower, since they remain less densely populated. Also vulnerable are the

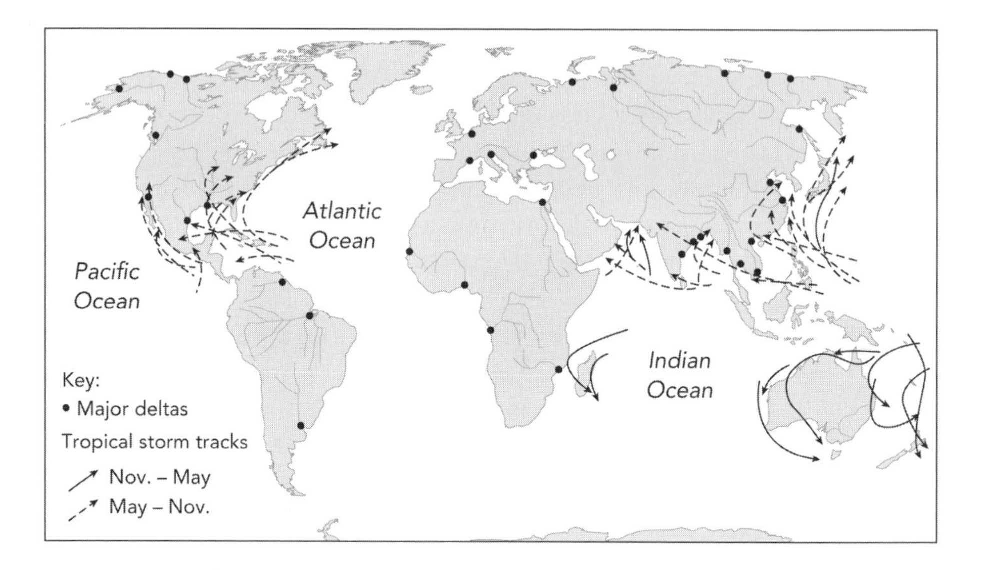

Figure 8.7 Major deltas of the world (black circles). Tracks of tropical cyclones are also shown. (Redrawn from Gornitz, 2005, "Natural Hazards," in M. L. Schwartz, ed., *Encyclopedia of Coastal Science*, fig. N3, p. 682, with kind permission from Springer Science + Business Media B.V. 2005)

low-lying shores of eastern Sumatra and large sections of Borneo (Kalimatan), Indonesia, and the tropical coral islands of the Pacific and Indian Oceans.

Coastal Wetlands

Important coastal ecosystems could also be lost. Salt marshes are increasingly recognized as valuable habitat for migrating birds and other wildlife, including small mammals, reptiles, and fish, as well as affording recreational opportunities. Furthermore, a healthy salt marsh acts as a buffer zone, protecting against the high waves and surges generated by major storms.

Salt marshes (and tropical mangroves) occupy the intertidal zone, supporting plants that are well adapted to the daily rise and fall of the tides and are immersed in brackish water for at least part of each day. Extensive salt marshes occur in southern Louisiana, along the shores of the Chesapeake and Delaware Bays, coastal Georgia, the Netherlands, and also the Norfolk barrier coast and Severn Estuary, England.

Salt marshes grow upward through a process of vertical accretion of sediment and plant organic matter, with some surface lowering due to compaction as water is squeezed out of pore spaces. Rivers and high tides deliver sediments to the marshes; storms and ebb tides remove them. The accumulation

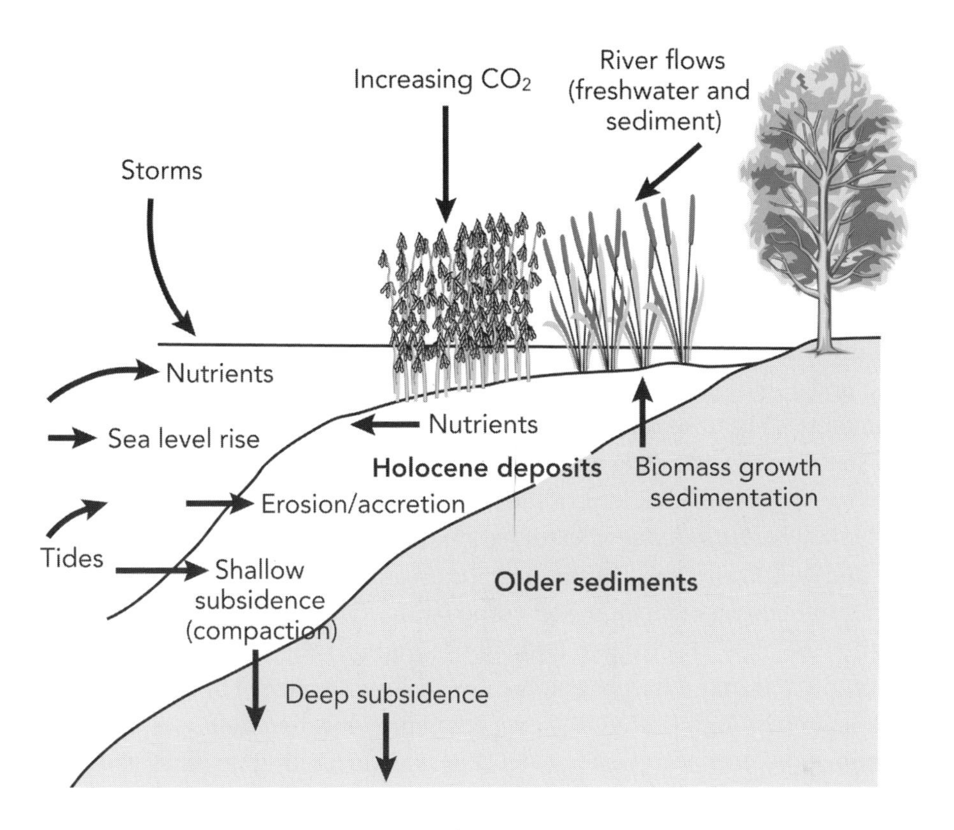

Figure 8.8 Processes that influence salt marsh development.

of organic matter from roots and stems helps to trap sediment and contributes to the upward accretion of the marsh. Salt marshes can generally withstand rising sea level, as long as the rate of vertical accretion exceeds the **relative (local) rate of sea level rise** and sediment compaction. This delicate balance is easily upset, for instance by decreasing the sediment input to the wetlands through damming of rivers and coastal protection schemes, as has happened in the Mississippi Delta (see above) and elsewhere (fig. 8.8).[27] Historic and ongoing land reclamation has significantly reduced the area of marshlands in many places (e.g., eastern England, the Netherlands, and the United States). Severe storms and hurricanes can wash away sediments and erode marshes, particularly along their edges. The ability of coastal wetlands to survive a rising sea level will depend on the balance among these various factors.

Tropical and subtropical mangroves occupy an ecological niche analogous to temperate-zone salt marshes. While some mangroves grow as far north as southern Florida and Louisiana, impressive mangrove stands have developed in the Niger Delta, Bangladesh, Guiana, Malaysia, and Indonesia. Mangroves are widely distributed, but at least 35 percent have been extensively cleared within the last few decades for rice paddies, aquaculture, and wood.[28] Widespread mangrove clearing in the tropics could expose the interior to damaging storm surges.

Coastal wetland ecosystems are finely attuned to sea level and sediment supply. Could they survive a more rapid rise in sea level? Louisiana may represent an exceptional situation, where natural geologic subsidence has been exacerbated by human activities. As a result, the relative sea level rise of ~10 millimeters per year stands at triple the current global average. Between 1956 and 1990 Louisiana lost its coastal wetlands at rates of 60–100 square kilometers (24–40 square miles) per year.[29] In addition to Louisiana, some salt marshes are drowning in portions of the Chesapeake and Delaware Bays, and some scattered locations of the New York shores.[30] While salt marshes in the United States and elsewhere have generally kept up with present sea level rise and will likely do so in the near future, some marshes nonetheless lost elevation because of shallow soil compaction. Others have been caught in a "sudden dieback," resulting in rapid deterioration and loss. The exact cause is uncertain and may vary geographically. Different factors have been blamed, such as not enough sediments, overgrazing by local wildlife, pollution, and microbial diseases.[31] However, even the fastest-accreting marshes would not likely survive rates of sea level rise in excess of 10–12 millimeters per year—a high-end scenario that could play out if ice sheet melting accelerates toward the end of this century. Marshes unable to accrete upward fast enough would be most vulnerable to future sea level rise.

Theoretically, wetlands may have a way out. Not only can they expand vertically, but they can also expand landward. However, their ability to migrate landward depends on the local slope, sediment supply, relative sea level rise, and available land. A steep slope, artificial barriers such as seawalls or bulkheads, or high-density coastal development could impede this option. The future of salt marshes will then depend not only on sea level rise but also on geography and the degree of coastal development.

Coral Islands and Reefs

Many coral islands, like the Maldives, Kiribati, and Tuvalu, are also at high risk of inundation because they typically stand only a few meters above mean sea level. As with salt marshes and mangroves, the ability of corals to keep up with sea level rise depends on how fast they can grow upward. Coral reefs display three basic response modes to a rise in sea level: (1) "keep-up" reefs—those that can keep pace with rising sea level; (2) "catch-up" reefs—those that are left behind by a rapidly rising sea level but resume growth quickly once the rate of sea level slows down; and (3) "give-up" reefs—those that cannot catch up even after sea level stabilizes. Some studies place 20 millimeters per year as a maximum "catch-up/keep-up" threshold, while others set that value at no higher than 15 millimeters per year.[32] If rates of sea level rise exceed 10–12 millimeters per year near the end of the century, all but the fastest-growing corals may fall behind and drown.

However, sea level rise is not the only phenomenon threatening reef-building corals. At present, they more frequently face stresses from widespread bleaching, a condition linked to ocean warming and major ENSO events.[33] Acidification of the oceans can also interfere with normal, healthy growth. As carbon dioxide builds up in the atmosphere, close to a third of the gas is absorbed by the oceans, which therefore become slightly more acidic. Surface ocean pH has decreased by 0.1 unit since industrialization.[34] A doubling of atmospheric CO_2, likely by 2100 according to many climate model projections, would decrease the ocean pH from 8.1 to 7.9. Calcium carbonate, from which the hard exterior skeleton of coral is made, becomes more soluble as acidity increases. Laboratory experiments suggest that higher CO_2 concentrations could interfere with calcification of some marine organisms. While no widespread coral damage has been reported from the minor ocean acidification to date, changes near shallow submarine volcanic vents that outgas CO_2 may provide a taste of thing to come. Near the vents, the marine population has shifted from organisms with calcium carbonate exteriors (coral, sea urchins, coralline algae, mollusks) toward sea grass and other species that have more tolerance for CO_2.[35] If CO_2 concentrations continue to increase, an acidified ocean could add to the other environmental stresses that undermine normal coral growth, such as rising water temperatures and sea level. The damages from severe cyclones also take their toll. Blasting of coral reefs for building materials, overfishing, and collection of exotic specimens have further damaged reef ecosystems, weakening their ability to respond to a rising sea.

Saltwater Intrusion

As sea level rises, salt water will penetrate farther upstream and into estuaries, and also infiltrate coastal aquifers. The upstream encroachment of salt water will be analogous to that which occurs at present during extreme droughts, when river runoff dwindles. At times of drought, several U.S. cities, such as New York and Philadelphia, depend on freshwater supplies that are located slightly upstream from the average brackish water–freshwater boundary. Rising sea level would push the boundary farther upstream, eliminating this water resource. A higher salt level in coastal aquifers would also contaminate urban surface water supplies. For example, the main water supply for the city of Miami, Florida, comes from the Biscayne Aquifer, which is recharged by the nearby Everglades. Parts of the Everglades have elevations less than one meter above the level of mean high tide. Salinization of the aquifer would threaten Miami's chief water resource. Extensive water management plans already under way aim to prevent such an eventuality.[36]

Low coral islands, like those in the western Pacific, would be particularly hard hit by saltwater intrusion. A freshwater lens overlies salt water, since it is less dense. A rise in sea level would reduce the thickness of that lens, thus diminishing the available freshwater capacity. Freshwater storage would be further reduced by any coastal erosion accompanying sea level rise. On some islands, increased coastal flooding and excess pumping of freshwater have already led to upward migration of the saltwater-freshwater interface. Unlike inhabitants of the mainland, the islanders have precious few other freshwater resources upon which to draw.

In southeast Asian deltas, rice cultivation extends up to the intertidal zone. Advancing salinity has already begun to affect water for drinking and irrigation in Bangladesh's fertile coastal zone.[37] Although the increased salinity stems in part from sea level rise, curtailed river flow due to upstream dams has aggravated the problem.[38] As sea level climbs, brackish water will likely extend farther inland, making once fertile agricultural land unproductive.

Coastal Erosion

Erosion is the geological process whereby water, wind, ice, and waves wear down the land surface and move loose soil and rock debris from one place to another. More specifically, the main agents of coastal erosion are waves, and to a lesser extent, wind. These agents remove or transfer sand from the beach offshore, inland, or along the shore. The energy of a wave varies in direct proportion to the square of its height. Thus, if a storm doubles the height of the incoming waves, their energy, or ability to erode the shore, quadruples (see chapter 1). While the waves and winds of severe storms produce the greatest changes, the normal lapping of waves, daily tides, and seasonal fluctuations continuously mold and sculpt the outline of the coast.

Beaches

More than 70 percent of the world's beaches are eroding. Although the rising sea is an underlying cause of the worldwide prevalence of coastal erosion, it is only one of multiple contributing factors. The Australian geomorphologist Eric Bird lists 21 causes of beach erosion, including reduced river sediments because of upstream dams; reduced cliff erosion or sand derived from inland dunes, or the seafloor, due to changes in local climate or ocean currents; mining of beach sand; increased storminess and wave attack; trapping of sand behind artificial structures; fewer ice shelves; land subsidence; and sea level rise.[39]

Beaches consist of loose, unconsolidated sediments ranging in size from fine sand to pebbles, cobbles, occasionally boulders, and varying amounts of broken shells. Storms wash in sand from the seafloor, longshore currents transport and deposit sand or pebbles eroded from nearby cliffs, rivers carry sand to the shore, and sand is often artificially added to beaches to replace losses from erosion. The erodibility of a beach, all other things being equal, depends on sediment size—the finer the particles, the easier it is for waves to wash them away. Thus, sand-sized grains or larger pebbles and cobbles remain on the beaches, while finer-grained silt and mud tend to deposit farther offshore. However, sediment size is just one factor determining how fast a beach erodes; wave height, storminess, relative sea level rise, and fresh sediment inputs also affect the erosion rate.

Around one-eighth of the world's coastline consists of barriers and barrier island beaches—elongated, low-lying, usually sandy landforms that partially enclose bodies of water, such as bays or lagoons, or are entirely surrounded by water. They are most well developed along the U.S. Atlantic and Gulf Coasts, and also along parts of the coasts of Australia, South Africa, and eastern South America. Barrier islands are landforms that are particularly vulnerable to the effects of sea level rise.

In many places, barrier islands are already responding to sea level rise. For example, along the U.S. Atlantic Coast, 75 percent of the shoreline away from spits, tidal inlets, and built structures has been eroding for the last 100–150 years.[40] During this period, barrier islands along the Louisiana coast, such as the Isles Dernieres and the Chandeleur Islands, have eroded, narrowed, and broken apart. A succession of severe hurricanes within the past decade[41] (as well as the effects of river engineering) have exacerbated the historically high regional erosion trends and have nearly wiped out the islands. Beaches have not had much chance to rebuild by natural processes before being attacked by the next hurricane. Are the disintegrating Louisiana islands a harbinger of the fate of barrier islands elsewhere as sea level rises?

Beach erosion is not limited to barrier islands. For example, in the UK, 67 percent of the eastern shoreline has been receding landward.[42] Many sections of low-lying Arctic coasts have also receded as a result of regional warming in addition to sea level rise.[43]

As sea level rises, the rate and extent of coastal erosion are likely to intensify, particularly on sandy beaches. The **Bruun Rule**, a widely applied, although flawed model for predicting how sea level rise would affect beaches, states that a typically concave-upward beach profile erodes sand from the upper beach and deposits it offshore, so as to maintain the same profile as the water level rises (fig. 8.9).[44] The Bruun Rule assumes equilibrium conditions—the shore neither gains nor loses sediment over a given time period and all sand lost from the upper beach deposits offshore. These ideal conditions are often not met. For example, longshore currents can deposit or remove sand from the beach. Heavy storms also sweep sand across the barrier beach, creating "washover" fans in the lagoon. The Bruun Rule ignores other processes that affect shoreline position, such as sand supply, complex patterns of offshore sediment transport, underlying bedrock geology, wave energy, and storm frequency. These complexities limit its applicability and predictive ability; therefore, some argue that it should be abandoned altogether.[45] Yet, in spite of its many shortcomings, it is still widely used to predict the shoreline's response to sea level rise because of its mathematical simplicity and the lack of a better model.

A general rule of thumb suggests that for every 1 meter rise in sea level, the shore retreats from less than 100 meters to more than several hundred meters inland.[46] The beach migrates upward and landward, retreating farther inland than it would for inundation alone. Thus, even a relatively minor sea level rise of 30 centimeters (1 foot) could move the shore back by 15–60

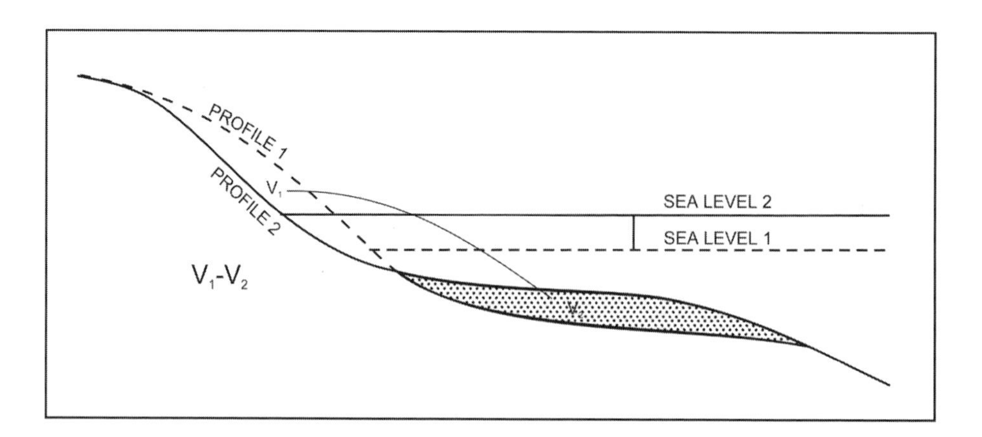

Figure 8.9 The Bruun Rule, showing the upward and landward shoreline displacement in response to sea level rise. The volume of sand deposited offshore is assumed to equal that lost from the beach.

meters (49–197 feet). However, the actual retreat rate will vary depending on local conditions.

As sea level rises, barrier islands would erode on the ocean side, with some sand being swept across the island toward the bay during large storms. Repeated episodes of sand creeping landward would gradually "roll" the island like a rug toward the mainland. But if sea level rises too fast, or not enough sand replaces that which has been lost, or the barrier island narrows beyond a critical threshold, it is in danger of breaking up, as Isle Derniere did in Louisiana. Sandy beaches occupying gentle slopes would also be gradually pushed landward. But small pocket beaches sandwiched between headlands or at the foot of cliffs would have nowhere to roll over. They would drown in place.

Bluffs and Cliffed Coasts

Three-quarters of the world's coasts consist of rocky cliffs.[47] Bluffs and cliffed coasts are undermined by the relentlessly pounding surf, and they too can experience erosion. Cliffs are steep, rocky slopes, whereas gentler bluff slopes are generally covered by loose debris, soil, or vegetation. Examples of well-known cliffed coasts include the white chalk cliffs of Dover and their counterparts across the English Channel in Normandy, France. Other cliffed coasts occur along segments of the Oregon coast, the Cliffs of Moher in County Clare, Ireland, the dark basalt cliffs south of Myrdalsjökull (glacier) in southern Iceland, and cliffs near Port Campbell, Victoria, Australia (fig. 8.10).

The heavy battering by waves laden with sand and gravelly fragments, especially during storms, eventually takes its toll. The cliff is gradually undercut by the wave action and grows progressively more unstable, until overlying rock masses break off and fall to the base. The pile of rocky debris at the base breaks down further and forms a narrow beach, or is removed by waves and currents. Cliffs or bluffs composed of softer formations, such as clays, sands, or other unconsolidated sediments, are cut back by repeated slumping, particularly after heavy rains. The erosion of cliffs and bluffs supplies sediments to the beach and nearshore. Softer, unconsolidated sediments, such as silts and clays, are much more easily eroded than hard, resistant rocks such as granite, quartzite, or basalt.

As the ocean level rises, the wave attack at the base of cliffs or bluffs strengthens and accelerates the erosion, at the same time supplying more rocky debris and sediment to the shore. The ultimate fate of this loose material—whether it is transported to the beach or offshore—is difficult to

Figure 8.10 Cliffed coast—the Twelve Apostles, near Victoria, Australia. The stacks, or erosional outliers, consisting of silty limestone, stand like pillars in the sea. They once were continuous with the cliffs on the mainland. (Photo: H. Apps; J. Clarke, 2009, "Cool Water Carbonates," in V. Gornitz, ed., *Encyclopedia of Paleoclimatology and Ancient Environments*, fig. C25, p. 139, with kind permission from Springer Science + Business Media B.V.)

predict at this time. However, as a general rule, rising waters will cause more cliff and bluff erosion, particularly in soft, unconsolidated sediments.

Arctic Coasts

Large portions of the polar Arctic coastlines may become unstable under mounting temperatures and ocean levels. The extent of Arctic sea ice has decreased dramatically during the summer in recent decades, largely due to higher temperatures, but also partly due to changing wind patterns.[48] Sea ice, when attached to the shore, protects it from being eroded by high waves during large storm events. Removal of this protective ice cover for longer and longer periods during the summer and early fall seasons results in greater amounts of coastal erosion.[49] Another consequence of higher temperatures arises from thawing permafrost—permanently frozen ground. Inland, thawing permafrost has already expanded the number and size of small ponds and lakes and also led to buckling roads, collapsing building foundations, and "drunken" forests. Along Arctic shores, sediments and soils once firmly held together by solid ice begin to crumble and disintegrate, leading to more-rapid coastal erosion.[50] Inasmuch as the Arctic is already heating up faster than anywhere else on Earth, these problems are likely to worsen as the planet warms and sea level rises.

The Human Imprint

Louisiana's rapid shoreline retreat, described earlier, vividly illustrates the human imprint—how flood control efforts along the mighty Mississippi River, as well as underground extraction of hydrocarbons and water have inadvertently amplified natural land subsidence and erosion. Elsewhere, artificial structures intended to prevent or reduce erosion have instead frequently done the opposite (fig. 8.11). Breakwaters are structures that protrude into the water in order to protect harbors against heavy waves. **Groins** are structures built perpendicular to the shore, often in series, constructed to reduce **longshore drift**.[51] Jetties, meant to stabilize barrier inlets or lagoon entrances, similarly project seaward. If improperly designed, these structures may instead intercept normal longshore drift, causing sand to accumulate on the updrift side and be eroded downdrift.[52] Seawalls are stone or concrete walls constructed to minimize flooding or overtopping by storm surge or high waves. However, the powerful wave energy often scours loose sand or sediment from beneath, ultimately undermining the structure. Rubble is often placed at the base of the seawall to dissipate this erosion. But a rising sea level will likely increase the amount of overtopping and bottom erosion associated with seawalls.

Assateague Island National Seashore, home to a herd of wild horses ("Chincoteague Ponies") and a rich variety of birds and fish, typifies the heightened beach erosion resulting from interception of longshore sand drift. Erosion on the northern end of Assateague Island accelerated after the construction of jetties in Ocean City, Maryland, to the north, which followed a major hurricane in 1933 that created an inlet separating Fenwick Island to the north from Assateague Island to the south.[53] A beach replenishment program for the northern part of the island began in 2002 and the second phase was initiated in 2004. Ongoing restoration efforts aim to replicate natural sand washover.

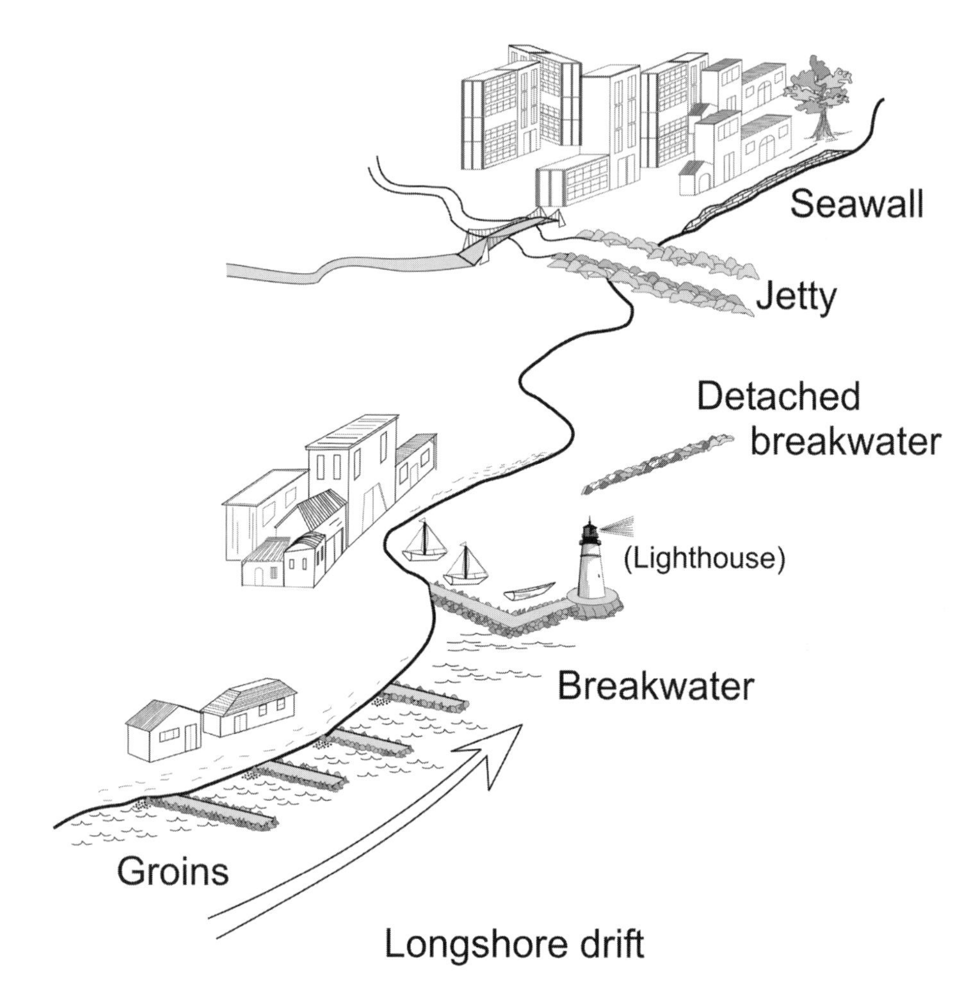

Figure 8.11 Schematic diagram of engineering structures designed to protect the coast from storms or erosion (groins, jetties, breakwaters, etc.). If improperly designed, they may have the opposite effect.

Erosion in Westhampton Beach, Long Island, New York, followed the stabilization of the Shinnecock Inlet farther east in 1942.[54] A series of 15 groins was built after flooding and erosion from a major nor'easter in March 1962. Another powerful nor'easter in December 1992 sliced through the barrier island in two locations, destroying around 60 homes.

Coastal erosion (and loss of wetlands as well) is widespread today, even in relatively "pristine" regions far from human settlements.[55] While the coast continuously adjusts to changes in the atmosphere, oceans, on land, and built-up areas, retreating in many places, advancing in others, an underlying common denominator is the rise in global sea level. The Bruun Rule predicts that future sea level rise could magnify the effects of coastal erosion more than a hundredfold. But beach erosion represents only one type of coastal hazard. Storms are another, often considerably more dangerous.

STORMS AND OTHER COASTAL HAZARDS

Fury from the Sea

Hurricane Katrina, rapidly gathering strength as it crossed southern Florida into the Gulf of Mexico, briefly reached Category 5 intensity before making its final landfall near the mouth of the Mississippi River as a Category 3 hurricane on August 29, 2005 (table 8.1). In its wake, Katrina left 80 percent of the city of New Orleans under as much as 15 feet of water, inflicted more than 1,500 fatalities, and became the most destructive and costliest U.S. natural disaster to date, causing $81 billion (2005 U.S. dollars) in damages (fig. 8.12). As tragic as the aftermath was, it came as no surprise to scientists. Half a year before Hurricane Katrina struck Louisiana, Shea Penland, a coastal geologist, described New Orleans as "a disaster in the making. New Orleans defenses would simply crumble if a truly enormous storm lingered over the city for too long."[56] Crumble they did! Levees designed to withstand a strike by a Category 3 hurricane failed in approximately 50 locations throughout the greater metropolitan area, inundating most of the city. Compounding the previously mentioned factors putting New Orleans at risk—delta subsidence, severe shore erosion and land loss, insufficient sediments—much of the city sits below sea level. Worse yet, most of the deaths and destruction resulted not so much from wind or rain damage as from flooding caused by multiple

Table 8.1 The Saffir/Simpson Hurricane Scale.

Category	Typical Hurricane Characteristics			
	Winds (mph)	Barometric Pressure (mbar)	Surge (ft)	Damage
1	74–95	>979	4–5	Minimal
2	96–110	965–979	6–8	Moderate
3	111–130	945–964	9–12	Extensive
4	131–155	920–944	13–18	Extreme
5	>155	<920	>18	Catastrophic

Source: NOAA National Weather Service.

Figure 8.12 New Orleans after Hurricane Katrina. (Photo: Commander Mark Moran, Lt. Phil Eastman, and Lt. Dave Demers, NOAA Aviation Weather Center).

levee failures, hours after the eye of the hurricane had passed to the north. The levee failures occurred partly from overtopping of surge waters, but also because of instability of the foundation soils and weakness in the levee fill.[57] Furthermore, levees had not been designed to withstand subsidence and sea level rise subsequent to construction, so at some locations the barriers were lower than their intended protection level.

The deadliest U.S. disaster ever, however, was "Isaac's Storm," or the Galveston Hurricane of 1900, which struck on September 8, 1900, wiping out the bustling city of Galveston and killing more than 8,000 people.[58] At its peak, winds blew 120 miles per hour and a surge of 15–20 feet swept over the barrier island. Survivors clung to floating timbers as their homes washed away. Meteorologist Isaac Monroe Cline of the U.S. Weather Service, noting the rising swell and strengthening winds, realized that a severe storm was approaching the island and managed to telegraph his superiors at the Washington bureau of an impending "disaster." Although later maligned by some for his previous disregard of the need for a seawall,[59] he issued timely storm warnings and worked tirelessly during the hurricane to shelter people at his home, until the entire building collapsed and they were cast adrift. Although he, his children, and brother survived, his wife drowned. The next day's edition of the *Galveston News* simply listed the names of the known dead.

In spite of this enormous tragedy, the city of Galveston decided to rebuild. The shore was defended by a seawall 3 miles long, 17 feet high, 16 feet wide at the base and 5 feet wide at the top. The entire city was raised up to 17 feet behind the wall by dredging sand from the bottom of the harbor and dumping the loads onto the streets. Individual houses were raised as well, at the homeowners' expense. Cornelia Dean, referring to the decision to rebuild Galveston, wrote (in 1999) in her book *Against the Tide*: "But Galveston had made a Faustian bargain, and it would pay the price."

More than a century later, it paid that price when Hurricane Ike roared across the barrier island on September 13, 2008. Packing winds of 110 miles (175 kilometers) per hour, Ike was ranked only as a Category 2 hurricane, yet it created a storm surge equivalent to that of a Category 5 storm, illustrating the inexactness of current models in anticipating the strength of storm surges. Because of its immense size, the hurricane devastated much of the coastline from Louisiana to near Corpus Christi, Texas, reserving the brunt of its fury for the Galveston area. Waves crashed over the 17-foot seawall, and the surge flooded most of the island, flattening many waterfront buildings on the eastern end of the island (nearest to the hurricane's track). The storm surge ebb also produced flooding on the island's backside. However, unlike 1900, the 2008 hurricane caused only ~20 deaths in the Galveston area (around 50 altogether, including adjacent states), thanks to weather satellite tracking, ample advance warnings, and a mass evacuation. Nevertheless, thousands of people stubbornly chose to remain in their homes on Galveston Island and in Houston.

These dramatic examples underline the awesome power of hurricanes. But hurricanes are not confined to the Caribbean basin nor to the Gulf and Atlantic Coasts. Tropical cyclones periodically rage across the Pacific and Indian Oceans. For example, the coast of Bangladesh, on the delta of the Ganges-Brahmaputra River, is especially vulnerable to storm surges because of its low elevation, high population density, and storm tracks directed toward the apex of the Bay of Bengal. Nearly half a million people perished during a cyclone in 1970, and another 100,000 lost their lives in 1991. In nearby Myanmar, Cyclone Nargis wiped out entire villages and killed more than 138,000 people on May 2, 2008. The surge reached up to 5 meters (16.4 feet), plus an additional 2 meters (6.6 feet) of waves—levels comparable to those of Hurricane Katrina.[60] Since the approaching storm was tracked on weather satellite, many fatalities could have been avoided had timely warnings been issued and evacuation plans been better organized.

The destructiveness of coastal storms reaches beyond the tropics, however. In 1953, the "North Sea Flood," an extra-tropical cyclone, hit western Europe, drowning more than 1,800 people in the Netherlands and around 300 in the United Kingdom. Driven by high winds close to spring tide, the floodwaters of 3–5 meters overwhelmed dikes and levees in the Netherlands and inundated a large part of the country. The disaster led to the construction of the Delta Works—a massive engineering project aimed at protecting much of the coast (see chapter 10).

Hurricanes in a Warming World

Because of their high winds, torrential rainfall, and elevated surge levels, major tropical cyclones are of particular concern to coastal residents.[61] Although wind and rain damage can extend far inland, most of the deaths and property damage near the shore are caused by the surge and high waves. Are we already seeing an increase in hurricane activity? Have they become more destructive? Or do greater damages reflect increasing coastal development in harm's way? Will tropical cyclones and hurricanes increase in numbers and strengthen in a warmer world, thus exacerbating the hazards of seaside living? Answers to these questions gain urgency as the coasts are set on a collision course between expanding populations and the growing threat from the rising sea.

Claims that we are indeed seeing an upswing in the number of hurricanes, especially the most destructive ones, have generated a storm of counterclaims. Some studies note an increase in both Atlantic hurricane frequency and intensity within the past 30 years—a trend presumably linked to increasing sea surface temperatures and global warming.[62] However, some earlier storms may have gone undetected during the pre-satellite era.[63] Furthermore, a count of only landfalling hurricanes may not truly capture all hurricanes, including those that stay well out to sea. Another complication is that normal multi-decadal climate variations, such as the Atlantic Multidecadal Oscillation (AMO), may lead to periods of high hurricane activity alternating with quieter periods. A misleading trend can emerge, depending on the time period selected for the analysis. Do the last 30 "stormier" years, therefore, represent a true upward trend or just one part of the AMO cycle?

Seven of the 10 costliest U.S. hurricanes have occurred since 1980.[64] Have the costs increased because of greater hurricane strength or because of increasing coastal development? A fairly calm period along the U.S. East Coast during the 1960s and 1970s may have induced a false sense of complacency and encouraged an unprecedented spurt of high-rise condominium construction along large segments of the coast. However, once the data have been adjusted to account for changes in inflation, population, and income, most of the apparent upward trend in storminess disappears.[65]

Predicting future storm behavior, especially tropical cyclones, is tricky. At first glance, it may seem reasonable to expect that the number of hurricanes would increase as their "breeding ground"—the area of upper ocean water above 26°C—expands. However, warm ocean water alone does not make a hurricane. Hurricane development also depends upon vertical wind shear, the vertical atmospheric temperature gradient, and climate variations such as the El Niño-Southern Oscillation or the Atlantic Multidecadal Oscillation. For example, more Atlantic basin hurricanes occur during La Niña episodes and fewer during El Niño years.[66] Atlantic hurricanes are suppressed during La Niñas because the stronger vertical wind shear inhibits hurricane development, with the reverse holding true in El Niño years. Even more uncertain are future changes in hurricane tracks (where they would make landfall)—an important consideration for the increasingly urbanized coast. Will the likelihood of a hurricane striking a particular city, say New Orleans or Miami, change over time? Will northern cities become more frequent hurricane targets as oceans grow warmer? We cannot yet foresee how changes in these other factors would affect hurricane development.

The vulnerability of a community to storm flooding depends not only on the storm's meteorological characteristics but also on the land and offshore topography, timing of storm landfall relative to the tidal cycle, and structural strength of buildings and seawalls. Densely populated or highly urbanized areas that are especially vulnerable to storm surges include the Bay of Bengal, Yangon (Rangoon), Bangkok, the Pearl River Delta (China), Shanghai, greater London, New Orleans, Miami Beach and south Florida, Venice, and the New York City metropolitan area.

Because severe coastal storms already impact the coasts, their threat appears more immediate than sea level rise. However, even without a change in future storm behavior, a high storm surge coupled with rising sea level creates a recipe for greater flood risk, especially in urbanized coastal areas.

Even a small rise in sea level can increase the number of floods above a specified height, without any change in storminess. In Venice, the frequency of 110-centimeter-high tides (regarded by Venetians as "exceptional high water") has increased during the 20th century, due to the historic 25-centimeter relative sea level rise (including an anomalously spurt between 1930 and 1970, from groundwater overdraft). In the past 40 years since over-pumping ceased, high water levels have tended to recur every 3 months, on average, compared to a 2.2-year recurrence rate in the early 20th century.[67]

The Thames Barrier was built to protect London from powerful storm surges, like the devastating one in 1953. Completed in 1982 and officially dedicated by the Queen in May 1984, the Barrier has been closed with increasing frequency over its 28 years of operation—from four closures during the 1980s to 75 closures during the 2000s.[68] Of these closures, 64 percent were triggered by storm surges and 32 percent by river flooding. As with Venice, the rising sea has increased the number of high-surge events requiring Thames Barrier shut-downs.[69]

Storminess did not increase along the U.S. East Coast during the 20th century, yet coastal flooding has become more frequent in places like Charleston, South Carolina, and Atlantic City, New Jersey.[70] Because of the regional rise in sea level of 8–16 inches (20–40 centimeters) since 1900, the storm surge and waves are now superimposed on top of a higher ocean. By the 2080s, sea level in New York City could expect an additional 12–23 inches above the historic rise. Consequently, the flooding currently produced by a storm that has a chance of occurring only once a century would then recur, on average, once every 15–35 years.[71]

Tsunamis

We learn geology the morning after the earthquake, on ghastly diagrams of cloven mountains, upheaved plains, and the dry bed of the sea.

—RALPH WALDO EMERSON (1808–1882), 1860

Coastal residents also face hazards besides storm surges and sea level rise. Tsunamis are another. On March 11, 2011, a deadly wall of water devastated the northeast coast of Honshu island in Japan, killing more than 13,000 people. The overwhelming, onrushing water, penetrating far inland to elevations of more than 30 meters (>100 feet) in some places, destroyed nearly everything in its path. The tsunami was generated by a magnitude 9 undersea earthquake (among the five most powerful earthquakes ever) as the Pacific Plate slid beneath Honshu's underlying plate.

A few years earlier, on December 26, 2004, tourists sunbathing on idyllic tropical beaches in Phuket, Thailand, on a bright sunny day suddenly noticed a huge wave emerge from the sea and push relentlessly toward them. It was already too late for many of those desperate to reach higher ground. More than 230,000 people, tourists as well as local residents of northern Sumatra, Sri Lanka, eastern India, Thailand, and other countries bordering the Indian

Ocean, drowned that day in what may have been the deadliest tsunami in history. Areas where mangrove vegetation had been preserved suffered less damage and fewer fatalities.

On another disastrous day over a century ago, the small volcanic island of Krakatau (also spelled Krakatoa) in the Sundra Straits between Java and Sumatra, Indonesia, exploded violently with a force equivalent to 7,500–8,750 Hiroshima-sized atom bombs, generating tsunami waves that reached a height of around 40 meters (131 feet). The paroxysmal volcanic eruptions wiped out the island of Krakatau on August 26 and 27, 1883, and claimed 36,000 lives, mostly from the ensuing tsunami. The sonic boom from the explosion could be heard as far away as Australia, and atmospheric shock waves were picked up by barometers around the world.[72]

Japan is well acquainted with the force of tsunamis. Three centuries ago, in 1700, several fishing villages on the Pacific coast of Japan were swamped by tsunami waves up to 3 meters (~10 feet) high. While the effects of this event in Japan were puny in comparison to those of the aforementioned tsunamis, its importance lies in its origin—on the other side of the Pacific Ocean, off the northwest coast of North America. Submerged tree trunks and peat layers beneath beach sand, found miles inland from the ocean, sit in marshes from Oregon to Washington. Tree rings and radiocarbon dates show an age of around 300 years, consistent with the Japanese records. Although the Pacific Northwest has been seismically quiescent ever since, the earthquake danger remains ever-present, inasmuch as the Cascadia Subduction Zone lies off the Oregon and Washington coast.

Unlike sea level rise, tsunamis are not produced by climate change. Instead, they are primarily caused by waves set off by the vertical displacement of ocean water, triggered by strong earthquakes (for example, the 2011 Japan tsunami), major volcanic eruptions (Krakatoa, 1886), submarine landslides, and more rarely the impact of an extraterrestrial object (the Chicxulub impact, Yucatán, at the K-T boundary; see chapter 3).[73] Curiously, out on the open ocean, the waves generated by a tsunami are barely detectable—merely a few feet high. However, as the waves approach the shore, they slow down but gain considerable height, thereby amplifying their destructiveness. Most destructive tsunamis originate in the Pacific Ocean, due to the high incidence of earthquakes and volcanic eruptions along the "Ring of Fire." However, the Indian Ocean is not immune, as the world learned to its chagrin in 2004. The U.S. Atlantic Coast is also potentially at risk because of slope instabilities that could set off massive landslides.

Methane hydrate may connect tsunamis to climate change. This form of ice, which traps methane or carbon dioxide within cage-like crystal structures, is stable only under pressure at subfreezing temperatures. Enormous quantities of methane hydrate are believed to be locked up in Arctic permafrost and in marine sediments on the continental slope and rise. They are already being eyed as a novel source of hydrocarbons for an energy-hungry world, but concerns exist about our ability to extract this resource safely, without explosively releasing methane.[74] The Storrega Slide, an enormous underwater landslide off the coast of Norway, generated a very large tsunami that impacted the coastlines bordering the Norwegian Sea and the North Sea roughly 8,100 years ago (chapter 5). The tsunami left deposits up to 20 meters high in some locations. While the trigger for the Storrega Slide is uncertain (e.g., earthquake or sediment instability), the slump overlies an area of known methane hydrate deposits and likely released large volumes of methane, which some speculate may have helped end the 8,200-year cold spell.[75]

As increasing temperatures slowly penetrate deeper into the oceans, the hydrates now solidly encased in unconsolidated marine sediments may begin to thaw and decompose. Methane plumes have been spotted in the Arctic Ocean west of Svalbard, largely derived from dissociating methane hydrate.[76] The Arctic Ocean in that region has warmed by 1°C over the last 30 years. As more solid methane hydrate decomposes, the destabilized soft sediments could slump downhill, setting off massive submarine landslides that release still deeper buried hydrates in a chain reaction, thus generating potentially destructive tsunamis.

LIVING DANGEROUSLY

The world's shorelines can be inherently dangerous places to live even today, because of their rapidly changeable nature. Yet the coasts are on a collision course between growing populations, on the one hand, and the destructiveness of major storms and sea level rise, on the other. Hurricanes, typhoons, and even severe non-tropical cyclones already produce heavy flooding and destruction. Coastal erosion is widespread and if not reined in (e.g., by "defensive" measures) will expose more people to damaging surges and waves. Many low-lying areas are sinking and losing land to the sea. Although natural and anthropogenic processes may largely mask the effects of sea level rise at present, coastal hazards will only intensify as the level continues to climb, especially toward the uppermost projections. Chapter 9 examines the growing vulnerability of highly populated rural and urban areas. It also investigates ways of adapting to the rising seas.

Coping with the Rising Waters 9

WHAT'S AT STAKE—IMPACTS ON LAND AND PEOPLE

Acqua Alta in Venice

Loud, piercing siren blasts signal approaching "exceptionally high water" (*acqua alta*, in Italian) when the tide reaches or exceeds 140 centimeters (55 inches) above the 1897 reference datum. Movable raised plank walkways are rapidly installed across the plazas, shopkeepers and hotel owners place sandbags across their doorways, and people don hip-high boots (fig. 9.1). The high tide rises above the canals and sweeps across the Piazza San Marco and down the narrow streets and alleys of this historic, art-filled city. A tide of 140 centimeters or more submerges more than half the city. The most recent such event (at the time of writing) occurred on Christmas Day, 2009 (at 145 centimeters). However, the record was set on November 4, 1966, at 194 centimeters (76 inches), flooding more than 80 percent of Venice! This severe flood inspired the MOSE project (**Mo**dulo **S**perimentale **E**lettromeccanico, or Experimental Electromechanical Module), also named after the biblical Moses, which was designed to protect Venice and its surroundings with an effective sea defense system.

Figure 9.1 *Acqua alta* in Venice. The flooded Piazza San Marco with raised walkways. (Photo: JoMa, Sept. 18, 2009, Wikimedia Commons.)

Acqua alta events occur when southeasterly sirocco or northeasterly bora winds combine with astronomical high tides to drive the waves in the narrow, relatively shallow Adriatic Sea toward Venice. The frequency of a "very intense" 110-centimeter tide (which triggers the alarm system) increased from once every 2 years to approximately 4 times per year during the 20th century.[1] Eustatic sea level rise alone did not push the waters higher. Land subsidence due to natural causes (tectonics, compaction, and sediment loading), groundwater mining (especially between 1930 and 1970), and artificial channels and jetties in the nearby industrial port of Porto Marghera added to the toll. Coastal scientist Laura Carbognin at the Institute of Marine Sciences in Venice and her colleagues foresee that a rise in sea level of a mere 20 centimeters (8 inches) could trigger the alarms for a 110-centimeter tide as often as 21 times a year. With a 50-centimeter (20-inch) rise, high tides would recur 250 times a year![2] The city would soon become unlivable.

Facing an Expanding Ocean

While the experience of Venice may seem unusual, even exceptional, today—a consequence of its unique geophysical setting, meteorology, and history—many coastal cities can expect to face a similar situation in the future. The first troubling manifestation of a rising sea will not be the gradual, barely perceptible increase in average water height but rather the more frequent high tides accompanying coastal storms that repeatedly flood low-lying neighborhoods. Formerly rare events will become commonplace and more damaging as the water laps ever higher and farther inland, even with no significant change in storminess.

How many people would confront the rising waters? How can we redesign our coastal cities and croplands to protect us from the encroaching sea? We start by estimating the number of people at risk for flooding or permanent inundation and then outline a number of strategies for coexisting with a higher ocean.

Even today, the coast is an inherently hazardous place, subject to the fury of wind, waves, and water. The surge and heavy waves from a major storm can re-sculpt the shoreline by scouring dune ridges, washing sand offshore, cutting new inlets, and redepositing sand on spits and elsewhere. Small houses can be swept out to sea and larger buildings seriously damaged. These risks will likely escalate as the ocean climbs relentlessly upward.

Furthermore, we are on a collision course between rapid worldwide coastal development and intensifying maritime hazards. Migration of people toward the coast has rapidly accelerated ever since the expansion of international trade during the colonial period, and especially in today's globalized economy. Two percent of the Earth's land area lies at or below 10 meters (33 feet), but holds 10 percent of the total population and 13 percent of the urban population.[3] Many of the world's largest, most populous cities sit along the coast and risk both permanent and episodic flooding of low-elevation neighborhoods. Thirteen out of the 20 most populated cities in the world are seaports, vital to international commerce. China exemplifies the seaward resettlement of population and urbanization, largely stimulated by its economic policies.[4]

The low topography of deltas makes them especially vulnerable to sea level rise (see fig. 8.7). Many deltas are also sinking because of geological and anthropogenic processes (e.g., see chapter 8). As a result, 85 percent of deltas are temporarily affected by severe flooding covering more than 260,000 square kilometers.[5] Major deltaic cities include Dhaka, Shanghai, Guangzhou, Ho Chi Minh City (Saigon), Bangkok, and Rotterdam, many of which are affected by anomalously high rates of relative sea level rise (largely due to land subsidence). Examples include the Chao Phraya Delta, near

Bangkok, Thailand (13–150 millimeters per year) and the Yangtze Delta, near Shanghai, China (3–28 millimeters per year).[6] The fertility of deltaic soil suits them for growing essential food crops. Crops, such as rice particularly in Asia, are threatened not only by extreme coastal flooding but by increasing saltwater intrusion as seawater gradually encroaches farther up estuaries and rivers.

Establishing an accurate count of the global population and area exposed to future sea level rise is problematic because many regions lack sufficiently accurate topographic and demographic data and because of inconsistencies among data sets. Divergent definitions exist for the "high risk" zone and associated populations and areas. Therefore, results may not be directly comparable.

A common yardstick of inundation risk is the "1-in-100-year" (or just "100-year") flood, which has a probability of exceeding a certain height once a century, or in other words, a 1 percent chance of occurring in any given year. As of 2005, about 40 million urban inhabitants were exposed to a 1-in-100-year coastal flood.[7] Currently, the top 10 cities most vulnerable to such a flood, ranked by population, are Mumbai, Guangzhou, Shanghai, Ho Chi Minh City (Saigon), Calcutta, the New York City metropolitan region, Tokyo, Tianjin, Bangkok, and Dhaka. This urban population is largely concentrated in Asia. The top 10 cities ranked by assets (buildings, transport and utilities infrastructure, and other long-lived structures) are Miami, greater New York City, New Orleans, Osaka-Kobe, Tokyo, Amsterdam, Rotterdam, Nagoya, Virginia Beach, and Shanghai—reflecting the greater concentration of wealth in the developed nations.

The most vulnerable cities of the future also lie predominantly in Asia. If sea level rises by 0.5 meter by the 2070s, the top 10 exposed cities by population become Calcutta, Mumbai, Dhaka, Guangzhou, Ho Chi Minh City (Saigon), Shanghai, Bangkok, Rangoon, Miami, and Hai Phong.[8] Ranked by assets, the cities are Miami, Guangzhou, greater New York City, Calcutta, Shanghai, Mumbai, Tianjin, Tokyo, Hong Kong, and Bangkok, reflecting the projected growth in population and wealth in Asian cities.

In the United States (excluding the Great Lakes), 8.4 million people (or approximately 3 percent of the 2000 population) reside within the 1-in-100-year flood zone, or "Special Flood Hazard Area."[9] The largest population at risk is concentrated along the Atlantic Coast, with Florida as the most vulnerable state. Both the Gulf and the Pacific Coasts (including Alaska) face roughly the same degree of risk in terms of exposed area.

Another measure of inundation risk delineates the population living below a specified elevation. One recent study finds that a 5-meter rise would potentially affect 670 million people (10.8 percent) and submerge 5.4 million square kilometers.[10] A 10-meter rise would affect 871 million people and drown 6.3 million square kilometers. Another report places the global population in the coastal zone below 10 meters elevation at 634 million people over 2.7 million square kilometers.[11] While sea level is quite unlikely to approach this height within this century, severe storm surges could overtop it in some places. Therefore, the 10-meter elevation, as an extreme upper bound, allows a comfortable margin of safety for protective measures.

Given the large uncertainties in global-scale demographic and topographic data sets, these figures can provide only very rough estimates of vulnerability to sea level rise. Furthermore, these estimates are based on current coastal populations and do not consider future displacements of people, economic losses, or costs of coastal protection. A study that incorporates some of these factors outlines the consequences of a collapse of the West Antarctic Ice Sheet (WAIS). Assuming a tiny (yet non-zero) chance for a 5-meter (16-foot) sea level rise in 100 years due to WAIS collapse, as many as 410 million people would be exposed to sea level rise at high tide.[12] Under such dire circumstances, 350,000 people per year would be forced to migrate, displacing up to 15 million by 2130, in the worst possible case. Without coastal protection and loss of income in the affected zone, monetary losses could range between $2.4 and $3.3 trillion GDP. Nevertheless, up to a third of the world's coasts would be deemed worthy of saving, in spite of the high protection costs.

An MIT computer model analysis predicts future economic activity, greenhouse gas emissions, and resulting climate change to calculate capital losses, protection costs, and wetland losses.[13] For a linear rise in sea level to 2100, global costs incurred total $1.99 trillion (in 1995 U.S. dollars expressed as purchasing power parity, PPP). However, these costs represent an upper bound, since sea level is likely to rise in a nonlinear (quadratic) fashion, in which case most damages occur later in the century, postponing the bulk of the costs (fig. 9.2).[14]

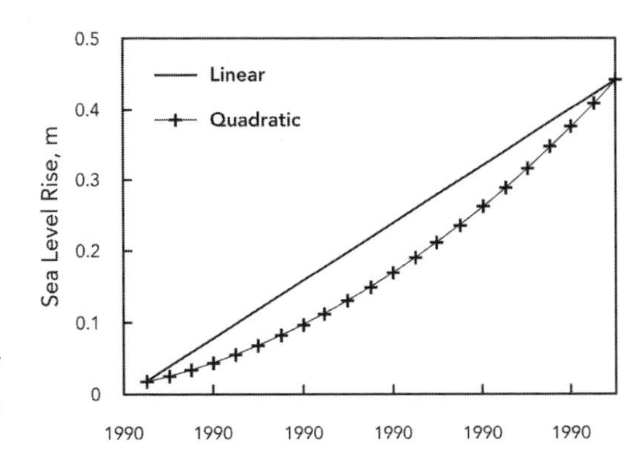

Figure 9.2 Exponential versus linear sea level rise. (After Sugiyama et al., 2008.)

LIVING WITH THE SEA—ADAPTING TO HIGHER WATER

As the situation in Venice shows, even a fairly small rise in relative sea level greatly amplifies damages from sporadic storm floods. While some more affluent, progressive communities or nations may be proactive, developing adaptation plans and measures far ahead of immediate danger to coastal residents, most towns or individuals will likely take action only after repeated incidents of severe coastal flooding or beach and wetland loss.

Coping with a rising sea can take many different forms. Some steps can be implemented rapidly, after the next major storm leaves damage and destruction in its wake, while others may require a large investment involving long-term planning and implementation.

In the context of sea level rise, adaptation means taking appropriate steps to minimize the threat of inundation. Adaptation can follow three basic pathways:

1. Protection
2. Accommodation
3. Retreat

Protection

Hard Solutions

Shoreline protection entails building structures or augmenting natural features to withstand current and anticipated shoreline retreat, storm surge, and sea level rise. It ranges from "armoring" the shoreline with "hard" defenses to "soft" defenses that mimic natural processes. Shoreline armoring is typically applied to defend substantial assets. Hard structures include **seawalls**, **bulkheads**, boulder ramparts (**revetments**, **riprap**), groins, **jetties**, and **breakwaters** (see fig. 8.11). The first three types of structures maintain the existing shoreline by preventing slumping or removal of soft, poorly consolidated sediments. The fortified shore withstands average surge and wave heights but still risks overtopping by extreme events. Groins and jetties extend outward and trap sand, widening the beach (but often intensifying erosion downdrift). Breakwaters shelter a harbor or beach from extreme wave action, but if poorly designed could also induce erosion.

Hard structures such as seawalls and revetments safeguard against flooding and storm surges, but may intensify erosion at the foot of the wall. Placement of rubble may help reduce this. Unless these engineered defenses are carefully adapted to local conditions, they may exacerbate coastal erosion downdrift of littoral currents, or undermine the embankment by washing out sediment from beneath the structure.

Other structures such as **dikes**, **tidal gates**, and **storm surge barriers** protect against extreme floods or permanent inundation. The high walls of dikes hold back the sea, but low-lying terrain behind the dike may need to be pumped dry. Tidal gates open and close with the tides, enabling water to drain out at low tide. Storm surge barriers close only during extreme surges or tides, allowing shipping to continue at other times. Inasmuch as these coastal defenses were designed for current sea level and storminess, existing structures would need to be retrofitted, for example by raising the height of seawalls, dikes, tidal gates, etc., or reinforcing them to resist stronger and higher waves. The design of new structures needs to accommodate a much higher water level (i.e., relative sea level rise plus storm surge, high tide, and waves).

Defending Holland from the Sea

Much of the southwestern part of the Netherlands is a low-lying delta largely built up by sediments deposited by three major rivers—the Rhine, the Meuse, and the Scheldt.[15] Ever since the Middle Ages, the Dutch have been protecting their nation from the ravages of the sea. In the 1930s, a dam—the Afsluitdijk—closed off the Zuiderzee (Zuider Sea), a large shallow embayment of the North Sea. This project, designed for sea defenses, land reclamation, and water drainage management, transformed the Zuiderzee into the IJsselmeer (IJssel Lake).

The disastrous North Sea flood of 1953 left 300,000 people homeless and caused 1,800 deaths.[16] That event prompted the development of the Delta Works, which consists of a series of dams, surge barriers, dikes, and sluices along the Rhine-Meuse-Scheldt delta, which closed off some of the estuarine outlets to the sea. The Dutch set the acceptable risk levels of dike failure as 1 in 10,000 years for North and South Holland, other high-risk areas at 1 in 4,000 years, and South Holland river flooding at 1 in 1,250 years. The first operational Delta Works storm barrier across the Hollandse IJssel River, completed in 1958, protects densely populated western Netherlands (the "Randstad") from flooding; it was later followed by the Eastern Schelde storm surge barrier in 1986. In addition, existing dams were raised and dikes reinforced. Completed in 1997, a movable barrier—the Maeslant Barrier—closes off the New Waterway (Nieuwe Waterweg) whenever a high storm surge threatens the city of Rotterdam, upstream, and the surrounding dikes (fig. 9.3).

Figure 9.3 Northern half of the Maeslant Barrier storm surge barrier. (Photo: Joop van Houdt, Sept. 8, 2010. Wikimedia Commons).

As impressive as these engineering works appear, they have been designed to withstand a sea level rise of just 20–50 centimeters (9–20 inches) per century.[17] As sea level rises, barrier closures will become more frequent. An 85-centimeter rise would force the Maeslant Barrier to close approximately 3 times a year; a 1.3-meter rise would lead to an average of 7 closures per year.

The 2008 Delta Committee report developed a long-term program to protect the nation's coast and interior. Existing or new storm surge barriers must withstand a worst-case scenario of regional sea level rise of 0.65–1.3 meters (2.1–4.3 feet) by 2100, and up to 2–4 meters (6.6–13 feet) by 2200.[18] Present flood protection standards in the diked areas must be raised tenfold by 2050.

Projects to reinforce weak links in the current dike system are well under way. One such weak link, for example, is the coastal resort town of Scheveningen, near The Hague, where a massive new dike 1 kilometer (0.6 miles) long and 12 meters (39 feet) high is under construction (fig. 9.4). A sweeping new boulevard will overtop most of the dike. Broad stairways ensure ready access to the beach, which is also being widened to dampen the power of

Figure 9.4 Raising the dikes and widening the beach, Scheveningen, the Netherlands. (Photo: V. Gornitz, 2010.)

the waves. Residents of buildings already lining the top of the dike enjoy a terrific seaside vista.

In addition to marine incursions, the Netherlands, like many other delta regions, faces river flooding as well. Winters are expected to grow wetter while summers become drier. Therefore, water management needs to defend against both inland and coastal flooding and also ensure an adequate water supply during dry spells. New developments beyond the dikes will be designed not to interfere with river discharge or lake water levels.

Other Hard Defenses

THE THAMES BARRIER, LONDON The North Sea storm that ravaged the Netherlands in 1953 also caused 300 deaths in the United Kingdom and pushed the British government into investigating means of defending London against future flooding. The Thames Barrier, essentially completed in 1982 and officially dedicated by Queen Elizabeth II in 1984, protects 125 square

Figure 9.5 The Thames Barrier, London. The gates close whenever water levels are predicted to exceed 4.85 meters (16 feet) in central London. (Photo: Arpingstone, Sept. 26, 2005, Wikimedia Commons.)

kilometers of central London from extreme tidal surges (fig. 9.5). The barrier spans 520 meters (1,700 feet) across the Thames and consists of four 61-meter (200-foot) and two nearly 30-meter (100-foot) navigable sections and several smaller non-navigable channels between a series of piers and abutments. Under normal conditions, the rotating, circular gates allow shipping to pass. A meteorological forecast of water levels exceeding 4.85 meters in central London due to a combination of high tides, storm surge, and river flow triggers an order to initiate a flood barrier closure. Hours before high water reaches London, warnings are issued to stop river traffic, close subsidiary gates, and advise other river users. The Thames Barrier remains closed until the tide recedes. The barrier also closes after heavy rains west of London produce floodwaters. River flooding caused a third of the closures up to 2009.[19]

Regional sea level has been rising at a rate close to 20 centimeters (8 inches) per century, due to glacial rebound and recent climate warming. Since completion, the Thames Barrier closed four times during the 1980s, 35 times in the 1990s, and 75 times between 2000 and 2010.[20] Because the risk of barrier failure increases with the number of closures, the maximum recommended limit is 50 barrier closures per year. Although the barrier was built to withstand projected sea level rise until around 2060–2070, recent sea level projections demonstrate the need for reinforcing the tidal defense system soon.

The Thames Estuary 2100 plan anticipates a sea level rise between 20 and 88 centimeters in the Thames estuary, based upon British Met Office Hadley Centre models. However, given the large uncertainty in future polar ice melt (see chapter 7), a sea level rise of up to 2 meters by 2100, although unlikely, cannot be ruled out. In a worst-case scenario, an extreme surge level of 0.7 meters could bring the water height to 2.7 meters at the entrance to the estuary. In preparation, the Thames Estuary 2100 plan recommends a three-stage flexible response approach:

1. 2010–2035. Maintain current flood defenses and undertake necessary improvements to the system beyond 2035. Monitor sea level and climate change and revise plans accordingly. Create new habitat areas for wetland migration.
2. 2035–2070. Reinforce and replace current defense network along the Thames River. Make final decision on building a new barrier (or select alternate options, e.g., reinforce and maintain current system) and begin planning stage. Continue to monitor sea level rise.
3. 2070–2100. Construct the new Thames Barrier (or alternate options). Modify defenses as needed to keep new barrier closures within operational limits.

MOSE, VENICE Recognizing the threat that more frequent *acqua alta* events pose to its irreplaceable artistic and historic heritage, the government-funded Venice Water Authority and Consorzio Venezia Nuova have decided to protect the surrounding lagoon by building the MOSE system of tidal gates. Construction began in 2003 and is expected to be completed by 2014. Other protective measures include raising quaysides and pavements and reinforcing the shore.

MOSE consists of a set of 78 mobile gates across the three tidal inlets into the Venetian Lagoon (fig. 9.6). When not in use, the gates fold flat on the seafloor and thus do not interfere with the normal flow of tidal currents or with shipping. Whenever tides above 1.1 meters are predicted, the gates are emptied of water using compressed air, which makes them rotate until they emerge from the water, temporarily blocking the high water from entering the lagoon. The gates are designed for a sea level rise of 60 centimeters, but can protect Venice from high water of up to 3 meters[21] when

Figure 9.6 The MOSE tidal gates, Venice. The gates swing up whenever tides exceed 1.1 meter (3.6 feet.) (Credits: Ministry of Infrastructure and Transport—Venice Water Authority, concessionary Consorzio Venezia Nuova image archive.)

raised above their optimum slope of 45°. The system is highly flexible. For an exceptional event, gates at all three inlets can be closed simultaneously, or alternatively, depending on forecast winds and tides, one or two gates can be closed independently.

The project has generated much controversy. Environmentalists express concern that by interfering with normal tidal flushing, the barriers might increase pollution or damage the fragile habitat of the lagoon wetlands, the largest in the Mediterranean. The authorities respond that environmental impacts will be minimal, since the barrier closures are infrequent and of extremely brief duration relative to the periods of open exchange with the sea. Effects would be comparable to those experienced in the lagoon each month during neap tides, or low water. The lagoon is a highly resilient environment and any adverse impacts from a given closure event would be erased by flushing over several tidal cycles. However, would the more frequent future *acqua alta* events affect the gates' performance and would environmental impacts still remain minimal? What if the sea climbs above the anticipated 60-centimeter level?

TIDAL BARRIERS, NEW YORK CITY The New York City metropolitan area ranks high among cities most exposed to coastal flooding.[22] In spite of its northerly location, tropical cyclones occasionally sweep past the city. In 1821, a hurricane produced a surge of 13 feet in 1 hour, flooding lower Manhattan as far north as Canal Street. In 1893, another hurricane flooded southern Brooklyn and Queens, wiping out a small barrier island off the Rockaways. Three major hurricanes pummeled adjacent Long Island and New Jersey during the 20th century: the "Long Island Express" (1938), Hurricane Donna (1960), and Hurricane Gloria (1985). Even winter storms (such as the December 1992 "nor'easter") cause extensive flooding in low-lying neighborhoods and seriously disrupt ground and air transportation.[23]

Future sea level rise would greatly increase urban flooding. Douglas Hill, consulting engineer, states: "New York City is presently planning to be flooded, not to prevent being flooded." The solution, according to Hill and his colleagues at Stony Brook University, is to build a set of three storm surge barriers at narrow points to protect the financial district of lower Manhattan and parts of adjacent Brooklyn, Queens, including LaGuardia Airport, and the Bronx (fig. 9.7).[24] Across the Hudson River, the barriers would shield Hoboken, Jersey City, and Newark, including Newark International Airport. A preliminary study, using advanced meteorological and hydrodynamic modeling to simulate an actual hurricane (Hurricane Floyd) and a nor'easter (December 1992), showed that all three barriers together could prevent flooding from storm surges. The flooding from rainfall runoff behind the barriers was deemed insignificant.[25]

Even if cost and environmental concerns were set aside, the barriers would not protect the entire city. Large sections of Staten Island, Brooklyn, and Queens, including JFK International Airport, would still be vulnerable. In addition, nearby communities on Long Island Sound and the barrier islands along the Atlantic shore lie beyond the scope of protection. Furthermore, the study did not investigate a worst-case scenario—a direct strike by a Category 3 or 4 hurricane plus future sea level rise at high tide (recall that the Dutch plan for the 1-in-10,000-year event, the British for the 1-in-1,000-year event).

As sea level rises, hard defenses would be periodically monitored, strengthened, and raised. The timing and extent of work would depend on how fast the sea rises, which varies geographically, as discussed previously. In London, for example, the defenses would be reinforced within the next

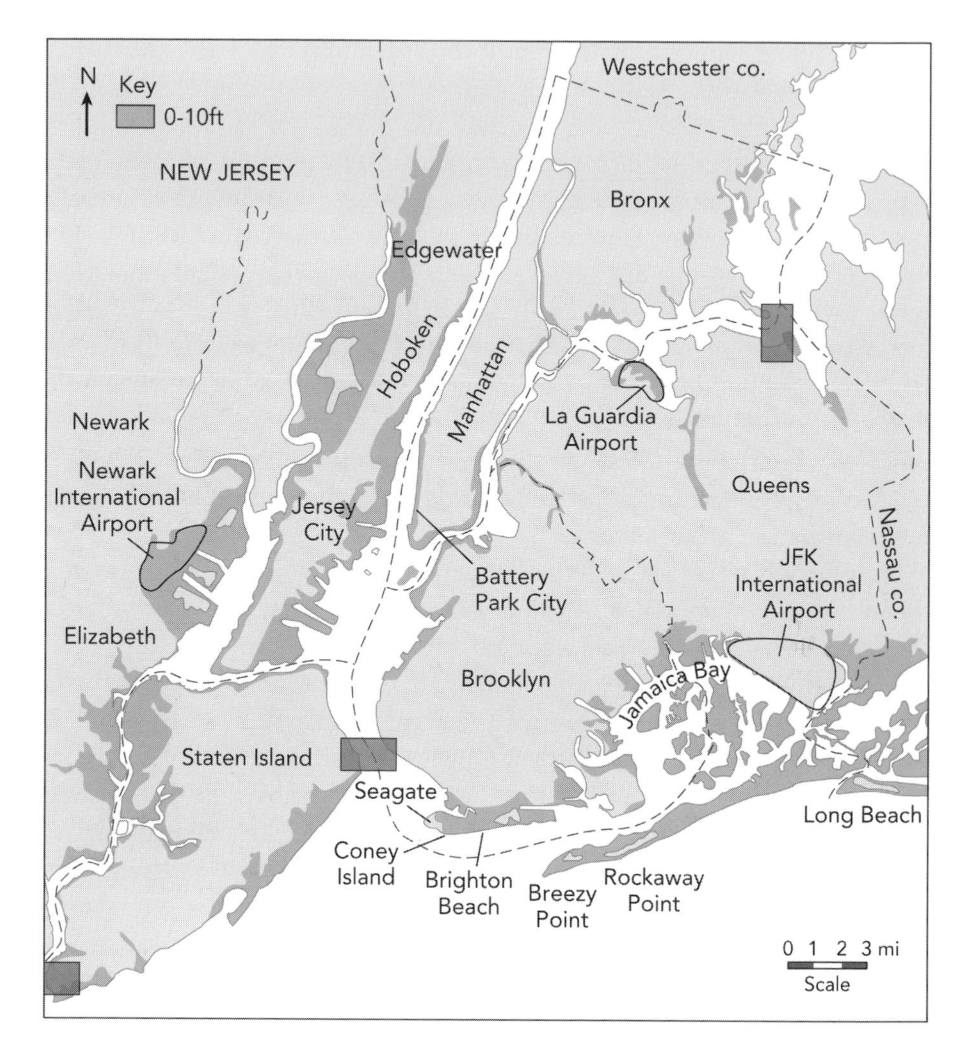

Figure 9.7 Location of proposed tidal barriers, New York City. (Hill, 2008; V. Gornitz et al., 2002.)

25–60 years, with the option to build a new barrier after that. The Netherlands is already planning to upgrade its sea defense system.

Soft Solutions

Soft defenses have become a preferred means of shore protection in many places, because of the negative impacts of hard stabilization on beaches. "Soft" defenses include **beach nourishment** and rehabilitation of dunes and coastal wetlands. Healthy beaches and salt marshes not only offer recreational opportunities and important habitat for a wide variety of wildlife, including fish, shellfish, waterfowl, and small amphibians, reptiles, and mammals, but also protect uplands against storm surges and high waves.

A broad beach with a high ridge of healthy dunes acts as the first line of defense against the sea's fury. Planting grasses and installing fences to trap sand stabilizes the dunes. Beach nourishment is another important method of shoreline protection. Sand added from offshore or inland sources can replace erosional losses and be used to widen and raise the beach. The U.S. East and Gulf Coasts have had a long history of erosion as a result of both geological and anthropogenic processes. Because of ongoing erosion, sand needs periodic replacement. The East Coast, from New York City to Key West, Florida, has undergone numerous beach nourishment projects since the 1920s. The number of these projects has accelerated since the 1970s, partly because of greater federal government involvement and partly because of a policy shift away from hard stabilization methods in favor of soft approaches. Between 1960 and 1996, approximately $1.3 billion (in 1996 dollars) was spent on beach stabilization along the East Coast.[26] Cumulative costs have totaled $2.4 billion nationally since the 1920s.[27]

Sea level rise will likely exacerbate erosion, causing sand replacement cycles to shorten and beach nourishment costs to increase. For example, one study suggests that beaches in the New York City metropolitan region would require up to 26 percent additional volume of sand, for a sea level rise of 60 centimeters to 1 meter by the latter half of this century.[28] A significant rise in sea level may ultimately make continued nourishment an option that is economically and physically unfeasible.

An extreme example of beach fill is to raise an entire barrier island, such as in Galveston. Following the deadly 1900 hurricane, survivors decided to raise the barrier island and remaining houses. More than a century later, on September 13, 2008, Hurricane Ike caused considerable property damage but relatively few fatalities on the island, because most of the population had been evacuated in time, thanks to satellite weather maps and a much improved warning system. The higher elevation and seawall saved Galveston from total destruction (but not the flooding) this time. Because of costs, such measures will be feasible only for fairly small, densely populated areas. Another question is whether Galveston would fare as well in 50 to 100 years, with higher seas, if another Hurricane Ike were to strike.

Tidal salt marshes generally keep pace with sea level rise, except for rapid subsidence (as in Louisiana and parts of Chesapeake Bay), low sediment supply, or altered natural biogeochemical cycles (e.g., Jamaica Bay, New York City). Because of the ecological services that salt marshes provide, their integrity should be maintained wherever feasible. Submerging salt marshes can be restored by adding sediment and replanting marsh grasses, sedges, or rushes to protect against erosion. Marsh replanting, often with native vegetation, can be combined with groins, breakwaters, or submerged stone sills to reinforce the marsh. Coastal uplands can be preserved as buffer zones for future landward salt marsh migration.

In tropical regions, intact mangroves, which shield the hinterland from the ravages of storm surges, should be protected from development and replanted where previously cleared.

"Living shorelines" is a form of coastal management that maintains or simulates natural processes.[29] Working with nature furthermore effectively safeguards the coast from future sea level rise. Strategic placement of plants, stone, sand, and other materials traps sediments and reduces wave energy, thereby cutting beach erosion and wetlands losses. Sand replenishment also extends the beach farther seaward. Newly restored dune ridges and beaches strengthen coastal defenses and provide bird and wildlife habitat, as well as recreational opportunities. For example, widening and extending the dune ridges seaward and replanting with native vegetation between The Hague and Hoek van Holland, along the Delfland coast, has created a whole new nature district (fig. 9.8).

Figure 9.8 Dune replanting and beach nourishment, south coast of Holland. (Photo: V. Gornitz, 2010.)

Accommodation

Around the world, more and more buildings crowd the beaches—small summer cottages or family-run motels replaced by mile after mile of wall-to-wall high-rise condos and large hotels. As sea level continues its relentless upward climb, coastal inhabitants will have to accommodate to a more aquatic existence. Where coastal floods are commonplace, buildings are frequently constructed on pilings or over piers (fig. 9.9). Yet powerful waves during major storms often topple many smaller houses or sweep them out to sea, stilts and all.

Providing generous space for water reduces flooding risks. One way is to create attractive "green" (well-vegetated) neighborhoods with natural drainage that enhances water infiltration into the ground. Another is to build underground garages that also store excess water at times of above-average river or sea level. The design of water-robust buildings helps minimize the impacts of extreme floods or rainfall. For example, garages or recreation facilities (e.g., gym, swimming pool, game room, or children's play space) can replace ground-floor apartments in multi-family dwellings. Second-story walkways can connect to adjacent raised streets or parks.

Figure 9.9 New homes on Westhampton Beach following the December 1992 nor'easter. (Photo: V. Gornitz, 1998.)

Floating buildings (like large houseboats) represent yet another mode of living with a higher water level. The Floating Pavilion, installed in the Stadshavens district of Rotterdam in June 2010, represents a prototype for future floating districts.[30]

Low-lying coastal cities often develop around a series of canals (the "Venetian" solution). While the best-known example is Venice, other cities like Amsterdam, Bruges, Copenhagen, Suzhou, China, and formerly Bangkok (many of its colorful canals have been filled and paved over in recent years, as the city has modernized) also boast extensive canal networks. Rotterdam plans to build additional moats and canals to store excess discharge at times of high water. The expanded canal system can also serve as a supplementary transportation corridor. Recalling its long maritime history, New York City reestablishes close ties between city dwellers and the water by providing additional open space and fresh air. In Manhattan, parks, footpaths, and bicycle pathways now line the waterfront.[31] Innovative harbor redevelopment proposals include creation of an artificial island archipelago using dredge material, or reconfiguring the shoreline by reducing the seaward gradient and creating new wetlands and parks.[32] These will also dampen wave energy and lessen surge impacts.

The Ganges-Brahmaputra Delta of Bangladesh is one of the large Asian deltas most vulnerable to sea level rise, because of its extent, low elevation, land subsidence, and large rural population. Exposed to both riverine and coastal flooding, the delta experiences increasing soil salinity. In order to adapt to adverse conditions, Bangladesh's Char Development and Settlement Project Phase III aims to provide protection from saltwater intrusion and flooding by building embankments, sluice gates, drainage channels, protective tree belts, and cyclone shelters, as well as improving local economic opportunities.[33] Replanting of mangroves on newly accreted land reduces erosion, stabilizes the soil, and shields against cyclone damage. In addition, farmers are encouraged to plant crops that are more salt-tolerant and to switch to shrimp farming.

Retreat

Defensive measures, even accommodation, may buy time and help us cope with near future sea level rise and storm surges. However, by offering an illusion of permanent protection against the rising sea, they encourage further development in inherently high-risk areas. It may ultimately become impossible to defend all heavily developed shorelines, particularly on barrier islands lined with high-rise condos outside of major urban centers. Beyond a certain point, repeated rebuilding after storms or even raising land may become too expensive or ineffective. Retreat may then become necessary. Even so, densely populated coastal mega-cities will probably continue to be defended at all costs. Yet foresighted prudent coastal management may avert some of the adverse consequences long before a rising sea renders many coastal areas uninhabitable. For example, sensible land zoning would limit housing density and building size and substitute "buffer zones" of parks, recreational facilities, or low-density development in high-risk areas.

A number of options exist to limit development in high-risk areas and to allow for eventual retreat. The government may acquire title to the property through eminent domain, or the property can be donated voluntarily to a conservation organization. Other measures include the creation of *erosion setbacks* and *easements* that establish buffer zones for coastal wetlands or beaches. *Buyout programs* reimburse shorefront landowners for abandoning property in high-risk zones. However, the public bears the cost.

Setbacks restrict shore construction based on erosion or elevation thresholds. In North Carolina, for example, single-family houses and small structures must be set back 30 times the historic average annual erosion rate (or a minimum of 60 feet). For larger structures (e.g., condos), the setback is 60 times the historic erosion rate. (For erosion rates above 3.5 feet per year, the setback is 30 times the average erosion rate plus 105 feet).[34] Regulations in Florida limit construction seaward of the 100-year flood zone and prohibit new construction seaward of the projected 30-year erosion zone. However, these regulations are poorly enforced. Residential units on eroding Rhode Island barriers must be set back 30 times the average annual erosion rate; commercial properties 60 times. Setbacks based on historic erosion rates presuppose a future continuation of those rates—unlikely, given anticipated sea level rise. Accurate historic erosion data may not always exist.

Because setbacks have been challenged in court as "takings,"[35] some other states, such as Maine, Massachusetts, Rhode Island, South Carolina, and Texas, have instead adopted variants of rolling easements.[36] *Rolling easements* allow landowners to remain on their property and develop it, but prevent armoring the shore. As nature takes its course unimpeded, the owner may eventually be forced to abandon the property. In *conservation easements*, a conservation organization such as the Nature Conservancy buys the right to prevent further development, but the landowner can still remain on the land. Given the voluntary nature of easements, adjacent land can still be armored, but the enhanced erosion downdrift may cancel any conservation benefits. Easements may not be very useful for shorelines that are already heavily developed.

Although designed for current conditions, these coastal management programs can be modified to accommodate higher future erosion rates and widened flood zones.

The ultimate step is *managed relocation*. Some coastal geoscientists, such as Orrin Pilkey and Robert Young, believe that repeated hard shoreline stabilization and beach nourishment will become untenable as the sea continues its relentless shoreward advance. They advocate a policy of pulling back from the shore and ending financial incentives to rebuild after storms.[37] "If you stay, you pay," they say.

Individual structures can be moved farther inland, as has been done most impressively for the historic Cape Hatteras Lighthouse, North Carolina, in 1999 (see chapter 8). More frequently, houses are moved some tens of meters landward within a given shorefront property, or other buildings threatened by imminent collapse due to coastal erosion are torn down.

Some countries are beginning to take landward relocation seriously. Great Britain now views "managed realignment" as a long-term planning tool, especially for estuaries and relatively undeveloped land.[38] The government recognizes that in the long term it will become uneconomic to defend many small communities along eroding coasts. "Realignment" entails moving existing coastal defenses to a new inland position. It is also a way to create space for salt marshes and tidal mudflats to migrate to as sea level rises. In England and Wales, 2.5 million people live in coastal areas less than 5 meters above the Ordnance Datum (the national datum to which heights above sea level are referred); 57 percent of the prime agricultural land lies below this elevation.[39] As may be expected, landowners and coastal managers largely oppose the concept. Generally, people have a low perception of hazard and risk until after damages occur. The unpopularity of setbacks and managed realignment has also been blamed on inertia, scientific uncertainties, a lack of public awareness, and top-down decision making.[40] Greater stakeholder involvement at the outset may, however, encourage increased public acceptance.

Corton, a small resort village (pop. 1,000) on the east coast of England, serves as an example of public attitudes toward managed relocation. The community sits atop partially vegetated cliffs of sandy boulder clay, sand, and gravels that are eroding in spite of a 1.5-meter-wide concrete seawall, revetments, and groins.[41] Several houses have already tumbled into the sea. When presented with several choices, residents vigorously opposed doing nothing to strengthen the shoreline or undergoing "managed realignment" and overwhelmingly favored "holding the line" against the sea for 50 years or longer. They feared potential loss of homes and businesses, also loss of strong historical family ties to the village. Unaware of the recent shift in coastal policy, they also expected the government to protect the shoreline against flooding and erosion. After considerable debate, a compromise was reached: the shore would be defended for another 20 years (from 2005) to give residents ample time to readjust, but in the long term, cliffs could retreat, allowing a more natural shoreline to emerge.

PREPARATIONS BEGIN

Preparing for sea level rise is much like buying insurance. We buy home insurance for potential damages due to fire, wind, theft, leaks, and other such natural or human disasters. The U.S. Federal Emergency Management Agency (FEMA) through its National Flood Insurance Program (NFIP) offers hurricane and flood insurance to property owners living within a "Special Flood Hazard Area" (i.e., the 1-in-100-year flood zone) in communities that comply with minimum federal flood protection standards.[42] NFIP adjusts its flood insurance rates annually, based on the latest hydrological data, including historic (but not future) sea level rise, among other factors.[43] Although designed for current hazards, NFIP's insurance rates could be modified to reflect the changing risks posed by enhanced coastal erosion and storm flooding from a rising sea level.

The optimal timing for implementing coastal adaptation plans remains to be determined. Acting too soon may unnecessarily increase present-day costs, if sea level rises less than anticipated; delaying too long may also cost more because of interim flood damages. Most shore protection measures, such as raising dikes or seawalls and expanding beach nourishment, can be implemented within a few years, so that a delay may not be critical. However, advance planning becomes more important for new buildings or infrastructure with longer life spans. New houses can be built upon higher pilings. Infrastructure can be "hardened" by raising seawalls, elevating roads, or building bridges higher as part of regular capital cycles, or retrofitted during regular maintenance operations. London's three-stage "flexible adaptation pathway" approach, previously discussed, adjusts to sea level rise uncertainties. New York City is also adopting a similar approach.

Several coastal cities have already begun to harden their sea defenses (the Thames Barrier in London and MOSE in Venice, for example). Other forward-thinking cities are beginning to prepare for a higher sea level, among them Rotterdam and New York City.

A Tale of Two Cities: Rotterdam and New York City

Rotterdam

> *All the Dutch towns are amphibious, but some are more watery than others.*
> —E. V. LUCAS, *A WANDERER IN HOLLAND*

Like many other major port cities, Rotterdam will be exposed to the consequences of sea level rise. Therefore it has embarked on an ambitious plan to "waterproof" the city, anticipating greater winter rainfall (causing additional riverine flooding) and a potential sea level rise of up to 1.3 meters by 2100 (table 9.1). Dutch water management has traditionally emphasized minimizing the coastal flood threat with dikes, storm surge barriers, and beach nourishment. Higher water levels increase overall maintenance costs, barrier closings, and shipping disruptions. Areas outside the dikes become more vulnerable to flooding. The Delta Committee therefore proposes to raise current flood risk standards by creating a "closable yet open" Rhine estuary through new movable flood barriers that would divert water in desired directions depending on hydraulic conditions. Other options include creating new connections within the existing network of canals and rivers or even entirely new waterways. Policy has also recently shifted toward working closely with the water. Protecting Rotterdam from sea level rise still entails reinforcing its water defenses—the dikes and the neighborhoods beyond the dikes. But it also means appropriate spatial planning, constructing water-adapted buildings, and allowing water to drain or to temporarily retain excess rainwater. Large areas of unembanked land designated for future urban expansion will need to be raised by 1–1.5 meters to meet the base flood level requirement of 3.9 meters above the Amsterdam Ordnance Datum.[44]

The Floating Pavilion is a harbinger of the floating districts planned for the Stadshavens section of Rotterdam. Floating homes exist in various places across the world. For example, thousands of people live in sampans in Hong Kong's harbor. In more-affluent settings, many houseboats crowd the shores of Lake Union, Seattle; Sausalito, California; and the 79th Street Marina along the Hudson River in New York City. Former barges now converted to comfortable homes line the canals of Amsterdam and Rotterdam. Rotterdam even boasts a floating hotel—the three-star H2otel, in historic Wijnhaven (Wine Harbor). Rotterdam envisions entire future floating residential districts, complete with office complexes and parks, and will expand its network of moats and canals.

Urbanization, by paving the ground with impermeable concrete or asphalt, prevents rainwater from infiltrating the soil and recharging the aqui-

Table 9.1 Rotterdam and New York City Compared

City	Rotterdam 2100 (rel. to 1990)	New York City 2080–2090 (rel. to 2000)
Sea Level Rise	0.65–1.3 m	0.3–0.58 m (central range) 0.67 m (max.)
"Rapid ice melt"	—	1.0–1.4 m
1-in-100-year flood height	2.8 m	2.9–3.3 m

Sources: Rotterdam data from Delta Commissie (2008); Aerts et al. (2009). New York City data from Horton et al. (2010).

fers. This results in street and basement flooding after heavy rains. In Rotterdam, excess rainwater can be temporarily stored beneath municipal parking or in water plazas that also serve as parks or playgrounds when dry. Planting "green roofs" also helps curb excess runoff.[45]

New York City

As they go about their daily tasks, most New Yorkers remain oblivious to the city's status as an island and a major seaport. The city boasts nearly 600 miles (970 kilometers) of shoreline, and four of its five boroughs are islands. However, the city's waterfront is undergoing a major transformation. Shipping, except for cruise ships, has largely moved to nearby Staten Island and Bayonne, New Jersey. Pedestrian walkways and bicycle paths now replace rotting piers and abandoned warehouses and factories. New high-rise apartment complexes sprout like mushrooms.

New York City recognizes the issues of global warming and sea level rise in its waterfront redevelopment plans. The city currently is among the 10 port cities most vulnerable to coastal flooding, in terms of population and assets.[46] By the 2070s, New York City will still remain among the top 10 port cities at risk, based on assets. Winter cyclones and hurricanes have flooded parts of the city in the past, most recently during the "nor'easter" of December 1992, discussed above.

The mayor's Office of Long-Term Planning and Sustainability manages city-owned infrastructure. Mayor Michael Bloomberg recently commissioned a study by experts from the NASA Goddard Institute for Space Studies, Columbia University, other regional universities, and the private sector to advise on climate change risks arising from changes in temperature, precipitation, and sea level change and to recommend adaptation strategies.[47]

The New York City Panel on Climate Change (NPCC) 2010 report projects future sea level rise based on seven Global Circulation Models (GCMs) and three greenhouse gas emissions scenarios (IPCC SRES A2, A1B, and B2). These projections, modified from IPCC methodology, include global contributions from thermal expansion and meltwater (glaciers, ice caps, and ice sheets), as well as local land subsidence, mainly due to glacial isostatic adjustments, and local changes in water height from sea temperature, salinity, and ocean currents. (Other factors affecting sea level, such as gravitational and rotational terms, were not included.) To simulate potential dynamic ice acceleration, an upper-limit, high-impact "Rapid Ice-Melt" scenario assumes that glaciers and ice sheets will melt at rates comparable to those of the Last Glacial Termination (chapter 5), when sea level climbed at an average rate of 0.39–0.47 inches (10–12 millimeters) per year. This scenario assumes that meltwater rises exponentially from the present mean ice melt rate of 1.1 centimeters per decade between 2000 and 2004, going to 2100. This term is added to the other three sea level terms, which remain unchanged.

The GCM-based projections show a sea level rise of 7–12 inches (18–30 centimeters)[48] by the 2050s and 12–23 inches (30–58 centimeters) by the 2080s (table 9.2). Sea level reaches ~41–55 inches (104–140 centimeters) by the 2080s in the "Rapid Ice-Melt" scenario.

The frequency, intensity, and duration of coastal flooding will likely increase along with a rising sea. The 100-year flood return curve (or "stage-frequency relationship") for New York City was calculated from a U.S. Army Corps of Engineers hydrodynamic model with both surge and tidal components.[49] Sea level rise reduces the 100-year return period to once in 15–35 years by the 2080s.

A higher average sea level would exacerbate street, basement, and sewer flooding and disrupt transportation more frequently. It would increase rates

Table 9.2 Sea level rise projections for New York City

	2020s	2050s	2080s
GCM-based scenarios	2–5 inches	7–12 inches	12–23 inches
	5.1–13 centimeters	18–30 centimeters	30–58 centimeters
"Rapid ice melt" scenario	5–10 inches	~19–29 inches	~41–55 inches
	13–25 centimeters	48–74 centimeters	104–140 centimeters

Source: Rosenzweig and Solecki (2010); Horton et al. (2010). Numbers represent sea level rise relative to the year 2000 for the mid-67 percent of the model projections.

of beach erosion, necessitating additional beach nourishment programs. Salt water would encroach farther into freshwater sources and potentially damage infrastructure.

The NPCC recommends that New York City begin to develop "flexible adaptation pathways" that can be adjusted periodically to the latest projections of sea level rise. Existing risk and hazard management strategies can then be revised as needed. The NPCC's Adaptation Assessment Guidebook (AAG) recommends that city agencies begin to prepare an inventory of infrastructure and assets at risk, link adaptation strategies to capital and rehabilitation cycles, and periodically monitor and reassess plans in response to newer climate information. In addition, the NPCC offers a general process for creating a set of climate change–related design and performance standards (climate protection levels, or CPLs). Most important is to update current 1-in-100-year flood zone maps (e.g., FEMA's maps) to incorporate future sea level rise and coastal flooding (table 9.2; fig. 9.10).

The New York City Department of City Planning has recently unveiled its Vision 2020: New York City Comprehensive Waterfront Plan for the revitalization of the waterfront.[50] The plan envisions enhanced public access to the waterfront and utilization of the waterways, as well as new economic development and residential construction. The delineation of the New York City Coastal Zone Boundary used in waterfront revitalization should also be updated to reflect the latest sea level rise projections.

REACTING TO THE RISING WATERS

In Norfolk, Virginia, sea level rise is no theoretical matter—it is already occurring! High spring tides regularly flood streets in some neighborhoods, forcing residents to re-park their cars away from the shore and detour around deep puddles. Norfolk, near the mouth of Chesapeake Bay, is surrounded on three sides by water. Natural subsidence plus settling and compaction of reclaimed marshland add to the rising water, making the relative sea level trend of 4.44 millimeters (0.17 inches) per year one of the highest along the East Coast. After extensive lobbying by local residents, the city decided to raise the worst-hit street by 46 centimeters (18 inches) and to readjust storm drain pipes to prevent street flooding. FEMA has also spent $144,000 to raise six houses, stimulating objections over high costs and the futility of endless countermeasures.[51] The mayor concedes that if the sea keeps rising, the city will eventually need to create "retreat zones," but those most affected (like their counterparts in Corton, England) strongly prefer "action at any cost."

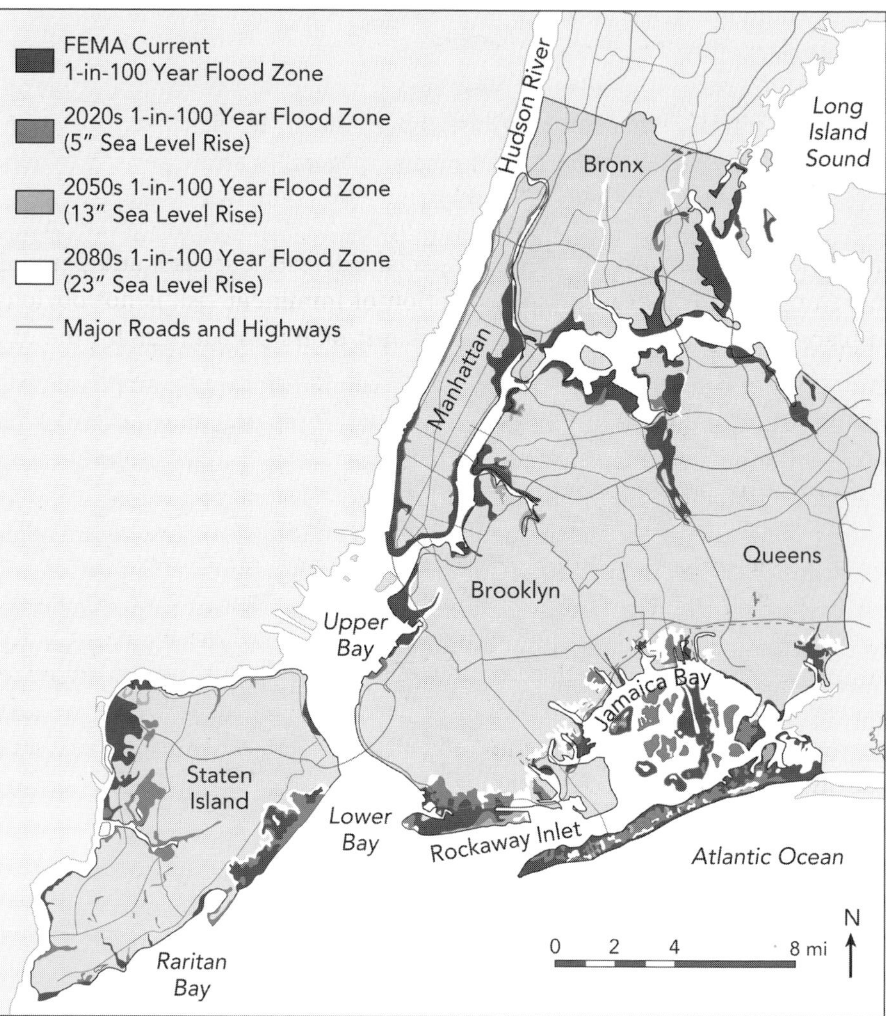

Figure 9.10 New York City FEMA 100-year-flood maps with sea level rise based on NPCC sea level rise projections. (Map by K. Grady, A. Marko, L. Patrick, W. Solecki, Climate Protection Level Workbook, in Rosenzweig and Solecki, 2010, fig. 3, p. 317.)

Meanwhile, the city will select its flood-mitigation projects more carefully and explore alternatives like inflatable dams or storm-surge floodgates.

In general, the response to the rising sea follows either of two divergent pathways. The first course entails staying put and holding the line for as long as possible. Coastal development continues (with minor restrictions) and the shore is defended by a mix of "hard" armoring, softer, more natural solutions, or accommodation by means of innovative architecture and design.

This approach works well for densely populated or already highly developed areas in the foreseeable future. However, it encourages further development in already high-risk areas and may not be able to withstand upper-end sea level rise scenarios.

The second pathway is a gradual rollback from the shore through a mix of land-use zoning, easements, erosion setbacks, and public land acquisition for wetland migration or parks, appropriate for sparsely populated or less-developed shorelines. At present, this approach may appear more contentious, not only because of clashing special interests but also, and perhaps more importantly, because the perception of imminent risk is not obvious to planners. Yet the prudent foresight that it offers prepares us for the day when defending the fort against the sea may become an unwinnable battle.

LOOKING AHEAD: HOW THE PAST INFORMS THE FUTURE

Having journeyed across the seas of time, in the final chapter we will review how past changes in climate and sea level may offer clues about future pathways. We also explore ways to reduce the potential marine menace. An important element of the solution involves mitigating (reducing) greenhouse gas emissions. Some plausible mitigation options will be briefly outlined. We also look beyond technological solutions to a change in perception that may help avert the most severe consequences.

> With the exception of massive volcanic events and asteroid impacts, no geological record exists for rises of greenhouse gases . . . at rates as high as the rate of 1 to 2 ppm CO_2/year since the second half of the 20th century. . . . The Earth climate is entering uncharted territory.
>
> —ANDREW GLIKSON, EARTH AND PALEO-CLIMATE SCIENCE, AUSTRALIAN NATIONAL UNIVERSITY

10 Charting a Future Course

THE PAST AS A GUIDE TO THE FUTURE

Over the eons, the ever-changing seas have danced to the daily rhythm of the tides, churned angrily in the midst of typhoons, and shape-shifted their basins to the languorous drift of tectonic plates. Time and again, the oceans have ebbed and swelled as the ice successively expanded and then slowly released its frigid grip across polar landmasses. What have we gleaned from the journey across the ages? This chapter explores the implications for the road ahead.

Records of sea level change grow dimmer as time recedes into the past. Nonetheless, different strands of geological evidence point to a high sea level stand in the late Cretaceous, 80–65 million years ago, followed by a long-drawn-out, yet uneven lowering. Once appreciable ice built up in Antarctica, around 34 million years ago, sea level changes have largely mirrored changes in climate. A series of glaciations succeeded the expansion of Northern Hemisphere ice sheets after around 2.75 million years ago. During the last 800,000 years, climate, ice volume, and sea level have varied in sync with carbon dioxide (CO_2) and methane (CH_4).[1]

Past sea levels and climates vastly different from those of today no longer surprise us. What truly differs now is the accelerating pace, its human

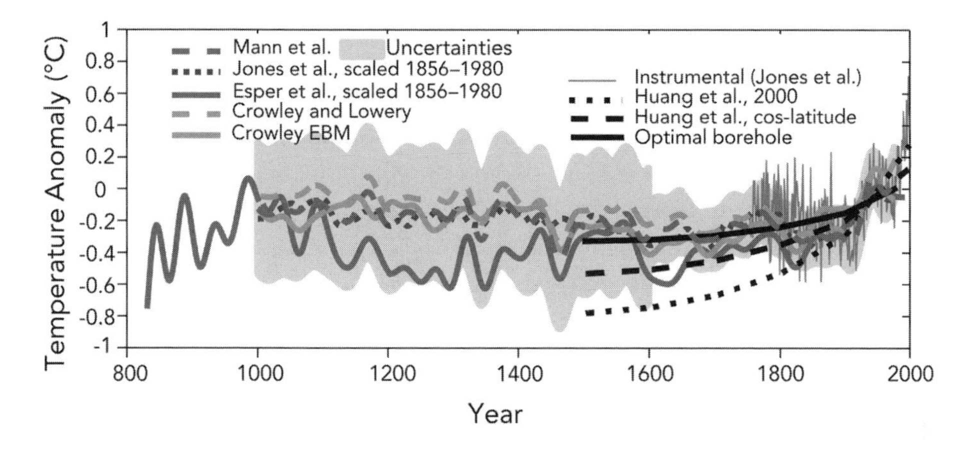

Figure 10.1 Northern Hemisphere temperatures for the last 1,000 years based on various climate proxy reconstructions. While considerable differences exist among various studies, the above average Medieval Warm Period and below average Little Ice Age are evident. The 20th-century warming stands out clearly. (Modified from M. E. Mann, 2009. "Climate Variability and Change, Last 1,000 Years." In *Encyclopedia of Paleoclimatology and Ancient Environments,* ed. V. Gornitz, fig. C56, p. 180, with kind permission from Springer Science + Business Media B.V.)

signature, and what that portends. Given the short instrumental record (circa 150 years) and modeling limitations, the archives of past climates and sea levels can serve as a guide for the future.

Michael Mann at Pennsylvania State University and his colleagues have reconstructed Northern Hemisphere temperatures of the last 1,300 years from paleoclimate records.[2] Temperatures rise slightly 1,000 years ago (the Medieval Warm Period), dip during the 15th to mid-19th centuries (the Little Ice Age), and climb steeply thereafter, forming a "hockey stick"–shaped curve (fig. 10.1). Although older temperatures are less certain, late-20th-century temperatures exceed the average for the last millennium. Controversy erupted over whether the present climate warming is indeed exceptional and over why recent temperatures derived from some Arctic tree rings deviated from the recent global trend.[3] While the cause of the deviant northern tree rings remains a mystery, many other independent climate proxies and instrumental records elsewhere corroborate the unusual late-20th-century warming, which thus appears real.

The following summarizes the case for a warming Earth:

1. The Earth's temperature is 33°C (59°F) higher than in the absence of naturally occurring greenhouse gases (i.e., primarily water vapor and CO_2).

2. In 1896 the Swedish chemist Svante A. Arrhenius (1859–1927) already predicted the warming capacity of CO_2 from fossil fuel combustion.

3. Carbon dioxide has increased steadily since at least 1958 (see fig. 6.3). Currently 100 parts per million (ppm) above preindustrial levels, it is growing by nearly 2 parts per million per year. Other greenhouse gases, such as methane and nitrous oxide, have been increasing as well.

4. Air bubbles encapsulated within Antarctic ice reveal that today's CO_2 (and methane, CH_4) levels are the highest in the last 800,000 years.[4]

5. Sea levels have closely tracked changes in ice volume and atmospheric greenhouse gases for at least the last half million years, and probably longer.[5]

6. The average global temperature increased by 0.74°C (1.33°F) between 1906 and 2005.[6] The 2000s have been the warmest decade since instrumental records have been kept.[7]

7. To account for the observed temperature trend since the 1960s requires anthropogenic greenhouse gas inputs (fig. 10.2).

8. Arctic summer sea ice has diminished rapidly since the 1970s and may disappear altogether within the next 50 years (chapter 6). The resulting ice-albedo feedback will amplify Arctic warming and cause more Greenland Ice Sheet melting, further raising sea level.

9. The world's glaciers have shrunk dramatically, adding around one millimeter per year to sea level (chapter 6). Himalayan glaciers are retreating, worrying countries such as Nepal and Bhutan that face increased landslides and flooding.[8] Matched early-20th-century and recent photographs of Himalayan glaciers dramatically document the rapid 20th-century ice recession,[9] also occurring in the Alps, the Rockies, and elsewhere (e.g., fig. 6.1).

10. The Greenland and Antarctic Ice Sheets have also thinned, raising sea level nearly one millimeter per year in recent years (see table 6.1). Ice streams accelerate as the grounding line retreats inland, fringing ice tongues or shelves melt from beneath, and surface melting lubricates basal ice flow.

11. Ever since the late 19th century, sea level has been rising 1–2 millimeters per year faster than during the previous few millennia (chapter 6).

12. Tide gauges register a 20th-century global mean sea level trend of 1.7–1.8 millimeters per year.[10] Satellite altimeters spot an even higher trend of ~3 millimeters per year since 1993.[11]

13. Although the IPCC foresees a sea level rise of only 18–59 centimeters by 2100, observed values already track the upper end of this decade's climate model forecasts. Sea levels correlated with historical temperatures suggest a rise of up to 1.4 meters by 2100.[12]

Could our actions inadvertently provoke unexpected, sudden climate changes like those of the last glaciation (chapter 4)? Could ice melt freshen

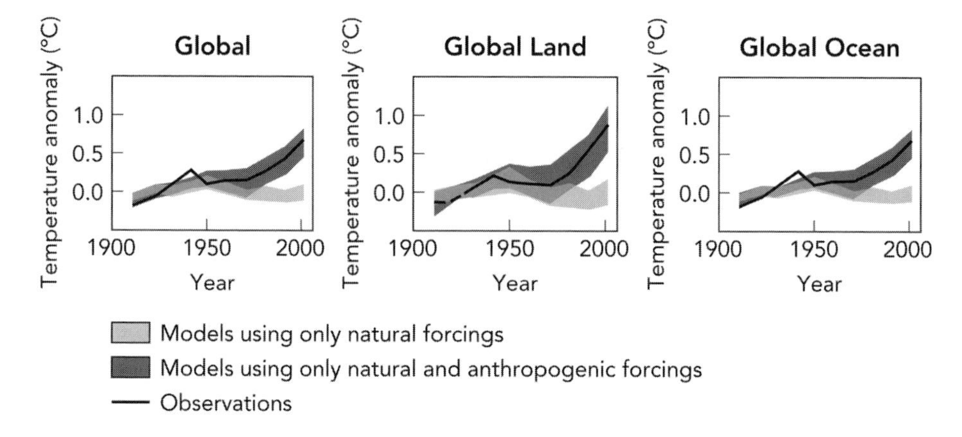

Figure 10.2 Natural versus anthropogenic factors contributing to 20th-century temperature rise. (Modified from IPCC, 2007a, Climate Change 2007: The Physical Science Basis. Working Group I Contribution to the Fourth Assessment Report of the Intergovernmental Panel on Climate Change, FAQ 9.2; Figure 1. Cambridge University Press. FAQ 9.2, fig. 1, p. 703.) Only those simulations that include both natural and human factors closely match observations (thick black line), especially since the 1960s.

the North Atlantic enough to slow North Atlantic Deep Water (NADW) sinking, as many computer models suggest? Such a slowdown would cool (or retard warming of) the North Atlantic, contrary to the rest of the overheating planet, yet simultaneously amplify sea levels along the East Coast of North America.[13] Recent observations are somewhat inconclusive.[14]

How fast would the ice sheets react? Ice sheets and glaciers appear highly sensitive to fairly minor air or ocean warming.[15] Although meltwater plays a small, yet growing role (see tables 6.1, 7.2), it could eventually elevate sea level by tens of meters, endangering hundreds of millions of people living along the coast. Is a near-future return to the high sea levels of the Last Interglacial—the Eemian—ca. 127,000–125,000 years ago, or even the mid-Pliocene (3.3–3.0 million years ago) realistic (chapters 3–5)?

BACK TO THE PLIOCENE?

Long a proponent of the concept of human-induced global warming, James Hansen, head of the NASA Goddard Institute for Space Studies, New York City, believes we are already at or very close to a point of no return. In order to avert a dangerous climate "tipping point" that would trigger a rapid sea level rise of several meters by the end of the century, he advocates a reduction in atmospheric CO_2 levels from the current 387 parts per million to no more than 350 parts per million.[16] Hansen reiterates the close association among past sea levels, climates, and CO_2.[17] Although carbon dioxide reached pre-industrial levels (~280–300 parts per million) during the Last Interglacial, polar summer temperatures were 3°C–5°C degrees warmer and sea level 4–6 meters (13.1–19.7 feet) higher. Greater solar insolation probably induced the balmier climate (amplified by feedbacks).

During the mid-Pliocene, the most recent period with warmth comparable to that expected by 2100, global temperatures averaged 2°C–3°C higher than present and CO_2 ranged between 350 and 450 parts per million. The Northern Hemisphere remained largely ice-free and the Antarctic Ice Sheet was smaller. Sea level soared by 25–30 meters (82–98 feet). Hansen, as a physicist, therefore finds it "almost inconceivable that BAU [business as usual] climate change would not yield a sea level change on the order of meters on the century timescale."[18]

Mid-Pliocene ocean circulation may have differed. Recent geophysical observations indicate a much lower underwater Greenland-Scotland Ridge (upon which Iceland sits) during the Pliocene, which allowed greater transport of warm water into the Arctic.[19] This could resolve the contradiction between paleoclimate data showing a much warmer North Atlantic with a stronger ocean conveyor system, and computer simulations that showed the reverse.[20] Although the analogy to the mid-Pliocene may not be exact, it nevertheless demonstrates the potential for much higher sea levels.

Some geologists believe that we have already entered a stage "without close parallel in any previous Quaternary interglacial."[21] They have even compared expected climate changes to the "hothouse climates" of the lower Jurassic, 180 million years ago, and the Paleocene-Eocene Thermal Maximum, 55 million years ago. Back then, however, in spite of similar CO_2 levels and temperatures, tectonic plates occupied vastly different positions and extensive polar ice caps were absent.

Could sea level rise attain peak rates of the Last Interglacial or the Last Glacial Termination (chapter 5) within this century? Disturbingly, ice melt appears to be increasing (chapter 7). Nonetheless, glaciers and ice sheets cannot discharge ice fast enough to raise sea level more than 1.5–2.0 meters within 100 years.[22] A more plausible upper bound by 2100 lies around 1 meter, based on an average deglacial rate of ~10 millimeters per year (in addition to thermal expansion and local subsidence, which could raise relative sea level to over a meter). But the story does not end in 2100.

Even though it may take many years to reach a "tipping point," the continuing atmospheric carbon dioxide buildup guarantees not only that the world will continue to heat up, but that it will remain hotter long after new inputs of carbon dioxide have ceased. While a return to a nearly ice-free "hothouse climate" appears remote within this century, it cannot be entirely ruled out in the future. Due to the ocean's high thermal inertia, surface heat takes a long time to reach the seafloor, thus raising sea level for many centuries. This "climate commitment" (chapter 7) is only part of the story. Once in the atmosphere, the added CO_2 is hard to eliminate, unlike methane and nitrous oxide, which have an average atmospheric lifetime of around 12 years and 110 years, respectively.[23] Carbon dioxide lingers in the atmosphere for thousands of years after fossil fuel combustion has ceased. While the oceans or biosphere takes up most of the anthropogenic CO_2 within a few centuries, a substantial portion (15–55 percent) may linger in the atmosphere for 1,000 years, and 11–14 percent could last for 10,000 years.[24] This persistent CO_2 keeps air temperatures high for centuries, possibly millennia, before equilibrium is restored.

"Business as usual" complacency is therefore ill-advised. The unprecedented increase in CO_2 and temperature, especially within recent decades, and uncertainties in future ice sheet behavior preclude completely excluding extreme scenarios. Changes that formerly unfolded over many centuries or millennia are now time-compressed into decades or centuries, introducing a large disequilibrium into the climate system. Modern to near-future CO_2 (and temperature) levels (comparable to the Last Interglacial and the mid-Pliocene) imply an eventual sea level rise of several meters (or more). Therefore, many climate experts urge constraining the global mean temperature rise to 2°C–2.5°C and CO_2 concentrations to within 350–450 parts per million to avert "dangerous anthropogenic interference with the climate system."[25] (It is currently 387 parts per million CO_2 and inching upward at ~2 parts per million per year).

The switch to carbon-neutral technologies may take many decades to significantly reduce greenhouse gas emissions and stabilize climate. While we procrastinate, the Earth continues to warm, more ice melts, and the sea climbs relentlessly upward.

SAILING INTO UNCHARTED WATERS

Most scientists and a growing proportion of the public now accept the mounting evidence for anthropogenic-induced climate change (summarized above), in spite of lingering questions about data quality and completeness, and computer model limitations. Many people, however, attribute it to natural cycles and go about their daily business as though nothing will change. Why? Probably because in most places, except for the Arctic, mountaintops, large deltas, or barrier islands, the observed changes (for example, a global mean temperature rise of 0.74°C (1.33°F) between 1906 and 2005)[26] are barely perceptible against the wide day-to-day swings in temperature, rainfall, heavy snowstorms, and other severe weather events. Most temperature and sea level projections show a gradual, nearly linear trend for several decades before a rapid acceleration later in the century (fig. 7.2). Although by that time climate change will become obvious to all, it may be too late to prevent serious impacts. A sharp disconnect exists between short election cycles, business planning for this year's or next year's profits, or even a human life span versus the much longer timescales over which climate evolves. Thus, climate change mitigation receives low priority.

Curbing Our Carbon Appetite

Retreating shorelines, submerging deltas, and drowning wetlands already foreshadow the future in many places. As the oceans lap ever higher and penetrate deeper inland, lengthening stretches of shoreline will shift landward or disappear underwater. Costs of heightened shoreline defenses and continual beach restoration could eventually become prohibitive. People will be forced to relocate farther inland, creating hundreds of thousands, if not millions, of coastal refugees. Will the adaptation strategies outlined in the previous chapter then suffice?

Preventing, or at least mitigating, the worst impacts of a potential waterworld requires striking at its ultimate source: global warming due to increasing anthropogenic atmospheric greenhouse gases. Averting a watery future demands that we curb our carbon appetite.[27]

Princeton University plant ecologist Stephen W. Pacala and engineer Robert Socolow recommend holding carbon dioxide levels to less than 500 parts per million by keeping emissions near the present level of 7 billion tons of carbon per year for the next 50 years.[28] They visualize limiting new CO_2 emissions over a 50-year period through seven equal "stabilization wedges," each of which represents an activity that starts at zero now and increases linearly up to the replacement of 1 billion tons of carbon per year (fig. 10.3). They recommend a set of 15 possible choices (others, not mentioned, could be substituted). One group of wedges includes improvements in energy **efficiency** and conservation. Others involve replacing coal with natural gas, capturing and storing waste CO_2 in geologic or biological sinks, and expanding both nuclear power generation and renewable energy such as wind, solar, or biofuels.

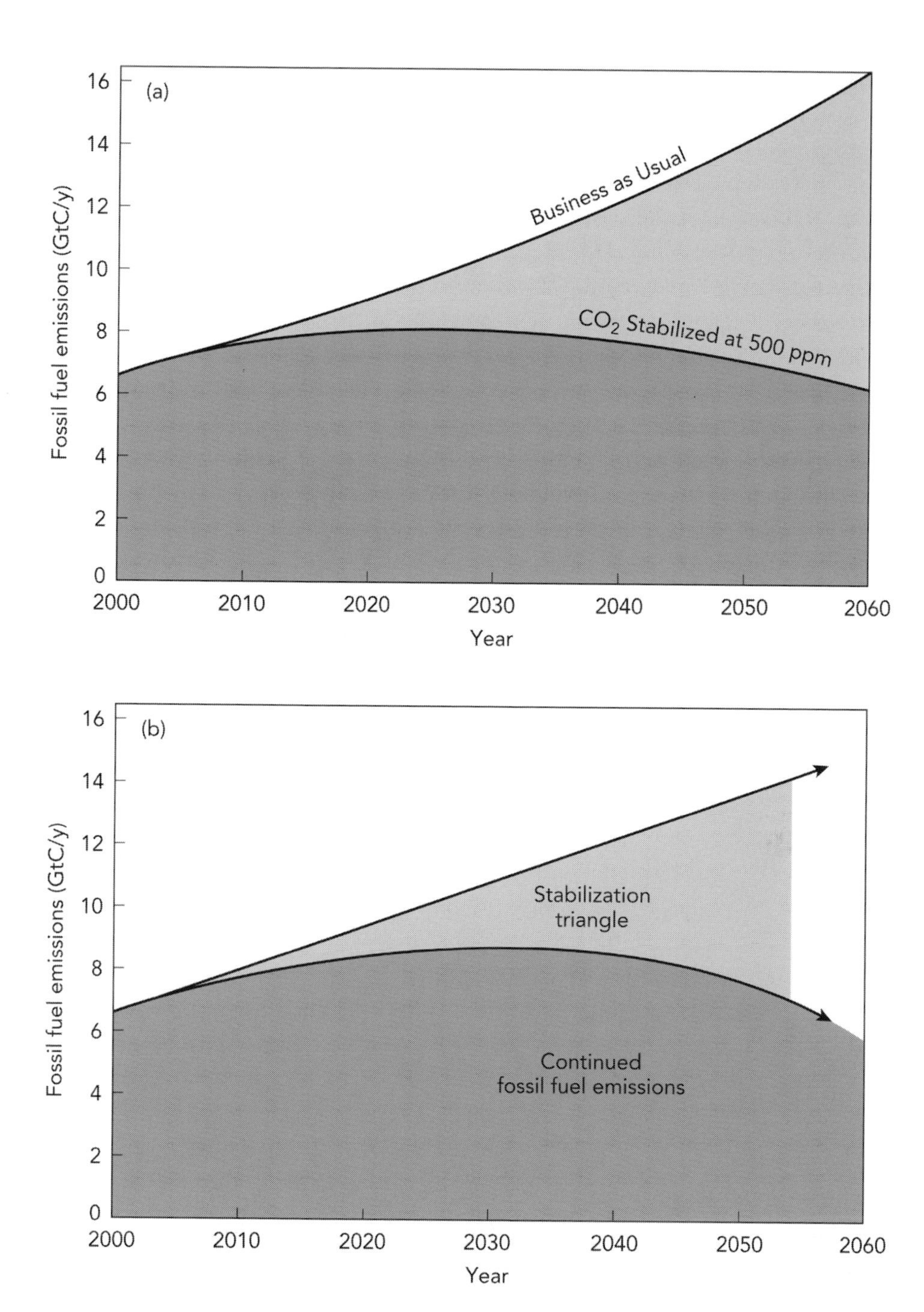

Figure 10.3 A. Growth in global CO_2 emissions to 2060 for Business as Usual (BAU) fossil fuel combustion (top curve.) Trajectory of CO_2 emissions with CO_2 stabilization at 500 parts per million (bottom curve). (1 GT C/y = 1 billion tons of carbon per year). B. Seven stabilization wedges to limit CO_2 emissions. Greenhouse gas emissions are assumed to remain constant at present levels for the next 50 years. New growth is accommodated by substituting a mix of energy efficiency, conservation, and alternative energy sources within the stabilization triangle. The downward arrow (bottom right) represents the need for additional future cuts in fossil fuel emissions. (After Pacala and Socolow, 2004, fig. 1.)

While not the main focus of this book, the next section briefly reviews some of these potential stabilization wedges.

Increased Energy Efficiency and Conservation

Gains in energy efficiency and conservation present opportunities for reducing CO_2 emissions and provide other benefits, such as reducing air pollution and energy costs. The average ~30–35 percent energy efficiency of coal-fired electric power plants can be improved by raising combustion temperatures and pressures. Co-generation (combined cycle) power plants, which produce both electricity and steam from the same fuel source, attain even higher efficiencies (up to 50–55 percent).

Additional gains stem from improved appliance (e.g., refrigerator) or lightbulb efficiencies. Although more expensive, compact fluorescent bulbs generally last 6–10 times longer and consume 75 percent less energy than incandescent bulbs (fig. 10.4). New U.S. lighting efficiency standards took effect nationwide January 1, 2012. The new 72-watt bulb, which replaces the 100-watt incandescent bulb, uses 28 percent less energy.

Mass transportation, especially in large urban agglomerations, is generally more energy-efficient than conventional automobiles, because buses, trams, and subways carry more people per mile for a given amount of fuel. Improved automobile engine design can also raise gas mileage. (Among the most energy-efficient vehicles today are the Honda Civic Hybrid, Toyota Prius Hybrid, Audi A3 TDI, Hyundai Elantra, Audi A3, and Lexus CT 200h).

Conservation reduces energy demand. Simple steps include turning off unused lights and lowering hallway lighting, cutting down on air-conditioning in summer and indoor heating in winter. Summer heat becomes more tolerable with open windows, large fans, drawn window shades, and lighter clothing. In winter, wearing heavy sweaters makes cooler rooms comfortable. Improved building design can also cut energy needs (see below). Improved energy efficiency is a key element of New York City's ambitious plaNYC 2030, which aims for a 30 percent cut in the city's greenhouse gas emissions by 2030.[29] A greater drive toward energy efficiency will target existing and new buildings (both public and private), as well as improved standards for appliances and electronics.

Environmental Architecture

Pueblo Bonito, Chaco Canyon, New Mexico, is a 1,000-year-old indigenous example of how buildings can be adapted to their environment—for example, through passive solar heating (fig. 10.5). This multistory pueblo, originally

Figure 10.4 Compact fluorescent bulb.

five stories high (the largest building in North America up to the 1880s), utilized a southern exposure to capture the sun's warmth during winter. The longest axis of the D-shaped pueblo complex faces south. In winter, its thick masonry walls absorb heat during the day and release it slowly during the night. Well-shaded thick stone walls maintain a relatively cool interior in summer.

Modern buildings can apply similar energy-conserving principles, such as increasing insulation or using appropriate construction materials. Insulation retains more heat, while cutting heating costs. Light-colored (more-reflective) materials or white roofs can cool buildings. Gardens or trees on roofs ("green roofs") also reduce the urban heat island effect[30] by providing insulation and lowering the need for air-conditioning.

Figure 10.5 Pueblo Bonito, Chaco Canyon National Historic Park, New Mexico. (Photo: Bob Adams Dec. 26, 2009. Wikimedia Commons.)

Urban Planning

Cities can avoid urban sprawl by developing compact communities and mixed land use, offering multiple transportation choices, and maintaining open green space. In addition to lowering their carbon footprint, inhabitants of "greener" cities enjoy an improved quality of life that features lower air pollution, convenient access to business, shopping, and entertainment, reduced commuting time, and more recreational opportunities.

Each city's unique geography shapes its future, as well as its historic development. Densely populated neighborhoods and an extensive mass-transit system make New York City the most environmentally efficient American city.[31] Improvements in New York's urban quality involve planting more street trees and reforesting parklands, painting "cool" white roofs, expanding public transportation, and revitalizing the waterfront. Urban planning in Rotterdam emphasizes water management, including innovative storage (i.e., "water plazas"), green roofs, floating buildings, and strengthening water defenses (chapter 9)

Decarbonization

Decarbonization of energy involves lowering CO_2 emissions by substituting fuels that are less carbon-intensive and expanding the use of renewable energy, nuclear energy, and carbon capture and storage.

Fuel Substitution

One type of fuel replaces a more carbon-intensive fuel. At a given energy efficiency, a switch from coal to oil cuts CO_2 emissions by around 26 percent, from oil to natural gas (methane) by 23.5 percent, and from coal to gas by 43 percent.[32] Natural gas also pollutes less.

Hydrogen is a clean, non-CO_2-emitting fuel that may eventually replace gasoline. It not only delivers more energy per gram than other fuels but is also independent of foreign imports. Yet producing hydrogen takes energy. Its chemical reactivity binds hydrogen to oxygen in water, H_2O, or to carbon in methane, CH_4 (and other hydrocarbons). At present, most hydrogen gas, H_2, is made from steam and catalysts that break down methane with up to 85 percent efficiency.[33] Water can also be split electrolytically into its constituents, which then re-combine in a fuel cell to produce electricity and water. Hydrogen still costs more than conventional fuels, and its manufacture from fossil fuels releases CO_2. Other drawbacks are the larger storage volume needed as compared with gasoline and the scarcity of refueling/recharging stations.

Hybrid vehicles represent a low-CO_2, energy-efficient alternative.[34] Hybrid cars carry an electric motor and an internal combustion engine, which runs on gasoline when the battery is depleted. As of 2009, 740,000 hybrid vehicles had been sold worldwide,[35] yet they still represent just a tiny share of the transportation sector.

Renewable Energy

Renewable energy derives from natural resources such as the sun, wind, water, geothermal energy, and tides. At present, renewable energy represents only 19 percent of the world's total energy consumption, expanding at rates of 10–60 percent annually between late 2004 and 2009.[36] Traditional biomass (wood or agricultural residues for heating and cooking) constitutes the largest share (13 percent), followed by hydropower (3.2 percent). Renewables now generate 18 percent of total global electrical power, of which 15 percent comes from hydropower.[37]

SOLAR ENERGY The Sun's radiation at the Earth's surface exceeds 10,000 times the current annual world energy consumption. The high annual insolation in low-latitude deserts and tropical to temperate regions provides ideal conditions for solar power development.[38] Yet, in spite of its huge potential, solar power still supplies just a tiny fraction of global energy needs.

Solar energy can be exploited in various ways. Passive solar architecture employs principles similar to those used at Pueblo Bonito: optimal building orientation, compactness (a low surface-area-to-volume ratio to cut energy loss), shading, favorable physical properties of construction materials, and good ventilation. Optimal orientation maximizes the duration of sunlight exposure and reduces the need for artificial illumination. Heat-retentive materials such as stone or cement absorb solar energy in daylight, cooling the building, and radiate stored heat at night. Solar buildings consume less conventional fuel for heating, cooling, ventilation, and lighting. Solar heating systems can also provide domestic hot water and heat swimming pools. Solar water heating is used in 70 million households worldwide, with China constituting 70 percent of the global total.[39]

Solar power plants and photovoltaic systems transform sunlight into electricity in various ways.[40] Concentrating solar power (CSP) systems typically employ mirrors or lenses to concentrate sunlight onto receivers that heat a fluid, which in turn drives a steam turbine and generator to produce electricity. One type of CSP system consists of parabolic troughs that focus sunlight onto fluid-filled tubes (usually water/steam) running along the main axis. Heated fluid flows to a turbine, where it generates electricity. The Solar Electric Generating Stations (SEGS) plants in California and Nevada Solar One near Boulder City, Nevada, use this type of technology (fig. 10.6).

Another CSP system employs fresnel reflectors—multiple flat or slightly curved mirrors that focus sunlight onto overhead receivers. Rotating dish systems collect sunlight with parabolic mirrors that direct it onto a receiver (for example, the Big Dish in Canberra, Australia). Solar power towers concentrate solar energy by means of numerous flat, sun-tracking reflectors, or heliostats, and focus the sunlight onto a receiver placed atop a tall tower (e.g., as in projects in California and Spain).

Photovoltaic (PV) devices convert sunlight directly into electrical current. Solar panels have been used on satellites ever since the advent of the Space Age. Other uses include power for domestic and commercial lights, appliances, machines, and pumps. Decentralized photovoltaic solar energy is

Figure 10.6 Solar Energy—Solar Power Generating Facility, NASA Kennedy Space Center, Florida. (Photo: NASA.)

economical in remote rural areas far from the electric grid, but more expensive than conventional electricity elsewhere.[41] While still constituting only a minute fraction of total energy use, photovoltaic installations are growing rapidly. In 2009 the three leading nations in solar photovoltaic capacity were Germany, Spain, and Japan, followed by the United States, Italy, South Korea, and other European Union members.[42]

Disadvantages include high initial costs, especially for photovoltaic systems, the large area needed for an efficient power system, and low or intermittent sunlight (e.g., on cloudy days or at night). Connecting the CSP or photovoltaic systems to the existing electrical grid overcomes intermittency. Off-grid CSP units store excess electricity using thermal storage systems or heated, molten salts, while off-grid PV systems can use rechargeable batteries.

BIOMASS Biomass derives from plants, agricultural and forestry residues, and municipal and industrial wastes. It can be used for fuel, power, or as a source of chemicals. For centuries, wood has traditionally been burned for heating and cooking. Although combustion releases CO_2, biomass is considered carbon neutral, since plants originally consumed CO_2 during photosynthesis.[43]

Biofuels are liquid or gaseous fuels manufactured from biomass. They include ethanol (alcohol) derived from corn and sugarcane, and biodiesel derived from vegetal oils and alcohol. Biofuels can replace fossil fuels in transportation. For example, global biofuel production in 2009 was equivalent to 5 percent of global gasoline consumption.[44] Brazil has pioneered ethanol production from sugarcane for autos since the 1970s. Its gasoline now contains up to 25 percent ethanol.[45] Ethanol furnished nearly 18 percent of the energy used in Brazilian transportation in 2008.[46]

Combustion of biomass is not without environmental impacts. Biomass releases particulates, nitrogen oxides, sulfates, carbon monoxide, and other pollutants into the atmosphere. Along with fossil fuels, it creates brown hazes that cover large parts of the world. Black carbon (or soot), formed by incomplete combustion of biomass and fossil fuels, adds to global warming by absorbing reflected solar radiation and decreasing the albedo, or brightness, of sea ice and glaciers.[47] Conversion of corn or sugarcane to vehicle fuel has been criticized, given rising food prices. However, other types of biomass can be substituted instead. Expansion of cropland for biofuel production may also exacerbate deforestation. (But Brazilian sugarcane occupies only a small percentage of arable land, far from Amazonia.)[48]

GEOTHERMAL ENERGY Geothermal energy stems from the Earth's internal heat released by radioactive decay of some minerals and stored in subterranean **magma** (molten rock) chambers or hot rock reservoirs. Hot springs arise from groundwater in contact with heated subsurface rocks. These underground heat sources (above 225°C, or 437°F) are tapped to heat buildings or to generate electricity. While the United States and the Philippines are the main producers of geothermal energy, this source supplies around 25 percent of Iceland's electricity. Other major producers of geothermal power include Indonesia, Mexico, Italy, and New Zealand.[49] The world's largest group of geothermal power plants is at the Geysers, in Northern California. Geothermal energy is also used for heating, with the maximum installed capacity in the United States and China. This resource furnishes 90 percent of Iceland's heating needs.

Geothermal energy generates few or no pollutants. (Minor releases of hydrogen sulfide or sulfur dioxide, and traces of noxious elements such as mercury, arsenic, antimony, and boron may occur.) It is also cost-competitive with coal power plants, making it among the cheapest energy alternatives. The Earth's vast stores of internal heat are considered a sustainable resource, though it can be depleted locally if extraction of steam or hot water exceeds recharge.

Geothermal energy is limited to geologically favorable locations, generally near major plate boundaries, where volcanic activity or deep-seated active faults allow the Earth's heat to escape to the surface. Elsewhere, deep drilling can reach hot rocks, but that approach is more expensive. Removal of subsurface hot water or steam may induce land subsidence and seismicity, as occurred in Basel, Switzerland, in 2009, causing cancellation of a planned geothermal systems project.

WIND ENERGY Wind power has grown rapidly, at an average 27 percent annually between late 2004 and 2009 (fig. 10.7).[50] Nations with the greatest wind power capacity in 2009 included the United States, China, Germany, and Spain. Wind delivers a sizable proportion of energy to Denmark (20 percent), Spain (14.3 percent), Portugal (11.4 percent), Ireland (9.3 percent), and Germany (6.3 percent). In the United States, wind power increased by 39 percent in 2009, and now supplies 2 percent of our electricity.[51] Iowa leads the United States, with 14 percent of its electricity derived from the wind, and Texas ranks

Figure 10.7 Wind turbines. (Photo: Joshua Winchell, U.S. FWS.)

second, with 5 percent.[52] Winds blowing across the High Plains of West Texas energize wind farms, many on depleted oil fields, and reinvigorate depressed towns. Texas is also investigating offshore wind turbine development. Meanwhile, in 2010, the U.S. Department of the Interior approved development of an offshore wind farm near Nantucket, Massachusetts, despite considerable local opposition. When fully operational, by late 2012, the project will be able to provide power to three-quarters of Cape Cod's residents.[53]

Suitable sites for wind farms strongly depend on average wind speeds, and like solar energy, they are subject to a highly intermittent supply. Integration into an existing power grid can help balance the uneven wind distribution. Concerns over dangers to birds and obstruction of landscapes can be overcome by careful placement of wind turbines at sites that avoid known nesting areas, avian flight paths, and scenic vistas.

TIDAL ENERGY Harnessing the tides could provide another source of clean energy. While most coastal areas experience two high and low tides daily, only a few places have a tidal range sufficiently large (more than 4 meters, or 13.1 feet) that it can be tapped for energy. Tidal energy technologies include barrages and turbines. The former are dam-like structures that capture tidal energy as water flows into and out of a bay or river. Turbines installed next to the barrage turn generators that produce electricity. However, tidal barrages may impact estuary ecology and sediment movement.

Tidal turbines, similar to wind turbines, operate underwater, driven by strong tidal currents. At present, the world's largest installation is the Rance Tidal Power Station along the Rance River in Brittany, France, with a capacity of 240 megawatts. Smaller plants operate in the Bay of Fundy, Nova Scotia; Kislaya Guba, Barents Sea, Russia; and China. Testing of the world's largest tidal turbine is under way in Orkney, Scotland.

In December 2010, Verdant Power applied to the Federal Regulatory Power Commission for a license to install up to 30 tidal power turbines near Roosevelt Island, in the East River, New York City, which would harness the extremely powerful currents flowing in that river (technically a tidal strait). If approved, this would be the first tidal power plant in the United States, and the world's first grid-connected tidal turbine system. The electricity would be used locally.[54]

HYDROPOWER Hydropower now supplies ~3.2 percent of the earth's energy needs, with China, the United States, Brazil, and Canada possessing the largest capacity. In addition to generating electricity, reservoirs also serve to control floods and provide water for cities and agriculture. China and India are either building or planning major new hydroelectric projects.

Although hydroelectric power is a clean and well-established source of energy, further large-scale expansion may soon reach a limit, after the best sites for major reservoirs are developed. Large reservoirs also create significant environmental and social impacts. Sediment trapped behind a reservoir exacerbates coastal erosion and deltaic subsidence downstream.[55] For example, the Three Gorges Dam, on the Yangtze River in China, will prevent all but 18 percent of the 1950s sediment flux from reaching the sea (fig. 10.7).[56] Siltation eventually clogs the reservoir and reduces its useful lifetime. Changes in flow regime often disrupt river ecology. Relocation of millions of people in populated areas, such as along the Yangtze River, creates social costs, such as a need for new homes and jobs.[57] The weight of water in a reservoir may trigger seismicity, as at Koynanagar, India, where a 6.3-magnitude earthquake in December 1967 killed nearly 200 people. The quake's epicenter was located near the Koyna Reservoir, in an otherwise aseismic region far removed from tectonic plate boundaries.

Figure 10.8 Three Gorges Dam, China. (ASTER, Terra Spacecraft, May 2006, Earth Observatory/NASA.)

Nuclear Energy

Nuclear energy furnishes around 6–7 percent of the world's energy needs and 15 percent of its electricity,[58] although this varies greatly by country. The United States, France, and Japan account for over half of the nuclear-generated electricity. As of 2004, the European Union obtained more than 30 percent of its electricity from nuclear energy, with France deriving most of its electricity from this source.[59]

Nuclear power creates no carbon dioxide (except for minor amounts in mining, transport, and processing of uranium fuel), it is comparatively energy-efficient, and it utilizes existing technology. However, concerns remain over reactor operating safety (e.g., the Fukushima Daiichi, Chernobyl, and Three Mile Island accidents), nuclear weapons proliferation, and permanent storage of nuclear wastes in non-seismic areas for thousands of years.

Following the catastrophic 9.0-magnitude earthquake and tsunami that struck northeast Japan on March 11, 2011, the Fukushima Daiichi Nuclear Power Plant (comprising six separate reactors) suffered the worst nuclear accident since Chernobyl, in 1986.[60] A nearly 15-meter (49-foot) wave unleashed by the quake easily overtopped the seawall, designed to withstand a 5.7-meter (19-foot) tsunami, causing an automatic shutdown of the reactors. Lacking any power for cooling, the reactors overheated, triggering explosions in four of them. Three suffered complete meltdowns. The Japanese government ordered an evacuation zone of a 20-kilometer (12-mile) radius. Two power plant employees died from "disaster conditions," but many other workers were exposed to unsafe radiation levels in their brave attempts to quench the fires.

Concerns about changing water table levels that could endanger the proposed Yucca Mountain nuclear waste repository in Nevada over its projected 10,000-year lifetime, or possible renewed faulting, have halted construction plans. Safety issues, as well as competitiveness relative to other energy sources, have limited growth of nuclear power.

Capturing and Sequestering CO_2

Coal-burning power plants generate roughly 45 percent of the world's electricity and discharge around 40 percent of all carbon dioxide emissions each year.[61] Such point sources make them good targets for trapping CO_2. The unwanted gas binds with chemicals called amines in an absorption tower. The amines are later heated and recovered, while the CO_2 is shipped elsewhere for storage. However, up to 30 percent of the plant's power output is wasted.[62]

At the Great Plains Synfuels Plant in North Dakota, currently the world's largest such installation, carbon dioxide released during the conversion of coal into syngas—a synthetic fuel, mainly hydrogen and carbon monoxide—is collected and piped 200 miles (~320 kilometers) north to the Weyburn oil field in Saskatchewan, Canada, for underground burial. Another method—oxyfuel combustion—burns crushed coal in pure oxygen instead of air, yielding carbon dioxide and water, which are then separated and the CO_2 captured. The Schwartze Pumpe plant, among the first such installations in Germany, sequesters its waste CO_2 in a nearby gas field, combining both steps.[63]

Once carbon dioxide has been captured, it must be permanently buried. Deep subsurface porous sedimentary rocks could accommodate between 1,700 and 11,000 gigatons of CO_2 worldwide (one gigaton = one billion metric tons).[64] Suitable rocks (e.g., deep saline formations and depleted oil/gas reservoirs) lie at depths below 800–1,000 meters (2,600–3,300 feet), sealed by an impermeable caprock layer far below aquifers. Under high pressure, CO_2 injected into deep strata dissolves in the brines occupying pore spaces. Enhanced oil recovery in depleted oil fields flushes out any residual oil with CO_2 under high pressure. The sequestered CO_2 essentially replaces the oil, which is recovered. Ideal sequestration sites lie near CO_2 sources and rock reservoirs with enough storage capacity to hold all CO_2 emitted during the power plant's lifetime (e.g., the Gulf Coast region).[65]

Alternatively, CO_2 can be solidified by reacting chemically with calcium or magnesium in **basalts** or **peridotites** to form carbonate minerals.[66] Although such reactions occur in nature, are they fast enough to effectively remove large quantities of CO_2? In a pilot study under way at the Hellisheidi geothermal plant in Iceland, steam and CO_2 are injected through wells down to 400–800 meters, where the heated fluid reacts with calcium in basalt to form calcite ($CaCO_3$)—the dominant mineral in limestone.[67] This technology, if successful, can be applied to other basaltic terrains. However, basalt's porosity is limited, and this rock type does not occur everywhere. At best it would supplement other sequestration efforts.

Offshore submarine reservoirs represent alternative sequestration sites. Since 1996, StatoilHydro has been injecting 1 million metric tons of CO_2 a year from the Sleipner natural gas field in the North Sea into a sandstone reservoir 1,000 meters under the seafloor.[68] Offshore sequestration does not disturb marine ecosystems, provides ample storage capacity, and avoids public opposition to land storage sites and legal disputes over whether pore space ownership lies with surface landowner or government. While carbon capture and sequestration represent a promising step toward a carbon-neutral future, numerous technical challenges, as well as large investment and operating costs, remain. Carbon sequestration could also conversely encourage more fossil fuel use.

Biological sinks (such as conservation tillage, reforestation, restoration of coastal wetlands and dune vegetation, expansion of parkland, urban green space, and green roofs) also sequester CO_2. In addition, they improve air quality, reduce soil erosion, protect shorelines, cool urban summers, and provide wildlife habitat.

Geoengineering

Some scientists promote **geoengineering** to save the Earth from overheating. The term generally applies to large-scale manipulation of the environment to neutralize greenhouse gas–induced warming. Most climate scientists were initially skeptical of such exotic ideas as creating "artificial volcanoes," "orbital sunshades," or "ocean iron fertilization." But in the 1980s, paleoclimatologist Wallace S. Broecker, at Columbia University, suggested spraying tons of sulfur dioxide into the stratosphere from high-flying jumbo jets to simulate the cooling effect of powerful volcanic eruptions.[69] The idea gained support after Nobel laureate Paul J. Crutzen, at the Max Planck Institute for Chemistry in Germany (discoverer of the ozone hole) later investigated the feasibility of the concept. More-recent modeling studies suggest that massive sulfur aerosol inputs could, in principle, largely counteract greenhouse gas warming.[70]

Another plan envisions a fleet of orbital screens or mirrors to cool the Earth by reflecting sunlight. Simpler interventions invoke light-colored roofs and pavements[71] or planting trees. Advocates view such "radiation management" as an "insurance policy" to buy time until other carbon fixes can be adopted. However, geoengineering does not solve the root cause of global warming; it merely masks the problem. Critics liken it to "climate change methadone," since it replaces one addiction (fossil fuels) with another (sulfur dioxide or orbiting sunscreens).[72] Once pumping sulfur into the atmospheric stops, global warming resumes big time. Furthermore, geoengineering cannot prevent other consequences of rising CO_2, such as ocean acidification and changes in precipitation patterns.

Although we are already inadvertently "engineering" nature, the feasibility of proposed geoengineering schemes has not been demonstrated, nor are all the side effects fully known. Artificial radiation management could affect cloud formation and plant growth in as yet unknown ways. Adverse unforeseen regional changes could also overshadow potential benefits.[73]

Other important challenges, in addition to effectiveness, costs, and who would pay, involve international decision making and oversight. Who would decide to undertake projects that affect the entire planet and that some nations might not willingly accept? Who would monitor projects to ensure reliability and safety? To date, international efforts to limit carbon emissions have fallen short of expectations. Would efforts at international cooperation on geoengineering follow a similar path?

Pricing Carbon Emissions

A cap-and-trade system (or emissions trading) uses the market system to limit greenhouse gas emissions. A central authority (national, regional, state, or provincial government) sets a cap, or limit, on the total allowable CO_2 emissions and then either distributes or sells allowances or permits, up to the cap, which is gradually lowered over time. Participating firms (generally electric power plants or carbon-intensive industries) buy permits up to their allotted limit, and if that amount proves insufficient, they must either curtail their emissions or buy additional permits from those who have spare credits. Individual participants decide how, or to what extent, they reduce their emissions. An oversupply of permits drives prices down while not cutting emissions, as occurred in the European Emissions Trading Scheme between 2005 and 2007.[74] It also disadvantages firms that have reduced emissions and banked their credits. On the other hand, too low a cap drives up the price of permits, impeding economic growth. "Safety valves" in some cap-and-trade schemes adjust to keep the value of allowances within specified limits. Some schemes allow polluters to buy **carbon offsets** that reduce emissions in a developing country by replanting trees or installing more-efficient power plants or factories.

Unlike a cap-and-trade system, under which industry pays to adapt to variable prices, an emissions tax tracks profits without constraining emissions. While the polluter risks less, the taxing agency is more exposed to market volatility.

Several trading programs already exist or will soon take effect. The largest is the 27-nation European Union Emissions Trading Scheme, affecting large power plants and heavy industry, which targets a 20 percent reduction in 1990 emissions levels by 2020. In the United States, the mandatory Regional Greenhouse Gas Initiative (RGGI), comprising 10 northeastern and mid-Atlantic states, plans a 10 percent emission cutback by power plants by 2018. Canada and seven western states, including California—its largest single member—form the Western Climate Initiative (WCI), which aims for a 15 percent reduction below 2005 levels by 2020. The Midwestern Regional Greenhouse Gas Reduction Accord, including six midwestern states and the Canadian province of Manitoba, plans to implement a multi-sector cap-and-trade mechanism to lessen greenhouse gas emissions.

Between 2003 and 2010, the pioneering Chicago Climate Exchange (CCX) served as a forum for trading greenhouse gas allowances and offsets, reducing greenhouse gas emissions by 700 million tons.[75] In July 2010, carbon emission trading was replaced by an emissions offset program. Ongoing and start-up regional programs support continuation of the related European Climate Exchange and Chicago Climate Futures Exchange (CCFE),[76] although the recent recession has dampened carbon trading.[77]

The coming global energy transformation involves all economic sectors, but waving farewell to fossils fuels still faces significant technological and economic impediments. Keeping greenhouse gas emissions below the "dangerous" 2°C level requires more than the seven "wedges" originally proposed by Pacala and Socolow.[78] It entails a major research and development effort akin to the Manhattan Project or landing men on the Moon. Will we successfully meet this challenge? Carbon-neutral energy is growing rapidly and now constitutes roughly a fifth of total energy consumption.[79] Aside from technical and business issues, the hitherto subtle effects of global warming inhibit incentives for change.

Averting Dangerous Consequences of Climate Change

Will actions by individual cities, regions, and nations (e.g., New York City, the Northeast and mid-Atlantic states, the European Union) to curb emissions suffice as long as others spew out CO_2? The international community recognizes climate change as a serious and potentially irreversible threat to the well-being of humanity and the planet. In 1992, 154 signatory nations to the United Nations Framework Convention on Climate Change (UNFCCC) voluntarily agreed to stabilize their greenhouse gas emissions to 1990 levels by 2000.[80]

In the subsequent 1997 Kyoto Protocol to the UNFCCC, most industrialized countries, including the European Union and some advanced developing economies, accepted legally binding greenhouse gas emission reductions averaging 5.2 percent below 1990 levels by 2012.[81] Developing countries, including China, India, and Brazil, were exempted, in view of their historically low emissions and their need for economic development. In addition to emissions trading, the Kyoto Protocol also introduced several other flexible mechanisms, such as the Clean Development Mechanism (CDM) and Joint Implementation (JI). The CDM enables a country to earn saleable credits to offset its Kyoto Protocol commitment by implementing an accepted emissions-lowering project (for example, installation of solar panels, more-energy-efficient boilers, or reforestation) in a developing country. The JI mechanism grants a country credits toward its commitment by investing in a greenhouse-gas-reduction/removal project in another country that also needs to curb its emissions.

The UN Climate Change Conference in Copenhagen, Denmark, in December 2009 endorsed continuation of the Kyoto Protocol and urged keeping global temperature below 2°C, by implementing deep cuts in global emissions. Annex I parties (i.e., industrialized countries and "countries in transition"—North America, Europe, including eastern Europe, the Russian Federation, Japan, Australia, New Zealand) would "commit to economy-wide emissions targets" by 2020. China, for example, pledged emissions reductions of 40–45 percent by 2020; Brazil agreed to 36.1–38.9 percent; the United States to 17 percent. The Copenhagen Accord also established the Green Climate Fund to support mitigation programs in developing countries. Reforestation programs were included in an overall global emissions reduction strategy. However, delegates to the Copenhagen Conference merely agreed to "take note of" rather than to commit to a legally binding treaty. Thus the agreement only outlined long-term goals and voluntary mitigation efforts.

The December 2010 UN Climate Change Conference in Cancun, Mexico, similarly acknowledged the need for major cuts in greenhouse gas emissions in view of the seriousness of climate change. Citing more-recent scientific findings, it even recommended holding temperatures below 1.5°C and proposed cutting emissions by 25–40 percent below 1990 levels by 2020. It also proposed a $100 billion Green Climate Fund by 2020 to assist poor countries in climate change adaptation and mitigation, and reducing emissions from deforestation (REDD).

Delegates to the November 28 to December 11, 2011, UN Climate Change Conference in Durban, South Africa, agreed to finalize an international legally binding treaty that addresses global warming by 2015 and that takes effect in 2020. The Kyoto Protocol, which is due to end in 2012, will remain in effect in the interim. Further progress was made toward establishment of the Green Climate Fund, for which a management framework was created.

These international conferences illustrate attempts to limit greenhouse gas emissions globally. However, a treaty with real teeth should be legally binding and agreed upon by all parties, especially major greenhouse gas emitters, allocate required carbon reductions fairly, and share technical advances with developing nations. Participants need to overcome the numerous disagreements still thwarting a comprehensive treaty. Meanwhile, initiatives by individual countries and cities move in the right direction.

WELCOME TO THE ANTHROPOCENE

> Humans were inevitably going to be part of the fossil record. But the true meaning of the Anthropocene is that we have affected nearly every aspect of our environment—from a warming atmosphere to the bottom of an acidifying ocean.
>
> —*NEW YORK TIMES* EDITORIAL, FEBRUARY 28, 2011

Rising seas are just one symptom of a much larger unfolding environmental crisis—one manifestation of the multidimensional planetary changes now under way. We now live in the **Anthropocene** epoch—increasingly marked by the human touch.[82] Ever since humanity first learned to control fire, people have transformed the Earth's surface. The agricultural revolution further altered natural vegetation patterns. However, after the Industrial Revolution of the late 18th century, and increasingly so after the mid-20th century, people have become major environmental and geologic agents. We are reshaping the planet by literally moving mountains, diverting water flows, denuding forests, eroding soils, altering biogeochemical cycles (i.e., nitrogen and phosphorus overloading), acidifying the ocean, diminishing biodiversity, and changing the climate. Exponential population growth coupled with rapid economic and technological development drive this planetary transmutation, unparalleled in Earth's history. As demand rises, growing scarcities of food, water, and mineral resources will increasingly stress our fragile environment.

Climate change brings additional stresses. Although some agricultural regions may benefit from a longer growing season resulting from additional warmth or extra rainfall, other regions stand to become drier or even turn into dust bowls.[83] Crop yields may drop, unless more drought-resistant varieties can be developed in time. Elsewhere, soil fertility may decline due to erosion of topsoil. Groundwater mining may lower the water table enough to make pumping water for irrigation too costly (e.g., the Ogallala Aquifer in Oklahoma and Texas). Increasing saltwater encroachment due to sea level rise may render many fertile low-lying deltaic or coastal farmlands (e.g., the Sacramento–San Joaquin valley, California, the Mekong Delta, Vietnam, or the Ganges-Brahmaputra Delta, Bangladesh) increasingly unproductive.

To forestall these rapid planetary transformations from bringing civilization to the brink requires a solution to the impending climate and environmental crisis that transcends short-term fixes. It calls for a major paradigm shift. No longer can environmental and climatic consequences of industrial activities remain "externalities" outside of economic models, with nature a mere provider of resources to be exploited and extracted.

Nature works through the interactions of multiple complex processes. In the Anthropocene, the human imprint merges with Earth's system operations. The new worldview embraces consilience—the unity of knowledge. It balances the dominant emphasis on economic growth with information from the environmental and earth sciences, as well as traditional human values.

Toward a New Balance

The unprecedented changes of the past century create many challenges in maintaining a livable planet. Integrated assessment models (IAMs) provide a tool with which to study complex global-scale environmental and human interactions. IAMs input results from global climate models (GCMs) into a cost/benefit analysis of various policy options to limit greenhouse gas emissions. The models compare costs of conventional (coal, petroleum, natural gas) and unconventional energy resources (renewables), their relative energy efficiencies, CO_2 emissions, and resulting environmental impacts. However, economic calculations often make simplistic (and subjective) assumptions of weights assigned to future versus present outcomes, mitigation benefits against costs, and technological changes.[84]

A more comprehensive systems framework simulates complex interconnected biogeophysical and socioeconomic dynamics.[85] In one such model, IMAGE 2.4, data on population, land use, energy utilization, agricultural activity, and economic activity determine greenhouse gas emissions that impact climate, oceans, and several biogeochemical cycles (e.g., carbon, nitrogen, and water).[86] The model also simulates other environmental stresses (e.g., soil degradation, water stress, biodiversity loses, and air/water pollution). Model impacts affect policy choices, which in turn feed back to the initial socioeconomic drivers, thus completing the loop. Several international studies, including the IPCC (2007) report, use versions of this model.

While such models yield useful insights into future energy choices, a deeper understanding of the complex Earth-Human system involves further multidisciplinary research.[87] Some objectives include the need to: (1) improve simulations of human interactions with biogeophysical and hydrological processes, (2) acquire more detailed information on natural and developed Earth systems, (3) anticipate and manage major global environmental disruptions, (4) develop technological and societal innovations that enhance global sustainability, and (5) adapt institutions, economics, and policies accordingly. The last two steps may represent the greatest challenge of all, since they call for a new vision of a livable environment where nature coexists comfortably with people.

Working with Nature

> Something there is, whose veiled creation was
> Before the earth or sky began to be . . .
> A name for it is "Way" . . .
> Man conforms to the earth;
> The earth conforms to the sky;
> The sky conforms to the Way;
> The Way conforms to its own nature.
>
> LAO TZU

Modern civilization relies on fossil fuels, as well as other essential minerals and metals. Coal and petroleum are used for electricity, heating, transportation, manufacturing, agriculture, clothing (synthetic fibers), drugs, and so on. But these non-renewable resources will not last indefinitely, as noted by geologist M. King Hubbert, in regard to petroleum.[88] At present, renewable energy constitutes only 3 percent of total global energy consumption.[89] Therefore, achieving sustainability involves not only embracing carbon-neutral energy but also cutting waste, recycling, reducing demand, and working more closely with nature.

Rebuilding dunes and restoring salt marshes and oyster reefs along "living shorelines" help buffer storm surges and sea level rise. For example, the Netherlands now integrates water safety with sustainability through "ecological engineering" along the shore, encouraging the natural interchange between wet and dry land. While dikes will still be raised and widened, so will the dunes and beaches. Expanded and interconnected coastal habitats provide migration corridors for animals and vegetation. Flood-resilient buildings adapt to changing water levels. Although unpopular in Holland (for historic reasons), yet being tried in Louisiana, removal of some levees would permit occasional floods to deposit fresh sediment in the delta and regenerate marshes.[90]

Reforestation sponges up excess atmospheric CO_2 while reestablishing habitat for wildlife. Good forest-management practices include selective logging, replanting, tailoring harvesting rates to local conditions, preservation of biodiversity and productivity. Reclamation of riparian vegetation and floodplains reduces bank erosion, reestablishes the river's natural flow, and provides increased flood protection. Sustainable farming reduces tillage, rotates crops, conserves water, recycles organic wastes, and maintains soil quality.

Development of renewable energy, as outlined above, can create entirely new industries and jobs (note parallels with computers and electronics). Other opportunities lie in fostering "smart" growth, such as "green" neighborhoods or cities served by low-carbon public transportation; designing adaptive architecture; manufacturing more-durable, less-wasteful goods; improving technology, productivity, and energy efficiency; providing more time for family or friends; and continuing education. Developing countries focus more on raising their basic standard of living, while attempting to avoid the pitfalls of environmental degradation.[91]

WHITHER THE RISING SEAS?

Our time travel has revealed the close interweaving of sea level and climate change, particularly since ice accumulated in Antarctica, 34 million years ago. Sea level has repeatedly swung up and down over multiple ice ages. Following the last ice age, sea level sprang upward ~120 meters, but then remained relatively stable for 7,000 years—until recently.

The changes already under way are unequaled in human history. Atmospheric CO_2 and CH_4 levels are now at their highest in 800,000 years. Northern Hemisphere temperatures are now likely the warmest in over a millennium. Sea level rose faster during the past century than during preceding millennia, and may have accelerated further during the last 20 years.

We are near a critical crossroad. Today's CO_2 level exceeds that of the Last Interglacial and now closely approaches mid-Pliocene levels. Sea levels many meters higher than present ones characterized both periods. Because of the ocean's large thermal inertia and long mixing time, and sluggish ice sheet response, the full climatic response lags many years behind the rise in CO_2. More importantly, because the anthropogenic CO_2 accumulated by 2100 (even if all emissions will have ceased) will need centuries or more to dissipate, warmer conditions will persist long enough for significant ice sheet melting. Thus we have not yet begun to experience the ultimate impacts of temperature and sea level rise.

To avoid a return to the Eemian, Pliocene, or even worse, the hothouse Eocene, with drastic consequences for the world's coastal communities, the transition toward a low carbon energy pathway should begin without further delay. Decisions and policies should integrate the latest climatic (and environmental) information in order to use nature's bounty wisely for humanity's well-being without destroying that bounty in the process. The choice is ours.

APPENDIX
Geologic Time Scale

CENOZOIC ERA

Quaternary Period	Holocene Epoch	<11,780 years
	Pleistocene Epoch	2.59 my
Neogene Period	Pliocene Epoch	5.33 my
	Miocene Epoch	23.0 my
Paleogene Period	Oligocene Epoch	33.9 my
	Eocene Epoch	55.8 my
	Paleocene Epoch	65.5 my

MESOZOIC ERA

Cretaceous Period	145.5 my

Jurassic Period	199.6 my
Triassic Period	251.0 my
PALEOZOIC ERA	
Permian Period	299.0 my
Carboniferous Period	359.2 my
Devonian Period	416.0 my
Silurian Period	443.7 my
Ordovician Period	488.3 my
Cambrian Period	542.0 my
PRECAMBRIAN ERA	
Proterozoic Eon	2,500 my
Archean Eon	4,000 my
Hadean Eon	~4,600 my
(formation of Earth)	

SOURCE: Dates based on International Stratigraphic Chart, International Commission on Stratigraphy, Sept. 2010. http://www.stratigraphy.org.

Notes

CHAPTER 1: THE EVER-CHANGING OCEAN

1. Gleick (1996); Shiklomanov (1997), p. 8.
2. United States Geological Survey (2006).
3. Sverdrup and Armbrust (2008), p. 100; Thurman (1997), p. 82.
4. Sverdrup and Armbrust (2008), p. 78.
5. Mathez and Webster (2004), p. 94; Sverdrup and Armbrust (2008), p. 105; Thurman (1997), p. 101.
6 See, for example, Mathez and Webster (2004), chap. 7, fig. 7.1.
7. For a more detailed explanation of the Coriolis effect, consult any standard oceanography text (e.g., Sverdrup and Armbrust (2008); Thurman (1997).
8. Sverdrup and Armbrust (2008), p. 231.
9. Seager (2006). This view of the diminished role of the Gulf Stream in generating a mild western European climate has been challenged by others; for example, see Rhines and Häkkinen (2003).
10. Broecker (1997).
11. Mathez and Webster (2004), p. 201.
12. Broecker (1997), p. 4.
13. Thurman (1997), p. 212.
14. E.g., Quadfasel (2005); Willis (2010).
15. Rahmstorf (2006).
16. Beckley et al. (2007).
17. Sverdrup and Armbrust (2008), p. 250.

18. Thurman, 1997, p. 241.

19. A fuller treatment of tidal forces is given by Pugh (2004), chap. 2. See also Sverdrup and Armbrust (2008), chap. 11, and Thurman (1997), chap. 10.

20. Pugh (2004), p. 31.

21. At summer solstice, around June 21, the Sun appears directly overhead at 23.5°N latitude along the Tropic of Cancer (Pugh, 2004, pp. 31, 40).

22. Gornitz (2005a) and citations therein.

23. Coch (1994).

24. Wood (1986).

25. Wood (1986).

26. Philander (1990); Cronin (2010), pp. 279–280.

27. Trenberth (1997).

28. http://www.pmel.noaa.gov/tao/elnino/el-nino-story.html.

29. McPhaden et al. (2006), p. 1741.

30. Ibid.

31. Tingstad and Smith (2007), p. 215.

32. Ryan and Noble (2002), p. 163.

33. Kosro (2002), pp. 347–348.

34. Strub and James (2002).

35. Other multi-annual climate anomalies that affect the North Atlantic Ocean are the Arctic Oscillation (AO), a pressure variation between the Arctic and mid-latitudes, and the Northern Annual Mode, which is very similar to the NAO.

36. Hurrell (1995).

37. Woodworth et al. (2007). The low air pressure can also produce a higher than average sea level, due to the "inverse barometer effect."

38. Woodworth et al. (2007), p. 942.

39. Maul and Hanson (1991).

CHAPTER 2: THE CAUSES AND DETECTION OF SEA LEVEL CHANGE

1. The Pliocene epoch lasted from 5.3 to 1.8 million years ago. (However, a recently revised geological timescale terminates the Pliocene at 2.59 million years; appendix.)

2. Dowsett and Cronin (1990).

3. E.g., Dwyer and Chandler (2009).

4. McCulloch and Esat (2000).

5. Peltier and Fairbanks (2006).

6. IPCC (2007a), chap. 5, p. 410.

7. The 19-year period corresponds approximately to the 18.8 lunar nodal cycle mentioned in chap. 1. The current tidal epoch used by NOAA runs from 1983 to 2001.

8. Douglas (2001), pp. 3–5; Pugh (2004), pp. 43–45.

9. Pugh (2004), p. 44.

10. See Glossary.

11. National Research Council (1990), pp. 6–16; Gornitz (2005b).

12. Peltier and Fairbanks (2006); Rohling et al. (2009).

13. Shiklomanov (1997).

14. IPCC (2007a), chap. 5, pp. 410, 415.

15. IPCC (2007b). Summary for Policy Makers; Technical Summary, pp. 58–66.
16. National Research Council (1990), p. 7.
17. Chao et al. (2008).
18. See Glossary.
19. IPCC (2007a), chap. 5, pp. 418–419; Milly et al. (2010).
20. Grotzinger et al. (2006), p. 25.
21. A. Cox (1973).
22. The oldest oceanic rocks are around 170 million years old, from the Pacific Ocean.
23. See Glossary.
24. The Tohoku (Sendai) earthquake is among the five most powerful recorded since 1900. However, on January 12, 2010, more than 200,000 people died in an earthquake (moment magnitude 7.0) that devastated Port-au-Prince, Haiti. Unlike earthquakes in Chile or Japan, the one in Haiti was produced by the lateral motion along a major fault zone (similar to the San Andreas Fault in California).
25. E.g., Miller et al. (2005), or Müller et al. (2008).
26. Müller et al. (2008).
27. Changes in seafloor spreading rates also affect the atmospheric concentrations of greenhouse gases such as carbon dioxide and sulfur dioxide, both of which are outgassed by volcanoes (including those not along plate boundaries). Both carbon dioxide and sea levels were high during past warm "greenhouse" climates, such as the Cretaceous period, and low during cold "icehouse" climates, such as the late Permian before the breakup of Pangaea. Changes in latitudinal position of continents and oceans also affected ancient climates. Snow and ice accumulate over large polar landmasses (e.g., Antarctica) and at high elevations, even in the tropics. Bright snow and ice reflect more of the Sun's energy back to space, whereas darker ocean water absorbs more solar energy. Thus, changes in the area and thickness of perennial ice alter the planet's heat balance, climate, and ultimately sea level. Since oceans absorb more heat than land does, changes in ocean versus continental area due to plate tectonics will affect the Earth's heat balance and its climate.

Plate tectonics also alters the patterns of ocean circulation and warm water transport from the tropics to the poles. In many places, water flows through fairly narrow constrictions, dubbed "ocean gateways," such as the Strait of Gibraltar or the Strait of Magellan at the southern tip of South America. If these gateways are closed by tectonic events, ocean circulation is disrupted or permanently changed (see chap. 3). See also DeConto, 2008.
28. Lambeck (2009), pp. 374–380; Peltier (2001).
29. E.g., Grotzinger et al. (2006), chap. 14.
30. Human water diversions and land use changes have also added to the subsidence (see chap. 8).
31. Ekman (1999).
32. Pugh (2004), chap. 1.
33. http://www.psmsl.org.
34. Bevis (2007).
35. For more mission details, see http://www.jpl.nasa.gov and http://www.cnes.fr.
36. Beckley et al. (2007).
37. Because of the orbital inclination, satellite coverage extends only 66°N and S. Thus no data are collected at higher latitudes near the poles.
38. Kearney (2009).

39. Radiocarbon dating employs the rate of radioactive decay of ^{14}C, produced by interactions of cosmic rays with atmospheric nitrogen. Carbon-14 has a half-life of 5,730 years (i.e., half of the ^{14}C initially present in a specimen disappears within 5,730 years, half of the remainder in the next 5,730 years, etc.). Living organisms incorporate ^{14}C into their cells but none after their death. Thus the radiocarbon age measures the time elapsed since the death of the organism. However, the atmospheric production of ^{14}C varies because of fluctuations in the amount of cosmic rays reaching the Earth. The radiocarbon age can be converted to the true age by measuring the $^{14}C/^{12}C$ ratio in tree rings from long-lived trees (like the bristlecone pine) that can be dated to the year by counting the number of tree rings, and by using other methods. ^{14}C has been calibrated to 40,000–50,000 years, with greatest accuracy to the last 11,000–12,000 years (Cronin, 2010, p. 35).

40. Chappell et al. (1996).

41. Isotope ratios are expressed in terms of the notation:

$$\delta^{a}B = (R_{sample}/R_{standard} - 1) * 1000$$

where *a* is the atomic weight of the heavier isotope, *B* is the chemical element, and *R* is the isotope ratio (heavy to light) in the sample and in a standard reference material. Values of δ are given in per mil, or parts per thousand ($^{o}/_{oo}$). Positive δ values occur when $R_{sample} > R_{standard}$; conversely, negative δ values result when $R_{sample} < R_{standard}$.

42. Ice cores from Antarctica and Greenland hold multifaceted archives of past climates that record variations in temperature, sea level, atmospheric circulation, and trace gas composition over glacial-interglacial cycles (e.g., Mayewski and White, 2002; see also chap. 4).

43. Cronin (2010), pp. 82–83.

44. King (1959), p. 6.

45. Beus and Morales (2003). Absolute ages from Mathis and Bowman (2006). http://www.stratigraphy.org.gssp.htm.

46. Müller et al. (2008).

47. Vail et al. (1977); Haq et al. (1987).

48. E.g., Hallam (1992); Christie-Blick et al. (1990).

49. Seismic sequence analysis can detect changes on the order of a million years or less, under favorable circumstances (Hallam, 1992, p. 34).

50. Miller et al. (2005). As sediments accumulate, they grow more compact as water initially trapped between mineral grains is gradually expelled. The weight of overlying sediments adds to the subsidence. Oceanic sediments also slowly subside with age, as they move away from hot, mid-ocean ridges. The original water depth can be estimated from the assemblage of fossil foraminifera. Past sea level changes can be reconstructed knowing these factors.

CHAPTER 3: PIERCING THE VEIL OF TIME

1. Müller et al. (2008). Other estimates of late Cretaceous sea level range from ~100 to 250 meters above present (e.g., Miller et al., 2005; Kominz et al., 2008).

2. Ananthaswamy (2008).

3. Alvarez (1997). Evidence for a catastrophic impact includes a worldwide iridium layer and the presence of shocked quartz and other minerals at the Cretaceous-Tertiary (K-T) boundary, 65 million years ago. Rare at the Earth's surface, iridium is more abundant in meteorites, along with nickel and cobalt. The global distribution of anomalously elevated iridium concentrations at the same time horizon argues for an extraterrestrial origin. Quartz (silicon dioxide) is a common mineral occurring in many different types of rocks. Under the extreme pressures of an impact event, it deforms internally along sets of intersecting planar surfaces, indicative of "shock metamorphism." The high-pressure forms of quartz—coesite and stishovite—have also been discovered at the K-T boundary, along with minuscule diamond grains, also commonly found at other impact sites.

4. Pierazzo (2009) and Rampino (2009) summarize climatic effects of the terminal Cretaceous impact.

5. Rampino (2009).

6. The exact cause of the mass extinction remains controversial (e.g., Schulte et al., 2010; Archibald et al., 2010; Courtillot and Fluteau, 2010). Some argue that volcanic inputs of sulfur dioxide and carbon dioxide into the atmosphere triggered short-lived cooling followed by a longer-lasting greenhouse effect and climate warming. Volcanism, on the other hand, would have led to a more gradual buildup of greenhouse gases, whereas the extinction occurred rather suddenly.

7. The Mesozoic era lasted between 251 and 65 million years. The succeeding Cenozoic era encompasses the Paleogene, Neogene, and Quaternary periods, spanning the last 65 million years. The older term "Tertiary" includes only the former two periods.

8. Zachos et al. (2001).

9. Moran et al. (2006); Ananthaswamy (2008).

10. Moran et al. (2006).

11. Huber (2008).

12. Moran et al. (2006).

13. Zachos et al. (2001); Pekar and Christie-Blick (2008).

14. Lewis et al. (2008).

15. Fox (2008).

16. The Tibetan Plateau strengthens the Asian monsoon. Although summertime temperatures in Tibet remain cool, it is nonetheless the warmest place on Earth at that elevation. This warm spot initiates continent-scale circulation analogous to land and sea breezes. During summer months, the zone of highest mean surface temperature shifts north, drawing moist air from the Indian Ocean, which replaces the hot, dry air rising over the Tibetan Plateau. As the moist air reaches the Himalayas, it rises, cools, and the moisture condenses, falling as rain over India and along the southern flanks of the Himalayas. By the time the storms move beyond this topographic barrier, the air has dried out and Tibet is almost a desert. This circulation system has intensified as the elevation difference between the plains at the foothills of the Himalayas and its lofty peaks has increased.

17. Raymo and Ruddiman (1992).

18. Kent and Muttoni (2008).

19. DeConto (2009).

20. Scher and Martin (2006).

21. Stickley et al. (2004).

22. Lyle et al. (2007).
23. DeConto and Pollard (2003).
24. Haug and Tiedemann (1998); Haug et al. (2004).
25. Miller et al. (2005); Katz et al. (2008).
26. See chap. 2, endnote 40.
27. See Glossary.
28. Müller et al. (2008).
29. Kominz et al. (2008).
30. Müller et al. (2008).
31. Lear et al. (2008).
32. Katz et al. (2008).
33. Katz et al. (2008).
34. E.g., Moran et al. (2006).
35. DeConto et al. (2008).
36. Miller et al. (2005); Kominz et al. (2008).
37. Naish et al. (2001).
38. E.g., Zachos et al. (2001); Shevenell et al. (2004).
39. Lewis et al. (2008).
40. Haug et al. (2004); Haug and Tiedemann (1998).
41. Haug et al. (2004).
42. Haug et al. (2004); Haug and Tiedemann (1998). The astronomical theory of the ice ages, originally proposed by Milutin Milankovitch, relates ice ages to periods of favorable configurations of the Earth's orbit, such as: (1) a low tilt angle of the rotational axis, (2) a near-circular orbit around the Sun, and (3) timing of the Earth's elliptical orbit to coincide with Northern Hemisphere summer solstice when Earth is farthest from the Sun.
43. Bartoli et al. (2005).
44. Lunt et al. (2008).
45. C. Karas et al. (2009).
46. This event has also been referred to as the Late (or Latest) Paleocene Thermal Maximum, or the Initial Eocene Thermal Maximum.
47. See chap. 2.
48. Zachos et al. (2001).
49. Katz et al. (1999); Thomas et al. (2002).
50. Katz et al. (1999).
51. Thomas et al. (2002).
52. Kennett et al. (2003).
53. Zachos et al. (2008).
54. Speijer and Morsi (2002).
55. Brinkhuis et al. (2006); Miller et al. (2005).
56. Brinkhuis et al. (2006).
57. Hsü (1972).
58. Hsü (1972); Rouchy and Caruso (2006).
59. Krijgsman et al. (1999).
60. Rouchy and Caruso (2006); Vidal et al. (2002).
61. Garcia-Castellanos (2009).
62. Vidal et al. (2002).
63. Dowsett et al. (2009).

64. Dowsett et al. (2009).
65. Schneider and Schneider (2010).
66. Dwyer and Chandler (2009).
67. Naish and Wilson (2009).

CHAPTER 4: WHEN THE MAMMOTHS ROAMED

1. J. Adams (2002).
2. The Quaternary, the most recent geologic period of the Cenozoic era, encompasses the last 2.588 million years. It formerly embraced the last 1.8 million years. The Quaternary consists of the Pleistocene epoch between 2.588 million and 11,780 years ago and the Holocene epoch from 11,780 years ago to the present. (Overview of Global Stratotype Sections and Points [GSSP's], compiled by Jim Ogg [2007] at http://www.stratigraphy.org/gssp.htm.)
3. A 41,000-year cycle had been observed somewhat earlier in mid-Pliocene marine sediments, but the older oscillations were much more muted (see chap. 3).
4. Clark et al. (2006); Lambeck et al. (2002a).
5. Glacials; ice ages—see Glossary.
6. Chappell et al. (1996); Lambeck et al. (2002a).
7. When averaged over thousands of years, rates of uplift on the Huon Peninsula (ranging from 2 to 4 meters per 1,000 years) can be assumed to remain constant. However, the actual uplift on a much shorter timescale proceeds in a much more jerky fashion—typically rising a meter or two with each large earthquake.
8. Chappell et al. (1996).
9. Cronin (1999), p. 157.
10. For the Quaternary, see Siddall et al. (2007); Lambeck et al. (2002a).
11. Cronin (2010), pp. 44, 82–83.
12. Clark et al. (2006).
13. Chappell (2002); Hearty et al. (2007).
14. Fairbanks (1989).
15. Siddall et al. (2007); Rohling et al. 2009).
16. Raynaud and Parrenin (2009); Brook (2008); Alley (2000); Mayewski and White (2002).
17. Siegenthaler et al. (2005); Brook (2008).
18. Cronin (1999), pp. 438–439.
19. Imbrie and Imbrie (1979/1986), pp. 69–75.
20. Imbrie and Imbrie (1979/1986), pp. 81–83.
21. Imbrie and Imbrie (1979/1986), pp. 84–86.
22. Milankovitch (1941/1969).
23. Imbrie and Imbrie (1979/1986), pp. 104–108.
24. Hays et al. (1976).
25. Maslin (2009).
26. Cheng et al. (2009).
27. See summary in Cronin (2010), pp. 135–139.
28. Raymo and Hubers (2008).
29. Clark et al. (2006); Bintanja and van der Wal (2008).

30. Bintanja and van der Wal (2008).

31. Brook (2008).

32. The first anatomically modern humans had evolved in East Africa by 200,000 years ago (Zimmer, 2005, p. 112).

33. Vaks et al. (2007).

34. Zimmer (2005), pp. 128–130.

35. Sea level may have also influenced even older waves of hominin migrations. For example, *Homo erectus* fossils that were 1.6–1.5 million years old were found in Sangiran, central Java, during an early Pleistocene glacial cycle and low sea level stand, when some of the Indonesian islands were probably connected to the south Asian mainland. Open woodland and savanna and grasslands fauna, flourishing in the region during the colder, drier climate, attracted *H. erectus*. Rising sea levels during interglacials, on the other hand, submerged wooded savanna lowlands, inhibiting migrations (Ciochon and Bettis, 2009).

36. Strictly speaking, the designations "Eemian" (Europe), "Sangamonian"(North America), and "Riss/Würm" (Alps) were defined in terms of vegetation, soil, and stratigraphic units, which are not of the same age everywhere. The eustatic sea level record provides a better reference for the last interglacial, since sea level change is virtually simultaneous around the world and it corresponds closely to climate change (see Müller, 2009).

37. IPCC (2007a); Otto-Bliesner et al. (2006); Overpeck et al. (2006), pp. 458–459.

38. Otto-Bliesner et al. (2006); Overpeck et al. (2006); Kopp et al. (2009).

39. Hearty et al. (2007).

40. Rohling et al. (2008).

41. Rohling et al. (2008).

42. Otto-Bliesner et al. (2006).

43. Otto-Bliesner et al. (2006).

44. Overpeck et al. (2006).

45. Kopp et al. (2009).

46. Lambeck et al. (2002a).

47. Cutler et al. (2003).

48. Hemming (2009).

49. Hemming (2009).

50. Maslin (2009).

51. Named after two glaciologists, Willy Dansgaard and Hans Oeschger.

52. E.g., Cronin (1999), pp. 221–236. These include changes in marine temperatures, sediment composition, and ecology, continental glacial advances and retreats, lake deposition, land vegetation (pollen), and Chinese loess.

53. Maslin (2009).

54. Bond et al. (2001); Almasi and Bond (2009).

55. Bond et al. (2001).

56. Chappell (2002).

57. Rohling et al. (2004).

58. IPCC (2007a), chap. 6, p. 450.

59. E.g., Fairbanks et al. (1989); Chappell et al. (1996); Siddall et al. (2003). See chapter 5 for additional information.

60. Some geologic evidence suggests that the Laurentide Ice Sheet may have been somewhat thinner, especially in the Canadian Arctic. A thinner ice sheet would have been more mobile and capable of reacting faster to climate change (e.g., see Heinrich events, Hemming, 2009; Bowen, 2009).

61. Bowen (2009)

62. Marshall (2009).

CHAPTER 5: THE GREAT ICE MELTDOWN AND RISING SEAS

1. Swanton (1909).

2. Fig. 6 in Peltier and Fairbanks (2006).

3. Fairbanks (1989).

4. E.g., see *The Epic of Gilgamesh*, trans. Maureen Gallery Kovacs (Stanford, Calif.: Stanford University Press, 1990) or *Gilgamesh*, trans. John Maier and John Gardner (New York: Vintage, 1981).

5. Several versions are summarized in Graham Hancock, *Underworld: The Mysterious Origins of Civilization* (New York: Three Rivers Press, Random House, 2002).

6. Ryan and Pitman (1998), p. 234.

7. Meltzer (2009), p. 34.

8. Keigwin et al. (2006).

9. Waters et al. (2011); Meltzer (2009), pp. 5, 92–93; Goebel et al. (2008). Many other proposed early sites have been challenged by archaeologists favoring the "Clovis first" hypothesis.

10. Meltzer (2009), pp. 129–130; Goebel et al. (2008).

11. Meltzer (2009), pp. 36–37; dates (in radiocarbon years) were converted to calendar years using his table 1, p. 9.

12. Clague (2009).

13. Clague (2009); Hetherington et al. (2004).

14. In addition to Meltzer (2009), interested readers may wish to consult: E. J. Dixon, *Bones, Boats, and Bison: Archeology and the First Colonization of Western North America* (Salt Lake City: University of Utah Press, 1999); J. E. Morrow and C. Gnecco, eds., *Paleoindian Archaeology: A Hemispheric Perspective* (Gainesville: University Press of Florida, 2006); J. Adovasio and J. Page, *The First Americans: In Pursuit of Archaeology's Greatest Mystery* (New York: Random House, 2002). T. D. Dillehay, *The Settlement of the Americas: A New Prehistory* (New York: Basic Books, 2000).

15. Morgan (2009).

16. Hoek (2009).

17. Steffensen et al. (2008).

18. Peteet (2009).

19. Baker (2009).

20. Tarasov and Peltier (2006); Tarasov and Peltier (2005). Geologic evidence also supports a flood along the Mackenzie River (Murton et al., 2010).

21. Lowell et al. (2005).

22. Bard et al. (2010).

23. Roberts (2009).

24. Thirty-five genera of large mammals became extinct in North America at the end of the Pleistocene (Meltzer, 2009, p. 44). They include the woolly mammoth, American mastodon, saber-toothed cat, horse, tapir, and several types of ground sloth. Others survived, such as bison, moose, elk, caribou, musk ox, and mountain sheep. Mammal extinctions also affected other continents, but fewer species died off.

25. Firestone et al. (2009). Fullerenes are members of a broad class of carbon molecules that can form hollow spheres, tubes, or sheets. The most common fullerene is C_{60}, or buckminsterfullerene, named after the inventor Buckminster Fuller (1895–1983). In C_{60}, the carbon atoms are connected at the corners of 20 hexagons and 12 pentagons shaped like a soccer ball, hence the nickname "buckyball." Fullerenes have been detected at known impact sites.

26. Kennett et al. (2009); West et al. (2008). These include cubic diamond (the usual kind), lonsdaleite (a hexagonal analog of diamond), and "n-diamond" (another type of cubic diamond; it forms at somewhat lower temperatures and pressures than the more common cubic type).

27. Alvarez (1997).

28. Melott et al. (2010). Above-average traces of ammonium and nitrate ions in the ice at the onset of the Younger Dryas may have been created by a comet crashing into the atmosphere.

29. Pinter and Ishman (2008).

30. Alvarez (1997).

31. Peltier and Fairbanks (2006).

32. Yokoyama et al. (2000); Clark et al. (2004); Hanebuth et al. (2009).

33. E.g., Peltier and Fairbanks (2006).

34. Peltier and Fairbanks (2006).

35. Stanford et al. (2006); Stanford et al. (2011).

36. Bard et al. (2010).

37. Tarasov and Peltier (2006).

38. E.g., Clark et al. (2002).

39. Bassett et al. (2005); Clark et al. (2002).

40. Leventer et al. (2006). Algae need sunlight for photosynthesis. If the surface of the fjord had been frozen at the time the sediments were deposited, the ice would have blocked the sun's rays.

41. Tarasov and Peltier (2006); Mackintosh et al. (2011).

42. Fairbanks (1989); Bard et al. (2010).

43. Leventer et al. (2006); Lowell et al. (2005); Tarasov and Peltier (2005).

44. Yu et al. (2010).

45. Carlson et al. (2008); Cronin et al. (2007); Liu and Milliman (2004).

46. Peltier and Fairbanks (2006).

47. Blanchon and Shaw (1995).

48. Törnqvist et al. (2004).

49. Cronin et al. (2007).

50. Yu et al. (2007).

51. Yu et al. (2007); Liu and Milliman (2004); Tooley (1989).

52. Alley and Ágústsdóttir (2005).

53. Lajeunesse and St. Onge (2008); Clarke et al. (2004); Barber et al. (1999).

54. Alley and Ágústsdóttir (2005).

55. Clarke et al. (2004); Törnqvist et al. (2004); Barber et al. (1999).
56. Yu et al. (2007).
57. Beget and Addison (2007).
58. Beget and Addison (2007).
59. The timing of meltwater pulse 1C needs to be narrowed down further. As things now stand, dates for the sea level spike range between 8,300 and 7,600 years ago (table 5.1), overlapping dates of the cold event. More likely, sea level rise would have simply accelerated once the climate warming resumed.
60. Ryan and Pitman (1998).
61. Ryan and Pitman (1998), pp. 155–160, 230–237.
62. Ryan (2007); Lericolais et al. (2009); Giosan et al. (2009).
63. Aksu et al. (2002).
64. Giosan et al. (2009).
65. E.g., Turney and Brown (2007).
66. Pross et al. (2009).
67. E.g., Dickinson (2004).
68. Mitrovica and Milne (2002).
69. E.g., Varekamp and Thomas (1998); Scott et al. (1995); Finkl (1995).

CHAPTER 6: THE MODERN SPEEDUP OF SEA LEVEL RISE

1. See Glossary. Recessional moraines are common in the Alps and other high mountain valleys. Jean Grove, in *The Little Ice Age*, describes the worldwide retreat of glaciers, including the Rhone glacier, since the mid-19th century.
2. Swiss Glacier Monitoring Network, http://glaciology.ethz.ch/messnetz/lengthvariation.html (last accessed June 30, 2009).
3. Past Variability of Canadian Glaciers. Canadian Cryospheric Information Network. http://www.socc.ca/cms/en/socc/glaciers/pastGlaciers.aspx (last accessed June 29, 2009). See also http://earthobservatory.nasa.gov/IOTD/view.php?id=7679.
4. Thompson (2009).
5. IPCC (2007b), Summary for Policymakers.
6. IPCC (2007a), chap. 3, pp. 241–253.
7. IPCC (2007b), Summary for Policymakers.
8. Long (2009); Pritchard et al. (2009); Shepherd and Wingham (2007); Velicogna and Wahr (2006).
9. Long (2009); van den Broeke et al. (2009).
10. Chen et al. (2009); Pritchard et al. (2009); Rignot et al. (2008).
11. Cazenave and Llovel (2010).
12. IPCC (2007b), Summary for Policymakers, p. 7.
13. IPCC (2007a), chap. 4, pp. 343–345.
14. IPCC (2007c), chap. 1, pp. 99–107.
15. IPCC (2007c), chap. 1, p. 94.
16. CO_2 and water, H_2O, form a weak acid, carbonic acid, H_2CO_3.
17. Carbon dioxide is consumed during photosynthesis in the Northern Hemisphere spring and summer; it is released during the winter season as the vegetation becomes dormant.

18. IPCC (2007b), Summary for Policymakers; IPCC (2007a), chap 2.

19. Greenhouses also retain heat because they act as a shelter from the wind and reduce heat loss by conduction and convection.

20. Brook (2008).

21. "NASA Research Finds 2010 Tied for Warmest Year on Record." *NASA Goddard Institute for Space Studies, Research News*. http://www.giss.nasa.gov/research/news/20110112/ (last accessed June 23, 2011).

22. IPCC (2007b), Summary for Policymakers; Mann et al. (2008).

23. Volcanoes emit various greenhouse gases, such as carbon dioxide and sulfur dioxide. Sulfur dioxide, although a greenhouse gas, reacts with water vapor in the atmosphere to produce sulfuric acid. The tiny acid droplets scatter sunlight and have a cooling effect. Depending on the latitude and prevailing wind patterns, a major volcanic eruption may cool the Earth by a few tenths of a degree Centigrade for several years. The Little Ice Age has been attributed to a period of minimal solar activity (i.e., the Maunder Minimum), but the effects of solar variability on present climate are believed to be low.

24. IPCC (2007a), chap. 9, pp. 702–703.

25. The trend has been corrected for glacial isostasy. Church and White (2011); IPCC (2007a, b).

26. Church and White (2011); Cazenave and Llovel (2010); Beckley et al. (2007).

27. Cazenave and Llovel (2010).

28. Church and White (2011).

29. Zemp et al. (2006).

30. Huss et al. (2010). Although Swiss glaciers have lost nearly half their volume since 1910, a significant portion derives from natural climate cycles, such as the decades-long Atlantic Multidecadal Oscillation (AMO), which affects climate patterns over much of the Northern Hemisphere. An increasing AMO index (especially since the 1980s) is associated with enhanced Alpine glacier melting.

31. IPCC (2007a), chap. 5, table 5.3.

32. Hock et al. (2009).

33. Cogley (2009); Meier et al. (2007).

34. Shepherd and Wingham (2007); Rignot et al. (2008); Witze (2008); van den Broeke et al. (2009).

35. Shepherd and Wingham (2007).

36. Rignot et al. (2011).

37. Rignot and Kanagaratnam (2006).

38. van den Broeke et al. (2009).

39. Tedesco, M. et al. (2011).

40. Mernild et al. (2009); Tedesco et al. (2008).

41. Khan et al. (2010).

42. Nettles and Ekström (2010).

43. Holland et al. (2008).

44. Rignot et al. (2010); Nick et al. (2009).

45. Steig et al. (2009).

46. Chen et al. (2009).

47. Steig et al. (2009).

48. De Angelis and Skvarca (2003); Rignot et al. (2004).

49. Cook et al. (2005).
50. Rignot et al. (2008).
51. Bamber et al. (2009).
52. Chen et al. (2009).
53. Naish et al. (2009).
54. Domingues et al. (2008).
55. Cazenave and Llovel (2010). See also table 6.2. The IPCC estimates that from 1993 to 2003 thermal expansion down to 3,000 meters raised sea level by 1.6 millimeters per year (2007a, chap. 5, p. 415).
56. Yin et al. (2009); Hu et al. (2009a).
57. Yin et al. (2009).
58. Church (2007); Quadfasel (2005); Bryden et al. (2005).
59. Milly et al. (2010).
60. Chao et al. (2008).
61. Cazenave and Llovel (2010).
62. IPCC (2007a); Domingues et al. (2008); Cazenave and Llovel (2010).
63. IPCC (2007a), p. 459.
64. Generally sediment cores from coastal salt marshes (e.g., see chap. 2).
65. Donnelly et al. (2004).
66. Gehrels et al. (2005).
67. Miller et al. (2009).
68. The high sea level trend in New Jersey may be attributed in part to sediment compaction and, for Atlantic City, groundwater withdrawal.
69. Gehrels et al. (2008).
70. Gornitz (1995); Shennan and Woodworth (1992).
71. Church and White (2011).
72. IPCC (2007a), chap. 4, p. 356.

CHAPTER 7: SEA LEVEL RISE ON A WARMING PLANET

1. http://www.ese.int/esaCP/SEMRAVANJTF_index_2.html; http://www.esa.int/esaCP/SEMD07EH1TF_index_2.html; http://nsidc.org/news/press/20090408_Wilkins.html.
2. Cook et al. (2005); Steig et al. (2009).
3. An ice shelf is a floating ice slab attached to the shoreline as a seaward extension of an ice sheet that is grounded (i.e., resting on) solid bedrock.
4. Cook et al. (2005); Rignot et al. (2004); De Angelis and Skvarca (2003).
5. Bamber (2009).
6. Vaughan (2008); Ivins (2009). See also chap. 6.
7. Wingham et al. (2009).
8. Holland et al. (2008).
9. Rignot et al. (2011); Pritchard et al. (2009).
10. Pritchard et al. (2009).
11. IPCC (2007a). See also previously, in chap. 6.
12. IPCC (2007a), pp. 812–822 and appendix 10.A.

13. Vermeer and Rahmstorf (2009); Horton et al. (2008); Rahmstorf et al. 2007.

14. Rignot et al. (2011); Pritchard et al. (2009); Shepherd and Wingham (2007); Velicogna and Wahr (2006).

15. Cazenave and Llovel (2010).

16. Rignot and Kanagaratnam (2006); Vaughan (2008); Wingham et al. (2009).

17. Vermeer and Rahmstorf (2009); Horton et al. (2008); Rahmstorf (2007).

18. Vermeer and Rahmstorf (2009).

19. Otto-Bliesner et al. (2006).

20. IPCC (2007); Kopp et al. (2009); see also chap. 4.

21. Otto-Bliesner et al. (2006).

22. Overpeck et al. (2006).

23. This calculation may be misleading in that it doesn't cover the entire sea level cycle from the end of the penultimate deglaciation to that of the last interglacial. Average rates of sea level rise over the length of the penultimate deglaciation resemble those of the latest deglaciation, lasting over 10,000 years, i.e., 10–13 millimeters per year (see Rohling et al. 1998).

24. Rohling et al. (2008); see also chap 4.

25. Otto-Bliesner et al. (2006).

26. Stanford et al. (2006; 2011); see also chap. 5.

27. Naish et al. (2009).

28. IPCC (2007a), p. 830.

29. Duplessy et al. (2007).

30. Grove (1988).

31. Meier et al. (2007).

32. Pfeffer et al. (2008).

33. Long (2009).

34. Semi-empirical calculations find a global sea level rise between 1 and 2 meters. Vermeer and Rahmstorf (2009); Rahmstorf et al. 2007; Horton et al. (2008). Recall, however, that local sea level can diverge widely from this—either higher or lower—for any of the reasons described in chaps. 2 and 6.

35. Hansen (2007).

36. Nick et al. (2009).

37. Holland et al. (2008).

38. Bell (2008a, 2008b); Kohler (2007).

39. Anderson (2007).

40. Pritchard et al. (2009).

41. Pritchard et al. (2009).

42. Rignot et al. (2008); Chen et al. (2009).

43. Duplessy et al. (2007).

44. Wingham et al. (2009).

45. Vaughan (2008).

46. Chap. 6; Church (2007); Bryden et al. (2005); Quadfasel (2005).

47. Yin et al. (2009); Hu et al. (2009a).

48. Hu et al. (2009a).

49. Thermal inertia is a measure of the responsiveness of a substance to variations in temperature. Materials with a high thermal inertia, such as water, will take longer to warm and cool as the temperature rises and falls during the diurnal cycle and over lon-

ger time periods (e.g., seasons). It is for this reason, for example, that ocean water at the beach is warmest in late August and early September, rather than in late June, at the time of the summer solstice. Similarly, the areal extent of Arctic sea ice reaches its minimum in September.

50. IPCC (2007a), pp. 822–823.
51. IPCC (2007a), pp. 828–829.
52. Zemp et al. (2006).
53. IPCC (2007a), p. 829.
54. IPCC (2007a), p. 829.
55. IPCC (2007a), p. 830.
56. IPCC (2007a), p. 830.
57. Pollard and DeConto (2009).
58. Pfeffer et al. (2008).

CHAPTER 8: SHORELINES AT RISK

1. After the ancestral Atlantic Ocean split apart and widened (chap. 2), ocean crust slowly cooled and subsided, enabling sediments to accumulate on the Atlantic coastal plain. However, some segments of the coastal plain sank more than others. For example, Chesapeake Bay lies within the Salisbury-Raritan Embayment. This long-term geological subsidence, to a small extent, may be adding to the above-average regional sea level rise.
2. Poag (1997); http://woodshole.er.usgs.gov/epubs/bolide/index.html).
3. Leatherman (1992).
4. Gibbons and Nicholls (2006).
5. Gibbons and Nicholls (2006).
6. Kearney and Stevenson (1991); also chap. 6.
7. Mills, Chung, and Hancock (2005).
8. Kearney et al. (2002).
9. Hawley (1949).
10. Between the 1850s and 1900, major hurricanes (3 or higher on the Saffir-Simpson Hurricane Scale [table 8.1]), struck the Savannah, Georgia–Charleston, South Carolina coastal region on September 8–9, 1854 (the "Great Carolina" hurricane); on August 25, 1885; on August 28, 1893 (the "Sea Islands" hurricane); and on October 13, 1893). Multiple tropical storms and weaker hurricanes traversed the region in the intervening years (Blake et al. 2007; http://www.nhc/noaa.gov/Deadliest_Costliest.shtml; http://weather.unisys.com/hurricanes/atlantic; http://www.hurricaneville.com/historic.html).
11. Chopra (2009).
12. Pilkey and Young (2009), p. 141.
13. Blum and Roberts (2009).
14. Martinez et al. (2006).
15. Penland et al. (2005).
16. "Climate Change in Kiribati." Office of the President, Government of the Republic of Kiribati, http://www.climate.gov.ki/Impact_on_coastal_areas.html and "Impact of Climate Change on Low Islands: Tarawa Atoll, Kiribati," chap. 4 in *Cities, Seas, and Storms: Managing Change in Pacific Island Economies*, vol. 4, *Summary Report*. 2000, World Bank: Washington, D.C. http://www.worldbank.org.

17. Allen (2004).

18. Bateman (2008).

19. Kwok and Rothrock (2009); Comiso et al. (2008).

20. Elsner et al. (2008).

21 IPCC (2007c), pp. 94, 235, 321. Bleaching is a whitening of the coral resulting from loss of symbiotic algae and/or their pigments.

22. Viles and Spencer (1995); IPCC (2007c), chap. 1, pp. 92–94; chap. 6, pp. 318–330.

23. LIDAR (Light Detection and Ranging) is similar to radar but operates at ultraviolet, visible, or near-infrared wavelengths. Mounted on an aircraft, in conjunction with Global Positioning System (GPS) and Inertial Measurement Unit (IMU), the LIDAR instrument determines surface elevations by sending laser pulses to a target and recording the time it takes for the pulse to return to the sensor. Operating at altitudes of 300–2,000 meters, LIDAR can achieve accuracies of a meter or less in the horizontal direction and approximately 15 centimeters vertically (e.g., see http://www.csc.noaa.gov/crs/rs_apps/sensors/lidar.html).

24. Titus (2005), pp. 838–839.

25. Gornitz (1991); IPCC (2007c), chap. 6, pp. 318–330.

26. A large percentage of the Netherlands lies below sea level, and the Dutch have constructed an elaborate system of dikes, polders, and levees to ensure that their country will never again be overwhelmed by disastrous floods like those of 1953.

27. See also Syvitski et al. (2005). Dams, in fact, retain enough water to reduce sea level rise by 0.5 millimeters per year (Chao et al. 2008).

28. Valiela et al. (2001).

29. Boesch et al. (1994).

30. Kearney et al. (2002); Hartig et al. (2002); Hartig and Gornitz (2005).

31. Cahoon et al. (2009), pp. 63–64.

32. Viles and Spencer (1995), pp. 217, 221–222.

33. IPCC (2007c), pp. 94, 235.

34. IPCC (2007c), p. 320. pH is a measure of acidity, more specifically defined as the negative logarithm of the H^+ ion concentration of the solution. A neutral solution has a pH of 7; lower values are increasingly acidic, higher values more alkaline. For comparison, the pH of stomach acid is 1–2, lemon juice 2, vinegar 2–3, distilled water 7, baking soda 8, surface seawater 8.1, milk of magnesia 10.5, and household ammonia 11.

35. Hall-Spencer et al. (2008).

36. Titus (2005), p. 843.

37. Chopra (2009).

38. Clarke (2003).

39. Bird (2008), p. 202.

40. Zhang et al. (2004).

41. E.g., Hurricanes Lili (2002, Cat. 1), Ivan (2004, Cat. 3), Katrina (2005, Cat. 3), Rita (2005, Cat. 3), Gustav (2008, Cat. 2), and Ike (2008, Cat. 2).

42. Taylor et al. (2004).

43. IPCC (2007c), p. 92.

44. Bruun (1962).

45. Cooper and Pilkey (2004).

46. Titus (2005), p. 839.

47. Bird (2008), p. 67.

48. Kwok and Rothrock (2009); Comiso et al. (2008).
49. E.g., Forbes et al. (2004).
50. Forbes (2005).
51. Sand or gravel moved along by currents that flow more or less parallel to the coast.
52. Bird (2008), p. 210.
53. http://www.nps.gov/asis/naturescience/upload/ProjectIntroduction.pdf; http://www.newworldencyclopedia.org/entry/Assateague_Island (last modified Sept. 1, 2009), http://www.epa.gov/climatechange/wycd/downloads/Cs_ches.pdf.
54. Gornitz et al. (2002).
55. IPCC (2007c), pp. 92–94.
56. Penland (2005).
57. Sills et al. (2008).
58. Dean (1999); Blake et al. (2007).
59. Lake (1999).
60. Fritz et al. (2009).
61. See chap. 1.
62. E.g., Elsner et al. (2008); Emanuel (2005); Webster et al. (2005).
63. Landsea (2007).
64. Adjusted for inflation to 2006 (Blake et al. 2007, table 3b).
65. Pielke et al. (2008).
66. E.g., Donnelly and Woodruff (2007). A curious reader may wonder about what effects ENSO, a Pacific Ocean–based phenomenon, has on Atlantic Ocean hurricanes. The trade winds hold the answer. During La Niña or neutral years, the trades carry low-pressure waves off the coast of West Africa toward the Caribbean, where higher ocean surface temperatures promote development into tropical storms and hurricanes. During an El Niño, however, as the easterly trade winds slacken, the westerly upper wind flow of the Walker cell strengthens (chap. 1). This sets up a vertical wind shear that tends to tear an incipient hurricane apart.
67. Carbognin et al. (2009).
68. The Environment Agency, 2010. http://www.environment-agency.gov.uk/research/library/publications/41065.aspx (last updated March 29, 2010). The Barrier is shut whenever a forecast predicts a combined surge, tide, and river flow above 4.85 meters at the London Bridge. River embankments in central London typically range from 5 to 6 meters above the Ordnance Datum Newlyn, ODN—the British national datum (see chap. 1 for a discussion of tidal datums); the Thames Barrier stands 7.2 meters above ODN.
69. Recognizing the need to prepare for climate change, the London Environment Agency has prepared a document—Thames Estuary 2100—outlining its plans to strengthen the tidal defense systems. See www.environment-agency.gov.uk .
70. Zhang et al. (2000).
71. The current New York City 100-year-flood height is 8.6 feet. Projected sea level rise increases this figure to 9.6–10.5 feet by the 2080s. See Horton et al. 2010.
72. Winchester (2003). The Krakatau eruption may have also put a damper on the start of the 20th-century speeded-up sea level rise, by cooling the upper ocean (e.g., see Gleckler et al. 2006). By contrast, any ocean cooling due to recent volcanic eruption has been largely offset by the anthropogenic warming trend.
73. Gornitz (2005b), p. 681.
74. Pearce (2009).

75. Beget and Addison (2007).

76. Westbrook et al. (2009).

CHAPTER 9: COPING WITH THE RISING WATERS

1. Carbognin et al. (2009).
2. Carbognin et al. (2009).
3. McGranahan et al. (2007).
4. McGranahan et al. (2007).
5. Syvitski et al. (2009).
6. Syvitski et al. (2009).
7. Hanson et al. (2011).
8. Hanson et al. (2011). The 0.5-meter sea level rise by 2070 is derived from Rahmstorf (2007) (see chap. 7 in this book). Rahmstorf finds that sea level could reach 1.4 meters above present levels by 2100.
9. As designated by FEMA; Crowell et al. (2010).
10. Usery et al. (2010).
11. McGranahan et al. (2007).
12. Nicholls et al. (2008). A more recent estimate of sea level rise due to WAIS collapse is 3 meters; see Bamber et al. (2009); see chap. 7.
13. Sugiyama et al. (2008).
14. After 2100, the quadratic rise in sea level will overtake the linear case. However, if protection for a 1-meter rise is installed prior to 2100, cumulative costs will be reduced, providing an incentive for advance foresight.
15. "The Netherlands" literally means "the low lands." Approximately 40 percent of the country lies below sea level and a much higher percentage is vulnerable to coastal or riverine flooding. The lowest areas are highly populated and encompass the bulk of the nation's economic activity.
16. Aerts (2009), pp. 32–33.
17. Delta Commissie (Committee) (2008).
18. Delta Commissie (Committee) (2008). The Delta Commissie upper-bound sea level rise scenario uses the highest IPCC emission scenario (A1F1); includes thermal expansion and meltwater components, recent trends in accelerated ice discharge, and local land subsidence; but excludes effects of gravitational changes.
19. Thames Barrier. Retrieved from http://en.wikipedia.org/wiki/Thames_Barrier (accessed October 19, 2010).
20. Thames Barrier. Retrieved from http://en.wikipedia.org/wiki/Thames_Barrier (accessed October 19, 2010). Thames Barrier closures—indicator two. The Environment Agency, http://www.environment-agency.gov.uk/research/library/publications/41065.aspx.
21. Sea level rise + tide + surge.
22. Hanson et al. (2011).
23. Horton et al. (2010); Gornitz et al. (2002).
24. Hill (2008).
25. Bowman et al. (2004).
26. Valverde et al. (1999).

27. Gornitz et al. (2002).
28. Gornitz et al. (2002).
29. Titus and Craghan (2009).
30. Aboutaleb (2009), pp. 24–25.
31. Most of the giant container ships entering the New York harbor now dock at Port Newark and Port Elizabeth. The few remaining Manhattan piers are chiefly used by cruise ships.
32. Nordenson et al. (2010).
33. Heering et al. (2010). Char is newly accreted land in the delta formed by sediments deposited by rivers. The Forest Department, Bangladesh imposes a 20-year period for planting and growth of mangroves on the char, but due to land scarcity, the new forests are often encroached upon and settled before the 20-year period has elapsed.
34. http://www.ncrec.state.nc/us/publications-bulletins/coastal.html/#informed.
35. If the property cannot be developed, or if construction is strictly limited by the setback, the owner can claim that it constitutes a "taking," i.e., government seizure of private property without due compensation.
36. E.g., Titus and Craghan (2009), pp. 95–96.
37. Chap. 9, "Sounding Retreat," in Pilkey and Young (2009).
38. De la Vega-Leinert and Nicholls (2008).
39. Rupp-Armstrong and Nicholls (2007).
40. De la Vega-Leinert and Nicholls (2008).
41. Tunstall and Tapsell (2007).
42. See Answers to Questions About the NFIP. FEMA F-084, March 2010, www.fema.gov/about/programs/nfip/index.shtm.
43. Titus and Neumann (2009), pp. 151–156.
44. Rijcken (2010); van Veelen (2010). The "base flood level" is a minimal safe level above which new buildings and infrastructure must be built to protect them from flooding.
45. Aerts (2009), pp. 77–81.
46. Hanson et al. (2011).
47. Horton et al. (2010).
48. Central range = middle 67 percent of values from the model-based distribution.
49. Gornitz, Couch, and Hartig (2002).
50. Vision 2020: New York City Comprehensive Waterfront Plan, http://www.nyc.gov/waterfront/.
51. Kaufman (2010).

CHAPTER 10: CHARTING A FUTURE COURSE

1. Rohling et al. (2009); Brook (2008).
2. Mann et al. (2008); Osborn and Briffa (2006).
3. Schiermeier (2010).
4. Brook (2008).
5. Rohling et al. (2009); Brook (2008).
6. IPCC (2007a).

7. http://www.giss.nasa.gov/research/news/20100121/; http://www.giss.nasa.gov/research/news/20110113; http://www.giss.nasa.gov/20110112.

8. Nayar (2009).

9. "River of Ice: Vanishing Glaciers of the Greater Himalaya," Asia Society, New York City (July 13–August 15, 2010). The exhibit compared historic and contemporary photographs by David Breashears.

10. IPCC (2007a).

11. Beckley et al. (2007); Cazenave and Llovel (2010).

12. See Rahmstorf et al. (2007); Horton et al. (2008).

13. See Yin et al. (2009); Hu et al. (2009a).

14. See Bryden (2005); Quadfasel (2005).

15. See Chen et al. (2009); Holland et al. (2008); Pritchard et al. (2009).

16. Hansen (2007); Hansen et al. (2008).

17. Rohling et al. (2009).

18. Hansen (2007).

19. Robinson (2011).

20. Dowsett et al. (2009). See also chaps. 3, 7.

21. Zalasiewicz et al. (2008). See also Ward (2010).

22. Pfeffer et al. (2008).

23. IPCC (2007a), p. 824.

24. Archer and Brovkin (2008); Gillette et al. (2011).

25. Rockström et al. (2009); Hansen et al. (2008).

26. IPCC (2007b): Summary for Policymakers.

27. Although carbon dioxide does not produce the strongest warming per molecule (methane surpasses it 80-fold), as the most abundant anthropogenic atmospheric greenhouse gas it is the main target of greenhouse gas reduction programs.

28. Pacala and Socolow (2004).

29. PlaNYC—A Greener, Greater New York, April 2007. The City of New York. Mayor, Michael R. Bloomberg. http://www.nyc.gov/html/planyc2030/downloads/pdf/full_report.pdf

30. "Urban heat island" refers to the higher temperatures, especially at night, of an urban area as compared to its suburban or rural surroundings. Replacement of natural vegetation, or croplands, with heat-retaining materials and waste heat from buildings cause temperatures to rise.

31. PlaNYC—A Greener, Greater New York, April 2007. The City of New York. Mayor, Michael R. Bloomberg. http://www.nyc.gov/html/planyc2030/downloads/pdf/full_report.pdf.

32. Pittock (2005), p. 170.

33. Service (2004).

34. Demirdöven and Deutch (2004).

35. http://en.wikipedia.org/wiki/Hybrid_electric_vehicle (last updated February 7, 2011).

36. Flavin (2008). REN21, Renewables 2010 Global Status Report, p. 15. http://www.ren21.net/Portals/97/documents/GSR/REN21/_GSR_2010_full_revised%2020Sept2010.pdf ; IPCC (2007d), p. 278.

37. REN21, Renewables 2010 Global Status Report, p. 16.

38. IPCC (2007d), p. 278.
39. Climate change mitigation. http://en.wikipedia.org/wiki/Climate_change_mitigation.
40. IPCC (2007d), p. 278; http://en.wikipedia.org/wiki/Solar_power (last updated January 24, 2011); http://www.ere.energy.gov/basics/renewable_energy (last updated June 2010).
41. IPCC (2007d), p. 279.
42. REN21, p. 19.
43. Pittock (2005), p. 177.
44. REN21, p. 53.
45. Pittock (2005), p. 177.
46. http://www.wikipedia.org/wiki/Ethanol_fuel_in_Brazil#cite_note-BEN2009–23 (last updated January 25, 2011).
47. Ramanathan and Carmichael (2008).
48. For further discussion on the pros and cons of biofuel production, see http://www.wikipedia.org/wiki/Ethanol_fuel_in_Brazil#cite_note-BEN2009–23.
49. REN21, pp. 20, 23–24.
50. REN21, pp. 16–18.
51. Kirshenbaum and Webber (2010).
52. REN21, p. 17.
53. Kirshenbaum and Webber (2010).
54. Kilgannon (2011).
55. Syvitski et al. (2009); Bohannon (2010); see also chap. 8.
56. Hu et al. (2009b).
57. The Three Gorges Dam will also drown important historic and cultural landmarks.
58. Pittock (2005), p. 172; http://en.wikipedia.org/wiki/Nuclear_power/ (last updated January 13, 2011).
59. Pittock (2005), p. 172; http://en.wikipedia.org/wiki/Nuclear_power/ (last updated January 13, 2011).
60. Fukushima Daiichi Nuclear Disaster. http://en.wikipedia.org/wiki/Fukushima_Daiichi_nuclear_disaster (last updated July 10, 2011).
61. Willyard (2009).
62. Willyard (2009).
63. Willyard (2009).
64. Orr (2009).
65. Orr (2009).
66. See glossary.
67. For more on the Hellisheidi project, see http://www/or.is/English/Projects/CarbFix/AbouttheProject/ and http://www.ldeo.columbia.edu/news-events/turning-co2-stone
68. Schrag (2009).
69. Violent volcanic eruptions inject sulfur dioxide (SO_2) into the atmosphere, where it reacts with water vapor to produce sulfuric acid. The tiny acid droplets scatter sunlight, and depending on latitude and prevailing wind directions, may cool the Earth by a few tenths of a degree Celsius for several years, as happened after the 1991 Mount Pinatubo eruption in the Philippines.
70. Morton (2007).

71. Although home to over half the world's population, urban areas cover only a small percentage of the total land area.

72. E.g., Gavin Schmidt, Goddard Institute for Space Studies climatologist. RealClimate, http://www.realclimate.org/index.php/archives/2008/08/climate-change-methadone/.

73. Wayman (2011).

74. Emissions trading. http://en.wikipedia.org/wiki/Emissions_trading (last updated December 10, 2010).

75. Gronewold (2011).

76. The CCFE trades in carbon emissions, acid rain, mandatory regional greenhouse gas, and renewable energy markets.

77. European carbon emissions trading halted for three weeks following a major cyber theft in January 2011. The security breach undermined investor confidence.

78. Hoffert (2010).

79. As of 2008. REN21, p. 15.

80. Up to 192 parties, as of December 2009. http://en.wikipedia.org/wiki/United_Nations_Framework_Convention_on_Climate_Change (last updated Feb. 22, 2011).

81. As of July 2010, 191 nations, but not the United States, had signed and ratified the Kyoto Protocol. http://en.wikipedia.org/wiki/Kyoto_Protocol (last updated February 17, 2011).

82. Crutzen (2002); Zalasiewicz et al. (2008).

83. IPCC (2007b).

84. Ackerman et al. (2009).

85. E.g., Reid et al. (2010); Costanza et al. (2007).

86. Kram and Stehfest (2006).

87. E.g., see Reid et al. (2010).

88. Hall and Day (2009). Offshore drilling is deeper, more expensive, and riskier (e.g., the Gulf of Mexico oil spill in spring 2010).

89. REN21, p. 15 (excluding hydropower and wood burning).

90. Inman (2010).

91. Wen Jiabao, premier of China (which has become the world's leading CO_2 emitter), announced plans to lower its economic growth to 7 percent a year and to cut its energy intensity (energy used per unit of economic output) by 16–17 percent over the next five years. Wen added that the environment should no longer be sacrificed for "reckless . . . unsustainable growth" (J. Qiu, "China Announces Energy-Savings Plans," *Nature*, March 4, 2011. http://www.nature.com/news/2011/110304/full/news.2011.137.html). However, total energy consumption continues to rise, and government pronouncements are frequently ignored or evaded.

GLOSSARY

abyssal plains Smooth, deep ocean floor covered by thin sediment layers and microscopic marine shell oozes.

acqua alta (Italian, high water) Storm surge combined with high tide that leads to flooding in Venice.

albedo The fraction (or percentage) of radiation striking a surface that is reflected by the surface.

Anthropocene The period, beginning approximately with the Industrial Revolution, when humans have increasingly transformed the Earth's surface.

aphelion Farthest annual distance of the Earth to the Sun (presently July 4).

apogee Farthest distance in the lunar orbit around the Earth every 27.6 days.

aquifer A permeable underground rock layer, or layers, through which water held within interconnected pores can flow. An aquifer (for example a permeable sandstone), can be sandwiched between two impermeable layers (called **aquicludes**—such as shale) that prevent the water from leaking away.

Arctic Oscillation (AO) A seesaw pattern of atmospheric pressure between northern polar and mid-latitudes, similar to the NAO and NAM (below). During the positive phase, low polar latitude pressure and high mid-latitude pressure steer mid-latitude storms north, increasing rainfall in Alaska and northern Europe. During the negative phase, the pressure pattern reverses and outbreaks of cold air tend to grip North America and Europe.

asthenosphere A weak zone within the upper mantle, ~100 to 200 kilometers (62–124 miles) beneath the lithosphere, where rocks deform by plastic flow under pressure. It is the source of magmas that emerge as lava flows at the mid-ocean ridges.

astronomical theory of ice ages The theory that ice ages are caused by cyclical variations in the Earth's orbit. Also referred to as the Milankovitch theory, after one of its first proponents. (See also **eccentricity**, **obliquity**, and **precession**.)

backstripping A method of reconstructing past **eustatic** sea level from marine sediments by correcting for the effects of basin subsidence, sediment compaction, and loading.

basalt A dark, fine-grained volcanic rock made up largely of the minerals plagioclase feldspar and pyroxene, with or without olivine.

beach nourishment Replacing sand removed from a beach by erosion by dredging offshore sand and dumping it on the beach. Because of continued erosion, periodic episodes of sand nourishment may be needed to maintain the beach.

benthic foraminifera Bottom-dwelling marine organisms.

binge-purge theory Growth of an ice sheet to a critical threshold (binge phase) resulting in instability and rapid disintegration (purge phase). The instability may be caused by enhanced basal melting, rapid flow over subglacial lakes, or weakened buttressing of ice shelves.

bipolar seesaw An out-of-phase relationship in millennial climate variability between the Northern and Southern Hemispheres. The cause has been attributed to changes in relative strengths of hemispheric ocean deepwater formation and circulation following massive iceberg discharges.

breakwater A structure, attached to or parallel to the shore, that shelters a harbor or beach from extreme wave action.

Bruun rule As sea level rises, a beach will attempt to maintain its profile by eroding sand or gravel from the upper beach and depositing it on the adjacent seafloor, assuming no lateral addition or loss of sand. The net effect is an upward and landward shift in shoreline. Although widely used, the Bruun rule does not always apply, since longshore currents may add or remove sand, and storms may cause sand to wash over the beach, into tidal inlets, or lagoons.

bulkhead and seawall The former is a vertical wall to hold the shore in place, usually in residential areas and around marinas. The latter is built higher and stronger, usually in urban areas. It is designed to withstand overtopping and wave action by storms.

carbon offset A reduction in carbon dioxide or other greenhouse gas emissions in one place intended to offset or compensate for greenhouse gas emissions elsewhere.

continental rise A gentle incline at the base of the continental slope composed of sediment accumulations.

continental shelf Shallow seaward extension of a continent, with average depths of 130–135 meters (430–443 feet).

continental slope The steep slope between the outer margin of the continental shelf and continental rise.

Coriolis effect The deflection of winds and surface ocean currents due to the Earth's rotation. The motion is to the right in the Northern Hemisphere and to the left in the Southern Hemisphere.

cryosphere The frozen portions of the Earth's surface, including glaciers, ice sheets, permafrost, snow, and sea ice.

Dansgaard-Oeschger (D-O) event A sudden rapid climate warming within decades during the last glacial period, followed by a slower return to cold conditions. These oscillations recurred approximately every 500–4,000 years.

diatom Unicellular algae living in the ocean, coastal lagoons, or freshwater lakes. Sensitive to various environmental parameters (such as temperature, salinity, water chemistry), they are used to reconstruct past climates and sea levels.

dike A high, impermeable earthen wall intended to protect the hinterland against flooding.

eccentricity The degree of deviation of the Earth's elliptical orbit from a circle. It varies over roughly 100,000- and 400,000-year cycles.

ecliptic Plane of the Earth's orbit around the Sun.

El Niño-Southern Oscillation (ENSO) A quasi-periodic oscillation of the ocean-atmosphere system in the tropical Pacific manifesting as changes in air pressure, sea surface temperature, and wind and rainfall patterns, recurring approximately every 3 to 7 years.

efficiency (of electric power plants) The ratio of electrical energy generated to the energy content of coal (or other fuel) input.

eustasy A global change in sea level. The magnitude of the sea level change may vary from place to place (see also **relative sea level change**).

evaporites Minerals such as halite (rock salt) or gypsum, formed by evaporation of seawater or brines in an enclosed basin.

evapotranspiration The release of water to the atmosphere by evaporation from soils and transpiration by plants. Water is lost from plants through leaf openings, or stomata, especially during the day.

fetch Distance over which a wave travels before reaching the shore.

foraminifera (forams) Single-celled organisms that live in the ocean, brackish water, and lakes. Marine forams can be either **planktonic** or **benthic**. Many species have shells of calcite (calcium carbonate). Variations in oxygen isotope ratios in the shells of marine forams are used to infer past sea levels and ocean temperatures.

gas hydrate (also **methane hydrate,** or **clathrate)** An open-lattice form of ice in which cage-like spaces can accommodate gas molecules, usually methane, or less commonly carbon dioxide, or low molecular weight hydrocarbons. Stable only under high pressure and low temperature. Occurs in nature beneath permafrost and buried marine sediments.

geoengineering Large-scale deliberate anthropogenic modification of the environment, generally intended to lessen or neutralize greenhouse gas–induced warming.

geoid A surface of equal gravitational potential, corresponding to mean sea level in the absence of winds, tides, and ocean currents.

geostrophic current A current in which the force exerted by the Coriolis effect is balanced by the force of the horizontal pressure gradient.

glacial erratic A boulder or rock transported by a glacier or ice sheet that has been deposited far from its source, after the ice had melted.

glacial-interglacial cycle A roughly 100,000-year cycle that includes a glacial period of prolonged cold (glaciation) followed by a warm period (interglacial), during which ice sheets retreated.

glacial isostasy Changes in land elevation resulting from the added (or decreased) load of ice masses.

greenhouse effect, anthropogenic Theory that atmospheric greenhouse gases (e,g,, carbon dioxide, methane, nitrous oxide) generated by fossil fuel combustion, deforestation, and industrial activity are warming the Earth more than naturally occurring greenhouse gases (such as water vapor and carbon dioxide) are. These latter gases heat the Earth by 33°C (59°F) above the temperature that would exist in their absence.

groin Shore-perpendicular structure, usually built in series, designed to trap sand moving along the shore. Erosion is often enhanced downdrift of the groin field.

grounding line Boundary separating a mass of ice resting on solid bedrock from an attached ice shelf that floats on the adjacent sea.

groundwater mining Pumping out more water from the ground than is recharged through rainfall.

Heinrich event An abrupt cold period ending with a sudden warming. Heinrich events are marked by massive deposits of **ice-rafted debris** shed by "armadas" of icebergs. At least six such events occurred within the last 70,000 years, recurring roughly every 7,000 years and lasting around 500 years.

hurricane A tropical cyclone with sustained winds of at least 119 kilometers (74 miles) per hour, central pressure >979 millibars (28.91 inches), and a surge of 1.2–1.5 meters (4–5 feet) on the Saffir-Simpson Hurricane Scale. Also called typhoon in the Pacific Ocean and cyclone in the Indian Ocean.

hydro-isostasy Changes in ocean elevation resulting from the addition (or loss) of water due to changes in the mass of ice sheets.

hydrological cycle Movement of water between ocean and land reservoirs (lakes, rivers, soil, groundwater, glaciers) by means of evaporation and precipitation (as rain, snow, or hail).

ice age Periods of ice accumulation at the poles. Ice has covered both poles for the last 2.8 million years. The Antarctic ice cores record at least 8 interglacial-glacial cycles within the past 800,000 years, during which the ice sheets retreated and expanded.

ice-rafted debris (IRD) Sediment or pebbles transported by floating ice and deposited in a body of water.

iceberg calving Process by which ice masses break off the mouth of a glacier into the sea. The resulting icebergs essentially become floating islands.

ice sheet A large ice mass covering an extensive area. The major ice sheets today are on Antarctica and Greenland.

ice shelf Floating ice, usually the marine extension of a glacier. Ice shelves weaken by calving and basal (bottom) melting.

Intergovernmental Panel on Climate Change (IPCC) United Nations Environment Programme (UNEP)– and World Meteorological Organization (WMO)–sponsored international group of scientists and technical experts convened to assess the current state of climate change and its environmental and socioeconomic impacts. The IPCC issued reports in 1990, 1995, 2001, and 2007. The next one is due in 2014.

interstadial A warm period within a glacial cycle.

intertidal zone The shore area between mean high tide and mean low tide.

inverse barometer effect A rise in water level caused by low atmospheric pressure. As a rule of thumb, a drop in air pressure of 1 millibar raises sea level by around 1 centimeter.

isostasy A state of gravitational equilibrium between the earth's **lithosphere** and **asthenosphere**, in which the different elevations of continents and oceans largely reflect differences in lithospheric thickness and density. Isostatic changes occur by erosion, sediment loading, volcanism, plate collision, or growth or retreat of ice sheets (e.g., **glacial isostasy**).

jetty A shore-perpendicular structure built on the sides of a barrier island inlet, river, or lagoon entrance in order to stabilize the inlet as an aid to navigation.

Last Glacial Maximum The period near the end of the last ice age when ice sheets and glaciers reached their maximum global extent—roughly between 23,000 and 19,000 years ago.

Last Glacial Termination The end of the **last ice age**, between 20,000–19,000 years ago and 7,000 years ago.

last ice age (or glaciation) The cold period following the end of the **last interglacial**, between ~116,000 and 20,000–19,000 years ago.

last interglacial The warm period preceding the **last ice age** (or glaciation) lasting from ~130,000 to 116,000 years ago. Also called the Eemian (Europe), or Sangamonian (U.S.).

lava Molten rock erupted at the surface by a volcano or along a volcanic fissure.

lipid A class of bio-molecules naturally occurring in fats, waxes, and certain vitamins (e.g., vitamins A, D, E, and K). The TEX_{86} molecule, used as a paleoclimate proxy, is a type of lipid that contains 86 carbon atoms. It has been found to be empirically correlated with sea surface temperature.

lithosphere The rigid, strong crust and upper mantle above the **asthenosphere**. It varies in thickness from several to 100 kilometers (62 miles) beneath the ocean and 100–150 kilometers (62–93 miles) beneath continents.

Little Ice Age A period from 1400 to 1850 marked by Northern Hemisphere temperatures slightly below the millennial average.

longshore drift (littoral drift) Current-driven flow of sediment nearly parallel to the shore.

lunar nodal cycle An 18.6-year tidal cycle caused by changes in the inclination of the Moon's orbit with respect to the **ecliptic**.

magma Molten rock generated by partial melting of the Earth's lower crust or upper mantle.

Medieval Warm Period A period of relative warmth roughly between 950 and 1100, when mean Northern Hemisphere temperatures were several tenths of a degree Celsius above the 2,000-year hemispheric average. The medieval warmth had not been exceeded until recent decades.

meltwater pulse Period of very rapid sea level rise during the **Last Glacial Termination** due to major discharges of meltwater.

Meriodional Overturning Circulation (MOC) North-south deep ocean circulation driven by density differences (temperature, salinity), winds, and tides.

metamorphism Changes in the physical characteristics and mineral composition of rocks caused by high temperatures and pressures during tectonic deformation.

mid-ocean ridge A lengthy, submerged volcanic mountain chain encircling the world's oceans. The mid-ocean ridges constitute the source of lavas that create new ocean floor and produce seafloor spreading.

moraine A ridge of rocky debris deposited by a glacier at its leading edge, or along the sides of the glacier valley. **Recessional moraines** are ridges of rock and sediment left by the glacier in successive stages as it retreats up the valley. A **terminal moraine** is one deposited at the outermost edge of the glacier or ice sheet.

moulin (French, mill) A network of channels through a glacier formed by meltwater.

nor'easter An extratropical cyclone, most active between November and March, that affects the Atlantic Coast of North America.

North Atlantic Deep Water (NADW) Part of the Atlantic **Meridional Overturning Circulation (MOC)** where warm northward-flowing surface currents become cold, dense, saline water that sinks and thereby drives ocean circulation (the latter sometimes called the global ocean conveyor system).

North Atlantic Oscillation (NAO) Climate variability in the North Atlantic region related to the atmospheric pressure difference between Iceland and Lisbon, Portugal (or the Azores). The Northern Annular Mode (NAM) and the **Arctic Oscillation (AO)** are similar types of climate oscillation.

obliquity The angle between the Earth's axis of rotation and the perpendicular to the ecliptic (the plane of the Earth's orbit around the Sun). It is presently 23.4°, but has varied between 22° and 25° over ~41,000 years. This angle affects the contrast between seasons.

oxygen isotope variations Changes in the proportions of ^{18}**O** and ^{16}**O** in marine carbonate shells mainly due to changes in ice volume and ocean temperatures. Also affected by changes in ocean chemistry, or biological activity.

Panama hypothesis A proposed explanation for the onset of North Hemisphere glaciation ~2.7 million years ago. According to one version of this hypothesis, the closure of the Isthmus of Panama initiated major changes in ocean circulation, which strengthened the Gulf Stream and the North Atlantic Deepwater Formation. This in turn led to a buildup of snow and ice in Greenland.

peridotite A coarse-grained deep-seated igneous rock consisting largely of olivine, with other minerals such as pyroxenes, amphiboles, or garnets. It is formed in the Earth's upper mantle.

perigee Closest distance in the lunar orbit around the Earth every 27.6 days

perihelion The closest annual distance of the Earth to the Sun (presently January 3).

period The time interval between successive crests or troughs of an ocean wave.

permafrost Permanently frozen ground, where temperatures remain below 0°C for two consecutive winters and the intervening summer.

photoelectric effect The ejection of electrons from matter by absorbing electromagnetic radiation, such as ultraviolet or visible light. This process forms the basis of photovoltaic cells.

planktonic Free-floating microscopic marine organisms (e.g., some foraminifera and diatoms).

plate tectonics The theory that the Earth's **lithosphere** is divided into a number of tectonic plates that move horizontally over a softer, plastically deforming **asthenosphere**. Plates move apart from the **mid-ocean ridge system**, leading to **seafloor spreading**, and converge at **subduction** zones.

polyculture In agriculture, growing a mixture of crops in a given field (to reduce disease or pest problems), in contrast to monoculture—planting a single crop only.

precession of the axis The force of the Moon and Sun on the Earth's equatorial bulge creates a wobble in the rotational axis that varies over a ~26,000 period.

precession of the equinoxes The change in timing of perihelion, when the Earth is closest to the Sun, with respect to the seasons, over a ~22,000-year cycle. Today, perihelion nearly coincides with the Northern Hemisphere winter solstice; 11,000 years ago, it coincided with the Northern Hemisphere summer solstice.

reference ellipsoid A flattened sphere approximating the Earth's shape that includes the equatorial bulge and polar flattening caused by rotation. It is used as a reference surface for measuring sea level.

relative (local) rate of sea level rise Spatial variations in the rate of sea level rise caused by ongoing glacial isostatic adjustments, tectonic movements, subsurface fluid extraction, or seasonal to interannual oceanographic processes (e.g., ENSO).

revetment or riprap A sloping embankment of loose rocks or concrete blocks intended to prevent shore erosion or flooding.

roches moutonnées Asymmetric rock outcrops sculpted by glacial erosion. Smoothly streamlined on the side of the oncoming glacier, they are shattered and jagged on the lee flank.

saltwater intrusion Replacement of freshwater by saline water in a coastal aquifer due to sea level rise, drought, or groundwater over-pumping.

seafloor spreading The process whereby new ocean floor is created at mid-ocean ridges, separating two tectonic plates and leading to an expansion of the seafloor.

sea ice Slabs of ice floating on the surface of the sea.

sequence boundary An **unconformity** (time gap) in the sequence of rock strata, generally interpreted as a fall in sea level and/or a cessation of sedimentation due to erosion or exposure to the elements.

sequence stratigraphy A method of interpreting stratigraphic sequences bounded by **unconformities,** or temporal breaks in the stratigraphic succession, in terms of rising and falling sea levels. Originally based on seismic reflection studies of marine sediments, the methodology now also includes analysis of rock strata sequences on land.

steric changes Changes in ocean water density (and ocean height) due to changes in temperature and/or salinity.

storm surge An increase in water level above that of the astronomical tide caused by high winds and low barometric pressure associated with a storm.

storm surge barrier A barrier or gate that is closed during a storm whenever the surge is expected to exceed a certain level.

subduction zone A zone where oceanic **lithosphere** (marine sediments, basaltic lavas and magmatic rocks) is thrust beneath less dense, more buoyant continents or volcanic island arcs.

swell Long-wavelength ocean waves, generated by distant storms, traveling great distances with little dissipation of energy.

thermal expansion Increase in ocean level due to a rise in water temperature.

thermal inertia A measure of the ability of a material to conduct and store heat. It denotes the ability of the material to store heat during the day and re-radiate it at night. Materials that have a high heat capacity, such as water, also generally have a high thermal inertia, showing only minor temperature changes during the diurnal cycle.

thermocline A diffuse thermal boundary between warmer upper ocean water and colder, deeper water.

thermohaline circulation Density-driven deep ocean circulation due to gradients in temperature and salinity.

tidal gates Barriers across streams that are opened or closed during the tidal cycle, allowing areas behind the gates to drain out during low tide.

tsunami Sea wave of long period triggered by a submarine earthquake, volcanic eruption, or massive underwater landslide. Fast-moving and barely noticeable in the ocean, these waves can reach tremendous heights (over 30 meters [100 feet]) at the shore.

unconformity The contact surface representing a time gap in the geologic sequence of rock layers. An unconformity is generally caused by a combination of land uplift, erosion, and absence of sedimentation.

wave height The difference in height (or amplitude) between the crest and trough of an ocean wave. Wave energy is proportional to the square of wave height, which determines its erosive power.

wavelength The distance from one crest (or trough) to the next crest (or trough) of an ocean wave.

wave run-up The forward rush of a wave advancing up the beach.

West Antarctic Ice Sheet (WAIS) Portion of the Antarctic Ice Sheet that lies west of the Transantarctic Mountains. It is potentially unstable because much of the ice sits below sea level and slopes down inland.

Bibliography

Aboutaleb, A. 2009. "Rising Water Level Approach Offers Opportunities." *Change Magazine: Water and Climate* 5 (1): 4–5.

Ackerman, F., DeCanio, S. J., Howarth, R. B., and Sheeran, K. 2009. "Limitations of Integrated Assessment Models of Climate Change." *Climatic Change* 95:297–315.

Adams, J., compiler. 2002. *Europe During the Last 150,000 Years.* http://www.esd.ornl.gov/projects/qen/nercEUROPE.html

Aerts, J., Major, D. C., Bowman, M. J., Dircke, P., and Marfai, M. A. 2009. *Connecting Delta Cities: Coastal Cities, Flood Risk Management, and Adaptation to Climate Change.* Amsterdam: Vrije Universiteit Press.

Aksu, A. E., Mudie, P. J., Rochon, A., Kaminski, M. A., Abrajano, T., and Yasar, D. 2002. "Persistent Holocene Outflow from the Black Sea to the Eastern Mediterranean Contradicts Noah's Flood Hypothesis." *GSA Today* 12 (5): 4–10.

Allen, L. 2004. "Will Tuvalu Disappear Beneath the Sea?" *Smithsonian Magazine* 35 (5): 44–53.

Alley, R. B. 2000. *The Two-Mile Ice Machine: Ice Cores, Abrupt Climate Change, and Our Future.* Princeton, N.J.: Princeton University Press.

Alley, R. B. and Ágústsdóttir, A. M. 2005. "The 8KY Event: Cause and Consequences of a Major Holocene Abrupt Climate Change." *Quaternary Science Reviews* 24:1123–1149.

Almasi, P. F. and Bond, G. C. 2009. "Sun-Climate Connections." In *Encyclopedia of Paleoclimatology and Ancient Environments,* ed. V. Gornitz, 929–935. Dordrecht: Springer.

Alvarez, W. 1997. *T. rex and the Crater of Doom.* Princeton, N.J.: Princeton University Press.

Ananthaswamy, A. 2008. "Once the South Pole Was Green . . . " *New Scientist*, June 21: 34–38.

Anderson, J. B. 2007. "Ice Sheet Stability and Sea-Level Rise." *Science* 315:1803–1804.

Archer, D. and Brovkin, V. 2008. "The Millennial Atmospheric Lifetime of Anthropogenic CO_2." *Climatic Change* 90:283–297.

Archibald, J. D. and 28 others. 2010. "Cretaceous Extinctions: Multiple Causes." *Science* 328:973.

Baker, V. "Glacial Megalakes." 2009. In *Encyclopedia of Paleoclimatology and Ancient Environments*, ed. V. Gornitz, 380–302. Dordrecht: Springer.

Bamber, J. L., Riva, R. E. M., Vermeersen, B. L. A., and LeBrocq, A. M. 2009. "Reassessment of the Potential Sea-Level Rise from a Collapse of the West Antarctic Ice Sheet." *Science* 324:901–903.

Barber, D. C., Dyke, A., Hillaire-Marcel, C., Jenning, A. E., Andrews, J. T., Kerwin, M. W., Bilodeau, G., McNeely, R., Southon, J., Morehead, M. D., and Gagnon, J.-M. 1999. "Forcing of the Cold Event of 8,200 Years Ago by Catastrophic Drainage of Laurentide Lakes." *Nature* 400: 344–348.

Bard, E., Hamelin, B., and Delanghe-Sabatier, D. 2010. "Deglacial Meltwater Pulse 1B and Younger Dryas Sea Levels Revisited with Boreholes at Tahiti." *Science* 327:1235–1237.

Bard, E., Hamelin, B., and Fairbanks, R. G. 1990. "U-Th Ages Obtained by Mass Spectrometry in Corals from Barbados: Sea Level During the Past 130,000 Years." *Nature* 346: 456–458.

Bartoli, G., Sarnthein, M., Weinelt, M., Erlenkeuser, H., Garbe-Schönberg, D., and Lea, D. W. 2005. "Final Closure of Panama and the Onset of Northern Hemisphere Glaciation." *Earth and Planet. Science Letters* 237:33–44.

Bassett, S. E., Milne, G. A., Mitrovica, J. X., and Clark, P. U. 2005. "Ice Sheet and Solid Earth Influences on Far-Field Sea-Level Histories." *Science* 309:925–928.

Bateman, S. 2008. "Sea Level Rise: Will the Maldives Disappear?" *RSIS Commentaries* 136/2008. S. Rajaratnam School of International Studies, Singapore, 3p.

Beckley, B. D., Lemoine, F. G., Luthcke, B., Ray, R. D., and Zelensky, N. P. 2007. "A Reassessment of Global and Regional Mean Sea Level Trends from TOPEX and Jason-1 Altimetry Based on Revised Reference Frame and Orbits." *Geophysical Research Letters* 34: L14608, doi:10. 1029/2007GL030002.

Beget, J. E. and Addison, J. A. 2007. "Methane Gas from the Storegga Submarine Landslide Linked to Early-Holocene Climate Change: A Speculative Hypothesis." *The Holocene* 17: 291–295.

Bell, R. E. 2008a. "The Unquiet Ice." *Scientific American* 298 (2): 60–67.

Bell, R. E. 2008b. "The Role of Subglacial Water in Ice-Sheet Mass Balance." *Nature Geoscience* 1:297–304.

Beus, S. S. and Morales, M., eds. 2003. *Grand Canyon Geology.* 2nd ed. New York: Oxford University Press.

Bevis, M. 2007. "Continuous GPS Positioning of Tide Gauges: Some Preliminary Considerations." *GLOSS Bulletin Editorial Issue No. 6.* http://uhslc. soest. hawaii. edu/uhslc/gpsd/bevis.html

Bindoff, N. L. and Willebrand, J. 2007. "Observations: Oceanic Climate Change and Sea Level." In *IPCC 2007. Climate Change 2007: The Physical Science Basis. Contribution of Working Group I to the Fourth Assessment Report of the Intergovernmental Panel on Climate Change*. S. Solomon et al., eds. Cambridge, UK and New York, NY, USA: Cambridge University Press, 385–428.

Bintanja, R. and van der Wal, R. S. W. 2008. "North American Ice-Sheet Dynamics and the Onset of 100,000-Year Glacial Cycles." *Nature* 454:869–872.

Bird, E. 2008. *Coastal Geomorphology: An Introduction*, 2nd ed. Chichester: John Wiley and Sons.

Blake, E. S., Rappaport, E. N., and Landsea, C. W. 2007. "The Deadliest, Costliest, and Most Intense United States Tropical Cyclones from 1851 to 2006 (and Other Frequently Requested Hurricane Facts)." *NOAA Technical Memorandum NWS TPC-5*. Miami: National Weather Service.

Blanchon, P. and Shaw, J. 1995. "Reef Drowning During the Last Deglaciation: Evidence for Catastrophic Sea-Level Rise and Ice-Sheet Collapse." *Geology* 23:4–8.

Blum, M. D. and Roberts, H. H. 2009. "Drowning of the Mississippi Delta due to Insufficient Sediment Supply and Global Sea Level Rise." *Nature Geoscience* 2:488–491.

Boesch, D. F., Josselyn, M. M., Mehta, A. J., Morris, J. T., Nuttle, W. K., Simenstad, C. A., and Swift, D. J. P. 1994. "Scientific Assessment of Coastal Wetland Loss, Restoration, and Management in Louisiana." *Journal of Coastal Research Special Issue* 20:1–89.

Bohannon, J. 2010. "The Nile Delta's Sinking Future." *Science* 327:1444–1447.

Bond, G., Kromer, B., Beer, J., Muscheler, R., Evans, M. N., Showers, W., Hoffmann, S., Lott-Bond, R., Hajdas, I., and Bonani, G. 2001. "Persistent Solar Influence on North Atlantic Climate During the Holocene." *Science* 294:2130–2136.

Bowen, D. Q. 2009. "Last Glacial Maximum." In *Encyclopedia of Paleoclimatology and Ancient Environments*, ed. V. Gornitz, 498–495. Dordrecht: Springer.

Bowman, M. J., Colle, B. A., Flood, R., Hill, D., Wilson, R. E., Buonaiuto. F., Cheng, P., and Zheng, Y. 2004. "Hydrological Feasibility of Storm Surge Barriers to Protect the Metropolitan New York–New Jersey Region: Final Report." *Marine Sciences Research Center Technical Report*, Stony Brook University, 28p.

Brinkhuis, H., Crouch, E. M., Bohaty, S., John, C. M., Reichart, G.-J., Schouten, S., Zachos, J. C., Dickens, G. R., Sinninghe Damsté, J. P., and Lotter, A. F. 2006. "Eustatic Sea Level Rise During the Paleocene-Eocene Thermal Maximum." In *Global Change During the Paleocene-Eocene Thermal Maximum.*, ed. A. Sluijs. Utrecht, The Netherlands: LLP Foundation.

Broecker, W. 1997. "Will Our Ride Into the Greenhouse Future Be a Smooth One?" *GSA Today* 7 (5): 1–7.

Brook, E. 2008. "Windows on the Greenhouse." *Nature* 453:291–292.

Bruun, P. 1962. "Sea Level Rise as a Cause of Shore Erosion." *Proceedings of the American Society of Civil Engineers, Journal of the Waterways and Harbor Division*. 88:117–130.

Bryden, H. L., Longworth, H. P., and Cunningham, S. A. 2005. "Slowing of the Atlantic Meridional Overturning Circulation at 25N." *Nature* 438:655–657.

Cahoon, D. R., Reed, D. J., Kolker, A. S., and Brinson, M. M. 2009. "Coastal Wetland Sustainability." In *Coastal Sensitivity to Sea Level Rise: A Focus on the Mid-Atlantic Region, U.S. Climate Change Science Program Synthesis and Assessment Product 4. 1*, ed. J. G. Titus et al. Washington, D.C.: U.S. Government Printing Office.

Carbognin, L., Teatini, P., Tomasin, A., and Tosi, L. 2009. "Global Change and Relative Sea Level Rise at Venice: What Impact in Terms of Flooding." *Climate Dynamics* DOI 10. 1007/s00382–009–0617–5.

Carlson, A. E., Legrande, A. N., Oppo, D. W., Came, R. E., Schmidt, G. A., Anslow, F. S., Licciardi, J. M., and Obbink, E. A. 2008. "Rapid Early Holocene Deglaciation of the Laurentide Ice Sheet." *Nature Geoscience* 1: 620–624.

Cazenave, A. and Llovel, W. 2010. "Contemporary Sea Level Rise." *Annual Reviews of Marine Science* 2:145–173.

Chao, B. F., Wu, Y. H., and Li, Y. S. 2008. "Impact of Artificial Reservoir Water Impoundment on Global Sea Level." *Science* 320:212–214.

Chappell, J. 2002. "Sea Level Changes Forced Ice Breakouts in the Last Glacial Cycle: New Results from Coral Terraces." *Quaternary Science Reviews* 21:1229–1240.

Chappell, J., Omura, A., Esat, T., McCulloch, M., Pandolfi, J., Ota, Y., and Pillans, B. 1996. "Reconciliation of Late Quaternary Sea Levels Derived from Coral Terraces at Huon Peninsula with Deep Sea Oxygen Isotope Records." *Earth and Planetary Science Letters* 141: 227–236.

Chen, J. L., Wilson, C. R., Blankenship, D., and Tabley, B. D. 2009. "Accelerated Antarctic Ice Loss from Satellite Gravity Measurements." *Nature Geoscience* 2: 859–862.

Cheng, H., Edwards, R. L., Broecker, W. S., Denton, G. H., Kong, X., Wang, Y., Zhang, R., and Wang, X. 2009. "Ice Age Terminations." *Science* 326:248–252.

Chopra, A. 2009. "Where Climate Change Hurts the Most." *U.S. News & World Report*, April 2009, pp. 52–53.

Christie-Blick, N., Mountain, G. S., and Miller, K. G. 1990. "Seismic Record of Sea-Level Change." In: *Sea Level Change.* National Research Council, 116–140. Washington, D.C.: National Academy Press.,

Church, J. A. 2007. "A Change in Circulation?" *Science* 317:908–909.

Church, J. A. and White, N. J. 2011. "Sea-Level Rise from the Late 19th to the Early 21st Century." *Surveys of Geophysics*, doi:10. 1007/s10712–011–9119–1.

Ciochon, R. L. and Bettis III, E. A. 2009. "Asian *Homo erectus* Converges in Time." *Nature* 458: 153–156.

Clague, J. J. 2009. "Cordilleran Ice Sheet." In *Encyclopedia of Paleoclimatology and Ancient Environments*, ed. V. Gornitz, 206–211. Dordrecht: Springer.

Clark, P. U., Archer, D., Pollard, D., Blum, J. D., Blum, J. D., Rial, J. A., Brovkin, V., Mix, A. C., Pisias, N. G., and Roy, M. 2006. "The Middle Pleistocene Transition: Characteristics, Mechanisms, and Implications for Long-Term Changes in Atmospheric pCO_2." *Quaternary Science Reviews* 25:3150–3184.

Clark, P. U., McCabe, A. M., Mix, A. C., and Weaver, A. J. 2004. "Rapid Rise of Sea Level 19,000 Years Ago and its Global Implications." *Science* 304:1141–1144.

Clark, P. U., Mitrovica, J. X., Milne, G.. A., and Tamisiea, M. E. 2002. "Sea-Level Fingerprinting as a Direct Test for the Source of Global Meltwater Pulse 1A." *Science*, 295:2438–2441.

Clark, P. U. and Mix, A. C. 2002. "Ice Sheets and Sea Level of the Last Glacial Maximum." *Quaternary Science Reviews* 21:1–7.

Clarke, G. K. C., Leverington, D. W., Teller, J. T., and Dyke, A. S. 2004. "Paleohydraulics of the Last Outburst Flood from Glacial Lake Agassiz and the 8200 BP Cold Event." *Quaternary Science Reviews* 23:389–407.

Clarke, T. 2003. "Delta Blues." *Nature* 422:254–256.

Coch, N. K. 1994. "Hurricane Hazards along the Northeastern Atlantic Coast of the United States." In *Coastal Hazards: Perception, Susceptibility and Mitigation. Journal of Coastal Research Special Issue No. 12*, pp. 115–147.

Cogley, J. G. 2009. "Geodetic and Direct Mass-Balance Measurements: Comparison and Joint Analysis." *Annals of Glaciology* 50:96–100.

Comiso, J. C., Parkinson, C. L., Gersten, R., and Stock, L. 2008. "Accelerated Decline in the Arctic Sea Ice Cover." *Geophysical Research Letters* 35:L01703, doi:10.1029/2007GL031972.

Cook, A. J., Fox, A. J., Vaughan, D. G.., and Ferrigno, J. G. 2005. "Retreating Glacier Fronts on the Antarctic Peninsula over the Past Half Century." *Science* 308:541–544.

Cooper, J. A. G. and Pilkey, O. H. 2004. "Sea-Level Rise and Shoreline Retreat: Time to Abandon the Bruun Rule." *Global and Planetary Change* 43:157–171.

Costanza, R., Leemans, R., Boumans, R., and Gaddis, E. 2007. "Integrated Global Models." In: *Sustainability or Collapse: An Integrated History and Future of People on Earth*, eds. Costanza, R., Graumlich, L. J., and Steffen, W., 417–446. Cambridge, MA: MIT Press.

Courtillot, V. and Fluteau, F. 2010. "Cretaceous Extinctions: The Volcanic Hyposthesis." *Science* 328:973–974.

Cox, A. 1973. *Plate Tectonics and Geomagnetic Reversal*. San Francisco: W. H. Freeman and Co.

Cronin, T. M. 1999. *Principles of Paleoclimatology*. New York: Columbia University Press.

Cronin, T. M. 2010. *Paleoclimates: Understanding Climate Change Past and Present*. New York: Columbia University Press.

Cronin, T. M., Vogt, P. R., Willard, D. A., Thunell, R., Halka, J., and Berke, M. 2007. "Rapid Sea Level Rise and Ice Sheet Response to 8,200-Year Climate Event." *Geophys. Res. Letters* 34:L20603.

Crowell, M., Coulton, K., Johnson, C., Westcott, J., Bellomo, D., Edelman, S., and Hirsch, E. 2010. "An Estimate of the U.S. Population Living in 100-Year Coastal Flood Hazard Areas." *Journal of Coastal Research* 26 (2): 201–211.

Crutzen, P. J. 2002. "Geology of Mankind." *Nature* 415:23.

Cutler, K. B., Edwards, R. L., Taylor, F. W., Cheng, H., Adkins, J., Gallup, C. D., Cutler, P. M., Burr, G. S., and Bloom, A. L. 2003. "Rapid Sea-Level Fall and Deep-Ocean Temperature Change Since the Last Interglacial Period." *Earth and Planetary Science Letters* 206: 253–271.

De Angelis, H. and Skvarca, P. 2003. "Glacier Surge After Ice Shelf Collapse." *Science* 299:1560–1562.

De la Vega-Leinert, A. C. and Nicholls, R. J. 2008. "Potential Implications of Sea-Level Rise for Great Britain." *Journal of Coastal Research* 24 (2): 342–357.

Dean, C. 1999. *Against the Tide: The Battle for America's Beaches*. New York: Columbia University Press.

DeConto, R. M. 2009. "Plate Tectonics and Climate Change." In *Encyclopedia of Paleoclimatology and Ancient Environments*, ed. V. Gornitz. Dordrecht: Springer.

DeConto, R. M. and Pollard, D. 2003. "Rapid Cenozoic glaciation of Antarctica Induced by Declining Atmospheric CO_2." *Nature* 421:245–249.

DeConto, R. M., Pollard, D., Wilson, P. A., Plike, H., Lear, C. H., and Pagani, M. 2008. "Thresholds for Cenozoic Bipolar Glaciation." *Nature* 455:652–656.

Delta Commissie. 2008. *Working Together with Water: A Living Land Builds for Its Future*. http://www.deltacommissie.com/doc/deltareport_full.pdf

Demirdöven, N. and Deutch, J. 2004. "Hybrid Cars Now, Fuel Cell Cars Later." *Science* 305: 974–977.

Dickens, G. R. and Forswall, C. 2009. "Methane Hydrates, Carbon Cycling, and Environmental Change." In *Encyclopedia of Paleoclimatology and Ancient Environments*, ed. V. Gornitz. Dordrecht: Springer.

Dickinson, W. R. 2004. "Impacts of Eustasy and Hydro-Isostasy on the Evolution and Landforms of Pacific Atolls." *Palaeogeography, Palaeoclimatology, Palaeoecology* 213:251–269.

Domingues, C. M., Church, J. A., White, N. J., Gleckler, P. J., Wijffels, S. E., Barker, P. M., and Dunn, J. R. 2008. "Improved Estimates of Upper-Ocean Warming and Multidecadal Sea-Level Rise." *Nature* 453:1090–1093.

Donnelly, J. P., Cleary, P., Newby, P., and Ettinger, R. 2004. "Coupling Instrumental and Geological Records of Sea-Level Change: Evidence from Southern New England of an Increase in the Rate of Sea-Level Rise in the Late 19th Century." *Geophysical Research Letters* 31:L05203 doi:10. 1029/2003GL018933.

Donnelly, J. P. and Woodruff, J. D. 2007. "Intense Hurricane Activity Over the Past 5,000 Years Controlled by El Niño and the West African Monsoon." *Nature* 447:465–468.

Douglas, B. 2001. "An Introduction to Sea Level." In *Sea Level Rise: History and Consequences*, ed. B. C. Douglas, M. S. Kearney, and S. P. Leatherman, 1–11. San Diego: Academic Press.

Dowsett, H. J., Chandler, M. A., and Robinson, M. M. 2009. "Surface Temperatures of the Mid-Pliocene North Atlantic Ocean: Implications for Future Climate." *Philosophical Transactions of the Royal Society* A 367:69–84.

Dowsett, H. J. and Cronin, T. M. 1990. "High Eustatic Sea Level During the Middle Pliocene: Evidence from the Southern U.S. Atlantic Coastal Plain." *Geology* 18: 435–438.

Duplessy, J. C., Roche, D. M., and Kageyama, D. M. 2007. "The Deep Ocean During the Last Interglacial." *Science* 316:89–91.

Dutton, A., Bard, E., Antonioli, F., Esat, T. M., Lambeck, K., and McCulloch, M. T. 2009. "Phasing and Amplitude of Sea-Level and Climate Change During the Penultimate Interglacial." *Nature Geoscience* 2:355–359.

Dwyer, G. S. and Chandler, M. A. 2009. "Mid-Pliocene Sea Level and Continental Ice Volume Based on Coupled Benthic Mg/Ca Paleotemperatures and Oxygen Isotopes." *Philosophical Transactions of the Royal Society* A 367:57–168.

Ekman, M. 1999. "Climate Changes Detected Through the World's Longest Sea Level Series." *Global and Planetary Change* 21:215–224.

Elsner, J. B., Kossin, J. P., and Jagger, T. H. 2008. "The Increasing Frequency of the Strongest Tropical Cyclones." *Nature* 455:92–95.

Emanuel, K. 2005. "Increasing Destructiveness of Tropical Cyclones Over the Past 300 Years." *Nature* 436:686–688.

Fairbanks, R. G. 1989. "A 17,000-Year Glacio-Eustatic Sea Level Record: Influence of Glacial Melting Rates on the Younger Dryas Event and Deep-Ocean Circulation." *Nature* 342:637–642.

Finkl, C. W. Jr., ed. 1995. "Holocene Cycles: Climate, Sea Levels, and Sedimentation." *Journal of Coastal Research Special Issue No. 17.* 402p.

Firestone, R. B. and 25 others. 2007. "Evidence for an Extraterrestrial Impact 12,900 Years Ago That Contributed to the Megafauna Extinctions and the Younger Dryas Cooling." *Proceedings of the National Academy of Sciences* 104:16016–16021.

Flavin, C. 2008. "Simulating a Clean Energy Revolution." In *Sudden and Disruptive Climate Change*, ed. M. C. McCracken, F. Moore, and J. C. Topping, Jr., 241–246. London: Earthscan.

Forbes, D. L. 2005. "Coastal Erosion." In *Encyclopedia of the Arctic*, ed. M. Nuttall. New York and London: Routledge.

Forbes, D. L., Parkes, G. S., Manson, G. K., and Ketch, L. A., 2004. "Storms and Shoreline Retreat in the Southern Gulf of St. Lawrence." *Marine Geology* 210:169–204.

Fox, D. 2008. "Freeze-Dried Findings Support a Tale of Two Ancient Climates." *Science* 320:1152–1154.

Fritz, H. M., Blount, C. D., Thwin, S., Thu, M. K., and Chan, N. 2009. "Cyclone Nargis Storm Surge in Myanmar." *Nature Geoscience* 2:448–449.

Garcia-Castellanos, D., Estrada, F., Jiménez-Munt, I., Gorini, C., Fernàndez, M., Vergés, J., and De Vicente, R. 2009. "Catastrophic Flood of the Mediterranean After the Messinian Salinity Crisis." *Nature* 462:778–782.

Gehrels, R. W., Kirby, J. R., Prokoph, A., Newnham, R. M., Achertberg, E. P., Evans, H., Black, S., and Scott, D. B. 2005. "Onset of Recent Rapid Sea-Level Rise in the Western Atlantic Ocean." *Quaternary Science Reviews* 24:2083–2100.

Gehrels, W. R., Hayward, B. W., Newnham, R. M., and Southall, K. E. 2008. "A 20th Century Acceleration of Sea-Level Rise in New Zealand." *Geophysical Research Letters* 35:L02717 doi:10.1029/2007GL032632, 5p.

Gibbons, S. J. and Nicholls, R. J. 2006. "Island Abandonment and Sea-Level Rise: An Historical Analog from the Chesapeake Bay, USA." *Global Environmental Change* 16:40–47.

Gillette, N. P., Arora, V. K., Zickfeld, K., Marshall, S. J., and Merryfield, W. J. 2011. "Ongoing Climate Change Following a Complete Cessation of Carbon Dioxide Emissions." *Nature Geoscience* 4:83–87.

Giosan, L., Filip, F., and Constatinescu, S. 2009. "Was the Black Sea Catastrophically Flooded in the Early Holocene?" *Quaternary Science Reviews* 28:1–6.

Gleckler, P. J., Wigley, T. M. L., Santer, B. D., Gregory, J. M., AchtaRao, K., and Taylor, K. E. 2006. "Krakatoa's Signature Persists in the Ocean." *Nature* 439:675.

Gleick, P. H. 1996. "Water Resources." In *Encyclopedia of Climate and Weather*, vol. 2, ed. S. H. Schneider, 817–823. New York: Oxford University Press.

Goebel, T., Waters, M. R., and O'Rourke, D. H. 2008. "The Late Pleistocene Dispersal of Modern Humans in the Americas." *Science* 319:1497–1502.

Gornitz, V. 1991. "Global Coastal Hazards from Future Sea Level Rise." *Global and Planetary Change* 89:379–398.

Gornitz, V. 1995. "A Comparison of Differences Between Recent and Late Holocene Sea-Level Trends from Eastern North America and Other Selected Regions." *Journal of Coastal Research* Special Issue No. 17:287–297.

Gornitz, V. 2001. "Impoundment, Groundwater Mining, and Other Hydrologic Transformations: Impacts on Global Sea Level Rise." In *Sea Level Rise: History and Consequences*, ed. B. Douglas, M. S. Kearney, and S. P. Leatherman, 97–119. San Diego: Academic Press,

Gornitz, V. 2005a. "Storm Surge." In *Encyclopedia of Coastal Science*, ed. M. L. Schwartz, 912–914. Dordrecht: Springer.

Gornitz, V. 2005b. "Eustasy." In *Encyclopedia of Coastal Science*, ed. M. L. Schwartz, 439–442. Dordrecht: Springer.

Gornitz, V. 2005c. "Natural Hazards." In *Encyclopedia of Coastal Science*, ed. M. L. Schwartz, 678–684. Dordrecht: Springer.

Gornitz, V. 2009. "Sea Level Change, Post-Glacial." In *Encyclopedia of Paleoclimatology and Ancient Environments*, ed. V. Gornitz, 887–893. Dordrecht: Springer.

Gornitz, V., Couch, C., and Hartig, E. K. 2002. "Impacts of Sea Level Rise in the New York City Metropolitan Area." *Global and Planetary Change* 32:61–88.

Gronewold, N. 2011. "Chicago Climate Exchange Closes Nation's First Cap-and-Trade System but Keeps Eye to the Future." *New York Times*, January 3.

Grotzinger, J., Jordan, T. H., Press, F., Siever, R. 2006. *Understanding Earth*. 5th ed. New York: W. H. Freeman and Co.

Grove, J. M. 1988. *The Little Ice Age*. London and New York: Routledge.

Hall, C. A. S. and Day, J. W., Jr. 2009. "Revisiting the Limits to Growth After Peak Oil." *American Scientist* 97:230–237.

Hallam, A. *Phanerozoic Sea-Level Changes*. 1992. New York: Columbia University Press.

Hall-Spencer, J. M., Rodolfo-Metalpa, R., Martin, S., Ransome, E., Fine, M., Turner, S. M., Rowley, S. J., Tedesco, D., and Buia, M.-C. 2008. "Volcanic Carbon Dioxide Vents Show Ecosystem Effects of Ocean Acidification." *Nature* 454:96–99.

Hancock, K. 2005. "Predictions of Relative Sea-Level Change and Shoreline Erosion Over the 21st Century on Tangier Island, Virginia." *Journal of Coastal Research* 21 (2): e36–e51.

Hanebuth, T., Stattegger, K., and Grootes, P. M. 2000. "Rapid Flooding of the Sunda Shelf: A Late-Glacial Sea-Level Record." *Science* 288:1033–1035.

Hanebuth, T. J. J., Stattegger, K., and Bojanowski, A. 2009. "Termination of the Last Glacial Maximum Sea-Level Lowstand: The Sunda-Shelf Data Revisited." *Global and Planetary Change* 66:76–84.

Hansen, J. 2007. "Scientific Reticence and Sea Level Rise." *Environmental Research Letters* 2:doi:10. 1088/1748–9326/2/2/024002.

Hansen, J., Sato, M., Kharecha, P., Beerling, D., Masson-Delmotte, V., Pagani, M., Raymo, M., Royer, D., and Zachos, J. C. 2008. "Target Atmospheric CO_2: Where Should Humanity Aim?" *Open Atmospheric Science Journal* 2:217–231.

Hanson, S., Nicholls, R., Ranger, N., Hallegatte, S., Corfee-Morlot, J., Herweijer, C., and Chateau, J. 2011. "A Global Ranking of Port Cities with High Exposure to Climate Extremes." *Climatic Change* 104:89–111.

Haq, B. U., Hardenbol, J., and Vail, P. R. 1987. "Chronology of Fluctuating Sea Levels Since the Triassic." *Science* 235:1156–1167.

Hartig, E. K. and Gornitz, V. 2005. "Salt Marsh Change, 1926–2003 at Marshlands Conservancy, New York." *Long Island Sound Research Conference Proceedings*.

Hartig, E. K., Gornitz, V., Kolker, A., Mushacke, F., and Fallon, D. 2002. "Anthropogenic and Climate-Change Impacts on Salt Marsh Morphology in Jamaica Bay, New York City." *Wetlands* 22:71–89.

Haug, G. H. and Tiedemann, R. 1998. "Effect of the Formation of the Isthmus of Panama on Atlantic Ocean Thermohaline Circulation." *Nature* 393:673–676.

Haug, G. H., Tiedemann, R., and Keigwin, L. D. 2004. "How the Isthmus of Panama Put Ice in the Arctic." *Oceanus* 42 (2): 1–4. (http://oceanusmag.whoi.edu/v42n2/haug.html)

Hawley, N. R. 1949. "The Old Rice Plantations in and around the Santee Experimental Forest." *Agricultural History* 23 (2): 86–91.

Hays, J. D., Imbrie, J., and Shackleton, N. J. 1976. "Variations in the Earth's Orbit: Pacemaker of the Ice Ages." *Science* 194:1121–1132.

Hearty, P. J., Hollin, J. T., Neumann, A. C., O'Leary, M. J., and McCulloch, M. 2007. "Global Sea-Level Fluctuations During the Last Interglaciation (MIS 5e)." *Quaternary Science Reviews* 26:2090–2112.

Heering, H., Firoz, R., and Khan, Z. H. 2010. "Climate Change Adaptation Measures in the Coastal Zone of Bangladesh." *Deltas in Times of Climate Change*. Rotterdam, September 29–October 1, 2010, PDD1.5–09 (Abstr.).

Hemming, S. 2009. "Heinrich Events." In *Encyclopedia of Paleoclimatology and Ancient Environments*, ed. V. Gornitz, 409–414. Dordrecht: Springer.

Hetherington, R., Barrie, J. V., MacLeod, R., and Wilson, M. 2004. "Quest for the Lost Land." *Geotimes*, February.

Hill, D. 2008. "Must New York City Have Its Own Katrina?" *Leadership and Management in Engineering* 8 (3): 132–138.

Hock, R., deWoul, M., Radic, V., and Dyurgerov, M. 2009. "Mountain Glaciers and Ice Caps Around Antarctica Make a Large Sea-Level Rise Contribution." *Geophysical Research Letters* 36:L07501 doi:10. 1029/2008GL037020.

Hoek, W. Z. 2009. "Bølling-Allerød Interstadial." In *Encyclopedia of Paleoclimatology and Ancient Environments*, ed. V. Gornitz, 100–103. Dordrecht: Springer.

Hoffert. M. L. 2010. "Farewell to Fossil Fuels?" *Science* 329:1292–1330.

Holland, D. M., Thomas, R. H., De Young, B., Ribergaard, M. H., and Lyberth, B. 2008. "Acceleration of Jakobshavn Isbrae Triggered by Warm Subsurface Ocean Water." *Nature Geoscience* 1:659–664.

Horton, R., Gornitz, V., and Bowman, M. 2010. "Climate Observations and Projections." 2010. In *Climate Change Adaptation in New York City: Building a Risk Management Response*, ed. C. Rosenzweig and W. Solecki, 41–85. *Annals of the New York Academy of Sciences*, 1196.

Horton, R., Herweijer, C., Rosenzweig, C., Liu, J., Gornitz, V., and Ruane, A. C. 2008. "Sea Level Rise Projections for Current Generations CGCMs Based on the Semi-Empirical Method." *Geophysical Research Letters* 35:L02715, doi:10. 1029/2007GL032486.

Howat, I. M., Joughin, I., and Scambos, T. E. 2007. "Rapid Changes in Ice Discharge from Greenland Outlet Glaciers." *Science* 315:1559–1561.

Hsü, K. J., 1972. "When the Mediterranean Dried Up." *Scientific American* 227 (6):26–36.

Hu, A., Meehl, G. A., Han, W., and Yin, J. 2009a. "Transient Response of the MOC and Climate to Potential Melting of the Greenland Ice Sheet in the 21st Century." *Geophysical Research Letters* 36, L10707, 6p.

Hu, B., Yang, Z., Wang, H., Sun, X., Bi, N., and Li, G. 2009b. "Sedimentation in the Three Gorges Dam and the Future Trend of the Changjiang (Yangtze River) Sediment Flux to the Sea." *Hydrology and Earth System Sciences* 13:2253–2264.

Huber, M. 2008. "A Hotter Greenhouse?" *Science* 321:353–354.

Hurrell, J. W. 1995. "Decadal Trends in the North Atlantic Oscillation: Regional Temperature and Precipitation." *Science* 269:676–679.

Huss, M., Hock, R., Bauder, A., and Funk, M. 2010. "100-Year Mass Changes in the Swiss Alps Linked to the Multidecadal Oscillation." *Geophysical Research Letters* 37:L10501 doi:10. 1029/2010GL042616.

Imbrie, J. and Imbrie, K. P. 1979/1986. *Ice Ages: Solving the Mystery*. Short Hills, N. J.: Enslow Publishers; 2nd ed. Cambridge, Mass.: Harvard University Press.

Inman, M. 2010. "Working with Water." *Nature Reports Climate Change*. April 6/doi:10.1028/climate. 2010 28. http://www.nature.com/climate/2010/1004/full/climate. 2010.28.html.

IPCC. 2007a. Solomon, S., Qin, D., Manning, M., Chen, Z., Marquis, M., Averyt, K. B., Tignor, M., and Miller, H. L., eds. *Climate Change 2007: The Physical Science Basis. Contribution of Working Group I to the Fourth Assessment Report of the Intergovernmental Panel on Climate Change.* Cambridge, UK and New York, NY, USA: Cambridge University Press. 996p.

IPCC. 2007b. Solomon, S., Qin, D., Manning, M., Chen, Z., Marquis, M., Averyt, K. B., Tignor, M., and Miller, H. L. "Summary for Policymakers." In *Climate Change 2007: The Physical Science Basis. Contribution of Working Group I to the Fourth Assessment Report of the Intergovernmental Panel on Climate Change.* Cambridge, UK and New York, NY, USA: Cambridge University Press. 996p.

IPCC. 2007c. Parry, M. L., Canziani, O. F., Palutikof, J. P., van den Linden, P. J., and Hanson, C. E., eds. *Climate Change 2007: Impacts, Adaptation, and Vulnerability. Contribution of Working Group II to the Fourth Assessment Report of the Intergovernmental Panel on Climate Change.* Cambridge, UK: Cambridge University Press. 976p.

IPCC, 2007d. Metz, B., Davidson, O. R., Bosch, P. R., Dave, R., and Meyer, L. A., eds. *Climate Change 2007: Mitigation of Climate Change. Contribution of Working Group III to the Fourth Assessment Report of the Intergovernmental Panel on Climate Change.* Cambridge, UK: Cambridge University Press.

Ivins, E. R. 2009. "Ice Sheet Stability and Sea Level." *Science* 324:888–889.

Karas, C., Nürnberg, D., Gupta, A. K., Tiedemann, R., Mohan, K., and Bickert, T. 2009. "Mid-Pliocene Climate Change Amplified by a Switch in Indonesian Subsurface Throughflow." *Nature Geoscience,* doi 10.1038/NGEO520.

Katz, M. E., Miller, K. G., Wright, J. D., Wade, B. S., Browning, J. V., Cramer, B. S., and Rosenthal, Y. 2008. "Stepwise Transition from the Eocene Greenhouse to the Oligocene Icehouse." *Nature Geoscience* 1:329–334.

Katz, M. E., Pak, D. K., Dickens, G. R., and Miller, K. G. 1999. "Source and Fate of Massive Carbon Input During the Latest Palocene Thermal Maximum." *Science* 286:1531–1533.

Kaufman, L. 2010. "Front-Line City Starts Tackling Rise in the Sea." *New York Times,* November 26.

Kearney, M. S. 2009. "Sea Level Indicators." In *Encyclopedia of Paleoclimatology and Ancient Environments*, ed. V. Gornitz, 899–902. Dordrecht: Springer.

Kearney, M. S., Rogers, A. S., Townshend, J. R. G., Rizzo, E., Stuetzer, D., Stevenson, J. C., and Sundborg, K. 2002. "Landsat Impagery Shows Decline of Coastal Marshes in Chesapeake and Delaware Bays." *EOS, Transactions of the American Geophysical Union* 83 (16): 173,177.

Kearney, M. S. and Stevenson, J. C. 1991. "Island Land Loss and Marsh Vertical Accretion Rate Evidence for Historical Sea-Level Changes in Chesapeake Bay." *Journal of Coastal Research* 7:403–415.

Keigwin, L. D., Donnelly, J. P., Cook, M. S., Driscoll, N. W., and Brigham-Grette, J. 2006. "Rapid Sea-Level Rise and Holocéne Climate in the Chukchi Sea." *Geology* 34: 861–864.

Kennett, D. J., Kennett, J. P., West, A., Mercer, C., Que Hee, S. S., Bement, L., Bunch,

T. E., Sellers, M., and Wolbach, W. S. 2009. "Nanodiamonds in the Younger Dryas Boundary Sediment Layer." *Science* 323:94.

Kennett, J. P., Cannariato, Kevin G., Hendy, Ingrid L., and Behl, Richard J. 2003. *Methane Hydrates in Quaternary Climate Change: The Clathrate Gun Hypothesis.* Washington, D.C.: American Geophysical Union.

Kent, D. V. and Muttoni, G. 2008. "Equatorial Convergence of India and Early Cenozoic Climate Trends." *Proceedings of the National Academy of Sciences*: doi10. 1073/pnas .0805382105, 6p.

Khan, S. A., Wahr, J., Bevis, M., Velicogna, I., and Kendrick, E. 2010. "Spread of Ice Mass Loss into Northwest Greenland Observed by GRACE and GPS." *Geophysical Research Letters* 37:L06501 doi:10. 1029/2010GL042460.

Kienast, M., Hanebuth, T. J. J., Pelejero, C., and Steinke, S. 2003. "Synchroneity of Meltwater Pulse 1a and the Bølling Warming: Evidence from the South China Sea. *Geology* 31:67–70.

Kilgannon, C. 2011. "Turning the East River's Flow Into Electricity." *New York Times*, January 3.

King, P. B. 1959. *The Evolution of North America*. Princeton, N. J.: Princeton University Press.

Kirshenbaum, S. R. and Webber, M. E. 2010. "A Tale of Two States: Offshore Wind in Texas and the Curious Case of Massachusetts." *Earth*, December, 37–39.

Kohler, K. 2007. "Lubricating Lakes." *Nature* 445:830–831.

Kominz, M. A. et al. 2008. "Late Cretaceous to Miocene Sea-Level Estimates from the New Jersey and Delaware Coastal Plain Coreholes: An Error Analysis." *Basin Research* 20: 211–226.

Kopp, R. E., Simons, F. J., Mitrovica, J. X., Maloof, A. C., and Oppenheimer, M. 2009. "Probabalistic Assessment of Sea Level During the Last Interglacial Stage." *Nature* 462:863–868.

Kosro, P. M. 2002. "A Poleward Jet and an Equatorward Undercurrent Observed off Oregon and Northern California During the 1997–1998 El Niño." *Progress in Oceanography* 54:343–360.

Kram, T. and Stehfest, E. 2006. "The IMAGE Model: History, Current Status and Prospects." In: *Integrated Modelling of Global Environmental Change: An Overview of IMAGE 2.4*, ed. A. F. Bouwman, T. Kram, and K. Klein Goldewijk, 7–23. Bilthoven, the Netherlands: Netherlands Environmental Assessment Agency (MNP).

Krijgsman, W., Hilgen, F. J., Raffi, I., Sierro, F. J., and Wilson, D. S. 1999. "Chronology, Causes, and Progression of the Messinian Salinity Crisis." *Nature* 400: 652–655.

Kwok, R. and Rothrock, D. A. 2009. "Decline in Arctic Sea Ice Thickness from Submarine and ICESat Records: 1958–2008." *Geophysical Research Letters* 36, L15501, doi:10. 1029/2009GL039035, 2009.

Lajeunesse, P. and St. Onge, G. 2008. "The Subglacial Origin of the Lake Agassiz-Ojibway Final Outburst Flood." *Nature Geoscience* 1:184–188.

Lake, E. 1999. *Isaac's Storm: A Man, a Time, and the Deadliest Hurricane in History.* New York: Crown Publishing Group.

Lambeck, K. 2009. "Glacial Isostasy." In *Encyclopedia of Paleoclimatology and Ancient Environments*, ed. V. Gornitz, 374–380. Dordrecht: Springer.

Lambeck, K., Esat, T. M., and Potter, E.-K. 2002a. "Links Between Climate and Sea Levels for the Past Three Million Years." *Nature* 419:199–206.

Lambeck, K., Yokoyama, Y., and Purcell, T. 2002b. "Into and Out of the Last Glacial Maximum: Sea-Level Change During Oxygen Isotope Stages 3 and 2." *Quaternary Science Reviews* 21:343–360.

Landsea, C. W. 2007. "Counting Atlantic Tropical Cyclones Back to 1900." *EOS, Transactions of the American Geophysical Union* 88 (18): 197, 202,

Lear, C. H., Bailey, T. R., Pearson, P. N., Coxall, H. K., and Rosenthal, Y. 2008. "Cooling and Ice Growth Across the Eocene-Oligocene Transition." *Geology* 36: 251–254.

Leatherman, S. P. 1992. "Coastal land Loss in the Chesapeake Bay Region: An Historical Analog Approach to Global Change Analysis." In *The Regions and Global Warming*, ed. J. Schmandt and J. Clarkson, 17–27. New York: Oxford University Press.

Lericolais, G., Bulois, C., Gillet, H., and Guichard, F. 2009. "High Frequency Sea Level Fluctuations Recorded in the Black Sea Since the LGM." *Global and Planetary Change* 66: 65–75.

Leventer, A., Domack, E., Dunbar, R., Pike, J., Stickley, C., Maddison, E., Brachfeld, S., Manley, P., and McClennan, C. 2006. "Marine Sediment Record from the East Antarctic Margin Reveals Dynamics of Ice Sheet Recession." *GSA Today* 16 (12): 4–10.

Lewis, A. R., et al. 2008. "Mid-Miocene Cooling and the Extinction of Tundra in Continental Antarctica." *Proceedings of the National Academy of Sciences* 105 (31): 10676–10680.

Liu, J. P. and Milliman, J. D. 2004. "Reconsidering Melt-water Pulses 1A and 1B: Global Impacts of Rapid Sea Level Rise." *Journal Ocean University of China* 3 (2): 183–190.

Long, A. J. 2009. "Back to the Future: Greenland's Contribution to Sea-Level Change." *GSA Today* 19 (6): 4–10.

Lowell, T. V., Fisher, T. G., Comer, G. C., Hajdas, I., Waterson, N., Glover, K., Loope, H. M., Schaefer, J. M., Rinterknecht, V., Broecker, W., Denton, G., and Teller, J. T. 2005. "Testing the Lake Agassiz Meltwater Trigger for the Younger Dryas." *EOS Transactions of the American Geophysical Union* 86:365, 372.

Lunt, D. L., Foster, G. L., Haywood, A. M., and Stone, E. J. 2008. "Late Pliocene Greenland Glaciation Controlled by a Decline in Atmospheric CO_2 Levels." *Nature* 454: 1102–1105.

Lyle, M., Gibbs, S., Moore, T. C., and Rea, D. K. 2007. "Late Oligocene Initiation of the Antarctic Circumpolar Current: Evidence from the South Pacific." *Geology* 35:691–694.

Mackintosh, A. and 11 others. 2011. "Retreat of the East Antarctic Ice Sheet During the Last Glacial Termination." *Nature Geoscience* 4:195–202

Mann, M. E., Zhang, Z., Hughes, M. K., Bradley, R. S., Miller, S. K., Rutherford, S., and Ni, F. 2008. "Proxy-Based Reconstructions of Hemispheric and Global Surface Temperature Variations Over the Past Two Millennia." *Proceedings of the National Academy of Sciences* 105 (36): 13252–13257.

Marshall, S. J. "Glaciations, Quaternary." 2009. In *Encyclopedia of Paleoclimatology and Ancient Environments*, ed. V. Gornitz, 389–393. Dordrecht: Springer.

Martinez, L. A., Penland, S. P., Cretini, F., O'Brien, S. P., Bethel, M., Guarisco, P. M., and Fearnel, S. 2006. "Barrier Island Comprehensive Monitoring Program Shoreline Change Analysis 1920"s-2005." *PIES Technical Report 01–2008*. Pontchartrain Institute for Environmental Sciences, University of New Orleans.

Maslin, M. 2009. "Quaternary Climate Transitions and Cycles." In *Encyclopedia of Paleoclimatology and Ancient Environments*, ed. V. Gornitz, 841–855. Dordrecht: Springer.

Mathez, E. A. and Webster, J. D. 2004. *The Earth Machine: The Science of a Dynamic Planet.* New York: Columbia University Press.
Mathis, A. and Bowman, C. 2006. "The Grand Age of Rocks: The Numeric Ages for Rocks Exposed Within Grand Canyon." *Nature and Science, Explore Geology, Grand Canyon.* National Park Service: U.S. Department of Interior. http://www.nature. nps .gov/geology/parks/grca/age/index.cfm.
Maul, G. A. and Hanson, K. 1991. "Interannual Coherence Between North Atlantic Atmospheric Surface Pressure and Composite Southern U.S.A. Sea Level." *Geophysical Research Letters* 18 (4): 653–656.
Mayewski, P. A. and White, F. 2002. *The Ice Chronicles: The Quest to Understand Global Climate Change.* Hanover, NH: University Press of New England.
McCulloch, M. and Esat, T. 2000. "The Coral Record of Last Interglacial Sea Level and Sea Surface Temperature." *Chemical Geology* 169:107–129.
McGranahan, G., Balk, D., and Anderson, B. 2007. "The Rising Tide: Assessing the Risks of Climate Change and Human Settlements in Low Elevation Coastal Zones." *Environment and Urbanization* 19 (1): 17–37.
McPhaden, M. J., Zebiak, S. E., and Glantz, M. H. 2006. "ENSO as an Integrating Concept in Earth Science." *Science* 314:1740–1745.
Meier, M. F., Dyurgerov, M. B., Rick, U. K., O'Neel, S., Pfeffer, W. T., Anderson, R. S., Anderson, S. P., and Glazovsky, A. F. 2007. "Glaciers Dominate Eustatic Sea-Level Rise in the 21st Century." *Science* 317:1064–1067.
Melott, A. L., Thomas, B. C., Dreschhoff, G., and Johnson, C. K. 2010. "Cometary Airbursts and Atmospheric Chemistry: Tunguska and a Candidate Younger Dryas Event." *Geology* 38:355–358.
Meltzer, D. 2009. *First Peoples in a New World: Colonizing Ice Age America.* Berkeley: University of California Press.
Mernild, S. H., Liston, G. E., Hiemstra, C. A., and Steffen, K. 2009. "Record 2007 Greenland Ice Sheet Surface Melt Extent and Runoff." *EOS* 90 (2): 13–14.
Milankovitch, M. 1941/1969. "Kanon der Erdstrahlung und Seine Andwendung auf der Eiszeitproblem." *Royal Serbian Academy Special Publication* 133. Belgrade. (English trans., 1969. "Canon of Insolation and Its Application to the Ice-Age Problem." Israel Program Scientific Translations. Washington, D.C.: U.S. Department of Commerce).
Miller, K. G., Kominz, M. A., Browning, J. V., Wright, J. D., Mountain, G. S., Katz, M. E., Sugarman, P. J., Cramer, B. S., Christie-Blick, N., and Pekar, S. F. 2005. "The Phanerozoic Record of Sea-Level Change." *Science* 310:1293–1298.
Miller, K. G., Sugarman, P. J., Browning, J. V., Horton, B. P., Stanley, A., Kahn, A., Uptegrove, J., and Aucott, M. 2009. "Sea-Level Rise in New Jersey Over the Past 5000 Years: Implications to Anthropogenic Changes." *Global and Planetary Change* 66:10–18.
Mills, W. B. Chung, C.-F., and Hancock, K. 2005. "Predictions of Relative Sea-Level Change and Shoreline Erosion Over the 21st Century on Tangier Island, Virginia." *Journal of Coastal Research* 21 (2): e36–e51.
Milly, P. D. C., Cazenave, A., Familglietti, J. S., Gornitz, V., Laval, K., Lettenmaier D. P., Sahagian, D. L., Wahr, J. M., and Wilson, C. R. 2010. "Terrestrial Water-Storage Contributions to Sea-Level Rise and Variability." In *Understanding Sea-Level Rise and Variability,* eds. J. A. Church, P. L. Woodworth, T. Aarup, and W. S. Wilson, 226–255. Chichester: Wiley-Blackwell.

Mitrovica, J. X. and Milne, G. A. 2002. "On the Origin of Late Holocene Sea-Level Highstands Within Equatorial Ocean Basins." *Quaternary Science Reviews* 21:2179–2190.

Moran, K. and 36 others. 2006. "The Cenozoic Palaeoenvironment of the Arctic Ocean." *Nature* 441:601–605.

Morgan, V. I. 2009. "Antarctic Cold Reversal." In *Encyclopedia of Paleoclimatology and Ancient Environments*, ed. V. Gornitz, 22–24. Dordrecht: Springer.

Morton, O. 2007. "Is This What It Takes to Save the World?" *Nature* 447:132–136.

Müller, R. D., Sdrolias, M., Gaina, C., Steinberger, B., and Heine, C. 2008. "Long-Term Sea-Level Fluctuations Driven by Ocean Basin Dynamics." *Science* 319:1357–1362.

Müller, U. C. 2009. "Eemian (Sangamonian) Interglacial." In *Encyclopedia of Paleoclimatology and Ancient Environments*, ed. V. Gornitz, 302–307. Dordrecht: Springer.

Murton, J. B., Bateman, M. D., Dallimore, S. R., Teller, J. T., and Yang, Z. 2010. "Identification of Younger Dryas Outburst Flood Path from Lake Agassiz to the Arctic Ocean." *Nature* 464:740–743.

Naish, T. and 55 others. 2009. "Obliquity-Paced Pliocene West Antarctic Ice Sheet Oscillations." *Nature* 458:322–328.

Naish, T. R. and Wilson, G. S. 2009. "Constraints on the Amplitude of Mid-Pliocene (3.6–2.4 Ma) Eustatic Sea-Level Fluctuations from the New Zealand Shallow-Marine Sediment Record." *Philosophical Transactions of the Royal Society A* 367:169–187.

Naish, T. R. and 32 others. 2001. "Orbitally Induced Oscillations in the East Antarctic Ice Sheet at the Oligocene/Miocene Boundary." *Nature* 413:719–723.

National Research Council. 1990. *Sea Level Change.* Washington, D.C.: National Academy Press.

Nayar, A. 2009. "When the Ice Melts." *Nature* 461:1042–1046.

Nettles, M. and Ekström, G. 2010. "Glacial Earthquakes in Greenland and Antarctica." *Annual Review of Earth and Planetary Sciences* 38:467–491.

Nicholls, R. J., Tol, R. S. J., and Valfeidis, A. T. 2008. "Global Estimates of the Impact of a Collapse of the West Antarctic Ice Sheet: An Application of FUND." *Climatic Change* 91:171–191.

Nick, F. M., Vieli, A., Howat, I. M., and Joughin, I. 2009. "Large-Scale Changes in Greenland Outlet Glacier Dynamics Triggered at the Terminus." *Nature Geoscience* 2:110–114.

Nordenson, G., Seavitt, C., and Yarinsky, A. 2010. *On the Water/Palisade Bay*. Ostfildern, Germany and New York City: Hatje Cantz Verlag and the Museum of Modern Art. 319pp.

Orr, F. M. 2009. "Onshore Geologic Storage of CO_2." *Science* 325:1656–1658.

Osborn, T. J. and Briffa, K. R. 2006. "The Spatial Extent of 20th Century Warmth in the Context of the Last 1200 Years." *Science* 311:841–844.

Otto-Bliesner, B. L., Marshall, S. J., Overpeck, J. T., Miller, G. H., Hu, A., and CAPE Last Interglacial Project Members. 2006. "Simulating Arctic Climate Warmth and Icefield Retreat in the Last Interglaciation." *Science* 311:1751–1753.

Overpeck, J. T., Otto-Bliesner, B. L., Miller, G. H., Muhs, D. R., Alley, R. B., and Kiehl, J. T. 2006. "Paleoclimatic Evidence for Future Ice-Sheet Instability and Rapid Sea-Level Rise." *Science* 311:1747–1750.

Pacala, S. and Socolow, R. 2004. "Stabilization Wedges: Solving the Climate Problem for the Next 50 Years with Current Technologies." *Science* 305:968–972.

Paillard, D. 2009. "Last Glacial Termination." In *Encyclopedia of Paleoclimatology and Ancient Environments*, ed. V. Gornitz, 495–498. Dordrecht: Springer.

Pearce, F. 2009. "Ice on Fire." *New Scientist*, June 27, pp. 30–33.

Pekar, S. F. and Christie-Blick, N. 2008. "Resolving Apparent Conflicts Between Oceanographic and Antarctic Climate Records and Evidence for a Decrease in pCO_2 During the Oligocene Through Early Miocene (34–16 Ma)." *Palaeogeography, Palaeoclimatology, Palaeoecology* 260:41–49.

Peltier, W. R. 2004. "Global Glacial Isostasy and the Surface of the Ice-Age Earth: The ICE-5G (VM2) Model and GRACE." *Annual Reviews in Earth and Planetary Sciences* 32:111–149.

Peltier, W. R. and Fairbanks, R. C. 2006. "Global Glacial Ice Volume and Last Glacial Maximum Duration from an Extended Barbados Sea Level Record." *Quaternary Science Reviews* 25: 3322–3337.

Penland, S. 2005. "Taming the River to Let in the Sea." *Natural History Magazine*, February.

Penland, S., Connor, P. F., Jr., Beall, A., Fearnley, S., and Williams, S. J. 2005. "Changes in Louisiana's Shoreline, 1855–2002." *Journal of Coastal Research Special Issue* 44:7–39.

Peteet, D. 2009. "Younger Dryas." In *Encyclopedia of Paleoclimatology and Ancient Environments*, ed. V. Gornitz, 993–996. Dordrecht: Springer.

Pfeffer, W. T., Harper, J. T., and O'Neel, S. 2008. "Kinematic Constraints on Glacier Contributions to 21st Century Sea-Level Rise." *Science* 321:1340–1343.

Philander, G. 1990. *El Niño, La Niña, and the Southern Oscillation*. San Diego, CA: Academic Press.

Pielke, R. A., Jr., Gratz, J., Landsea, C. W., Collins, D., Saunders, M. A., and Musulin, R. 2008. "Normalized Hurricane Damage in the United States, 1900–2005." *Natural Hazards Review* 9 (1): 29–42.

Pierazzo, E. 2009. "Cretaceous/Tertiary (K-T) Boundary Impact, Climate Effects." In *Encyclopedia of Paleoclimatology and Ancient Environments*, ed. V. Gornitz. Dordrecht: Springer.

Pilkey, O. H. and Young, R. 2009. *The Rising Sea*. Washington, D.C.: Island Press/Shearwater Books.

Pinter, N. and Ishman. S. E. 2008. "Impacts, Mega-Tsunami, and Other Extraordinary Claims." *GSA Today* 18 (1):37–38.

Pittock, A. B. 2005. *Climate Change: Turning Up the Heat*. CSIRO: Collingwood, Australia and London: Earthscan.

Poag, C. W. 1997. "The Chesapeake Bay Bolide Impact: A Convulsive Event in Atlantic Coastal Plain Evolution." *Sedimentary Geology* 108:45–90.

Pollard, D. and DeConto, R. M. 2009. "Modelling West Antarctic Ice Sheet Growth and Collapse Through the Past Five Million Years." *Nature* 458:329–333.

Pritchard, H. D., Arthern, R. J., Vaughan, D. G., and Edwards, L. A. 2009. "Extensive Dynamic Thinning on the Margins of the Greenland and Antarctic Ice Sheets." *Nature* 461:971–975.

Pross, J., Kotthoff, U., Peyron, O., Dormoy, I., Schmiedl, G., Kalaitzidis, and Smith, A. M. 2009. "Massive Perturbation in Terrestrial Ecosystems of the Eastern Mediterranean Region Associated with the 8. 2 kyr B. P. Climatic Event. *Geology* 37:887–890.

Pugh, D. 2004. *Changing Sea Levels: Effects of Tides, Weather, and Climate*. Cambridge, UK: Cambridge University Press.

Quadfasel, D. 2005. "The Atlantic Heat Conveyor Slows." *Nature* 438:565–566.

Rahmstorf, R. 2007. "A Semi-Empirical Approach to Projecting Future Sea-Level Rise." *Science* 315:368–370.

Rahmstorf, S. 2006. "Thermohaline Circulation." In *Encyclopedia of Quaternary Sciences*, ed. S. A. Elias. Elsevier: Amsterdam

Rahmstorf, S., Cazenave, A., Church, J. A., Hansen, J. E., Keeling, R. F., Parker, D. E., and Somerville, R. C. J. 2007. "Recent Climate Observations Compared to Projections." *Science* 316:709.

Ramanathan, V. and Carmichael, G. 2008. "Global and Regional Climate Changes Due to Black Carbon." *Nature Geoscience* 1:221–227.

Rampino, M. R. 2009. "Bolide Impacts and Climate." In *Encyclopedia of Paleoclimatology and Ancient Environments*, ed. V. Gornitz, 96–100. Dordrecht: Springer.

Raymo, M. E. and Hubers, P. 2008. "Unlocking the Mysteries of the Ice Ages." *Nature* 451:284–285.

Raymo, M. E. and Ruddiman, W. F. 1992. "Tectonic Forcing of Late Cenozoic Climate." *Nature* 359:117–122.

Raynaud, D. and Parrenin, F. 2009. "Ice Cores, Antarctica and Greenland." In *Encyclopedia of Paleoclimatology and Ancient Environments*, ed. V. Gornitz, 453–457. Dordrecht: Springer.

Reid, W. V., Chen, D., Goldfarb, L., Hackman, H., Lee, Y. T., Mokhele, K., Ostrom, E., Raivio, K., Rockström, J., Schellnhuber, H. J., and Whyte, A. 2010. "Earth System Science for Global Sustainability: Grand Challenges." *Science* 330:916–917.

REN21, Renewable Energy Policy Network for the 21st Century. 2010. Renewables 2010 Global Status Report. http://www.ren21. net/Portals/97/documents/GSR/REN21/_GSR_2010_full_revised%2020Sept2010.pdf

Rhines, P. B. and Häkkinen, S. 2003. "Is the Oceanic Heat Transport in the North Atlantic Irrelevant to the Climate in Europe?" *ASOF Newsletter*, Issue No. 1:13–17.

Rignot, E., Bamber, J. L., van den Broeke, M. R., Davis, C., Li, Y., van den Berg, W. J., and van MeijGaard, E. 2008. "Recent Antarctic ice mass loss from radar interferometry and regional climate modeling." *Nature Geoscience* 1:106–110.

Rignot, E., Box, J. E., Burgess, E., and Hanna, E. 2008. Mass Balance of the Greenland Ice Sheet From1958 to 2007. *Geophysical Research Letters* 35:L20502 doi:10. 1029/2008GL035417, 5p.

Rignot, E., Casassa, G., Gogineni, P., Krabill, W., Rivera, A., and Thomas, R. 2004. "Accelerated Ice Discharge from the Antarctic Peninsula Following the Collapse of Larsen B Ice Shelf." *Geophysical Research Letters* 31:L180401 doi:10. 1029/2004GL020697, 4p.

Rignot, E. and Kanagaratnam, P. 2006. "Changes in the Velocity Structure of the Greenland Ice Sheet." *Science* 311:986–990.

Rignot, E., Koppes, M., and Velicogna, I. 2010. "Rapid Submarine Melting of the Calving Faces of West Greenland Glaciers." *Nature Geoscience* 3:187–1191.

Rignot, E., Velicogna, I., van den Broeke, M. R., Monaghan, A., and Lenaerts, J. 2011. "Acceleration of the Contribution of the Greenland and Antarctic Ice Sheets to Sea Level Rise." *Geophysical Research Letters* 38:L05503, doi:10. 1029/2011GL046583, 5p.

Rijcken, T. 2010. "Rhine Estuary Closable But Open—An Integrated Systems Approach to Floodproofing the Rhine and Meuse Estuaries in the 21st Century." *Rotterdam's Climate Adaptation Research Summaries 2010*, pp. 10–11. www.rotterdamclimateinitiative.nl

Roberts, N. 2009. "Holocene Climates." In *Encyclopedia of Paleoclimatology and Ancient Environments*, ed. V. Gornitz, 438–442. Dordrecht: Springer.

Robinson, M. 2011. "Pliocene Climate Lessons." *American Scientist* 99:228–235.

Rockström, J. and 28 others. 2009. "A Safe Operating Space for Humanity." *Nature* 461:472–475.

Rohling, E. J., Fenton, M., Jorissen, F. J., Bertrand, P., Ganssen, G., and Caulet, J. P. 1998. "Magnitudes of Sea-Level Lowstands of the Past 500,000 Years." *Nature* 394:162–165.

Rohling, E. J., Grant, K., Bolshaw, M., Roberts, A. P., Siddall, M., Hemleben, Ch., and Kucera, M. 2009. "Antarctic Temperature and Global Sea Level Closely Coupled Over the Past Five Glacial Cycles." *Nature Geoscience* 2:500–504.

Rohling, E. J., Grant, K., Hemleben, Ch., Siddall, M., Hoogakker, B. A. A., Bolshaw, M., and Kucera, M. 2008. "High Rates of Sea-Level Rise During the Last Interglacial Period." *Nature GeoScience* 1:38–42.

Rohling, E. J., Marsh, R., Wells, N. C., Siddall, M., and Edwards, N. R. 2004. "Similar Meltwater Contributions to Glacial Sea Level Changes from Antarctic and Northern Ice Sheets." *Nature* 430:1016–1021.

Rona, P. A. 1995. "Tectonoeustasy and Phanerozoic Sea Levels." *Journal of Coastal Research Special Issue No.* 17: 269–277.

Rosenzweig, C. and Solecki, W., eds. 2010. "Climate Adaptation in New York City: Building a Risk Management Response." New York City Panel on Climate Change 2010 Report. *Annals of the New York Academy of Sciences* 1196. Boston: Blackwell Publishing.

Rouchy, J. M. and Caruso, A. 2006. "The Messinian Salinity Crisis in the Mediterranean Basin: A Reassessment of the Data and an Integrated Scenario." *Sedimentary Geology* 188–189:35–67.

Rupp-Armstrong, S. and Nicholls, R. J. 2007. "Coastal and Estuarine Retreat: A Comparison of the Application of Managed Realignment in England and Germany." *Journal of Coastal Research* 23 (6): 1418–1430.

Ryan, H. F. and Noble, M. 2002. "Sea Level Response to ENSO Along the Central California Coast: How the 1997–1998 Event Compares with the Historic Record." *Progress in Oceanography* 54.

Ryan, W. and Pitman, W. 1998. *Noah's Flood: The New Scientific Discoveries About the Event That Changed History*. New York: Simon and Schuster.

Ryan, W. B. F. 2007. "Status of the Black Sea Flood Hypothesis." In *The Black Sea Flood Question: Changes in Coastline, Climate, and Human Settlement*," ed. V. Yanko-Hornbah, A. S. Gilbert, N. Panin, and P. M. Dolukhanov, 63–88. Dordrecht: Springer.

Scher, H. D. and Martin, E. E. 2006. "Timing and Climatic Consequences of the Opening of Drake Passage." *Science* 312:428–430.

Schiermeier, Q. 2010. "The Real Holes in Climate Science." *Nature* 463:284–287.

Schneider, B. and Schneider, R. 2010. "Global Warmth with Little Extra CO_2." *Nature Geoscience* 3:6–7.

Schrag, D. P. 2009. "Storage of Carbon Dioxide in Offshore Sediments." *Science* 325: 1658–1659.

Schulte, P. and 39 others. 2010. "Cretaceous Extinctions: Multiple Causes—Response." *Science* 328: 975–976.

Scott, D. B., Brown, K., Collins, E. S., and Medioli, F. S. 1995. "A New Sea Level Curve from Nova Scotia: Evidence for a Rapid Acceleration of Sea-Level Rise in the Late Mid-Holocene." *Canadian Journal of Earth Science* 32:2071–2080.

Seager, R. 2006. "The Source of Europe's Mild Climate." *American Scientist* 94 (4): 334–341.

Service, R. F. 2004. "The Hydrogen Backlash." *Science* 305:958–961.

Shennan, I. and Woodworth, P. L. 1992. "A Comparison of Late Holocene and Twentieth-Century Sea-Level Trends for the UK and North Sea Region." *Geophysical Journal International* 106:96–105.

Shepherd, A. and Wingham, D. 2007. "Recent Sea-Level Contributions of the Antarctic and Greenland Ice Sheets." *Science* 315:1529–1532.

Shevenell, A. E., Kennett, J. P., and Lea, D. W. 2004. "Middle Miocene Southern Ocean Cooling and Antarctic Cryosphere Expansion." *Science* 305:766–1770.

Shiklomanov, I. A. 1997. *Comprehensive Assessment of the Freshwater Resources of the World.* Stockholm: World Meteorological Organization and Stockholm Environment Institute.

Siddall, M., Chappell, J., and Potter, E.-K. 2007. "Eustatic Sea Level During Past Interglacials." In *The Climate of Past Interglacials*, ed. F. Sirocko, M. Claussen, M.-F. Sanchez-Goni, and T. Litt, 75–92. Developments in Quaternary Science 7, series ed. J. Van der Meer. Amsterdam: Elsevier.

Siddall, M., Rohling, E. J., Almogi-Labin, A., Hemleben, Ch., Meischner, D., Schmelzer, I., and Smeed, D. A. 2003. "Sea-Level Fluctuations During the last Glacial Cycle." *Nature* 423: 853–858.

Siegenthaler, U. and 10 others. 2005. "Stable Carbon Cycle-Climate Relationship During the Late Pleistocene." *Science* 310:1313–1317.

Siegert, M. J. 2009. "Binge-Purge Cycles of Ice Sheet Dynamics." In *Encyclopedia of Paleoclimatology and Ancient Environments*, ed. V. Gornitz, 453–457. Dordrecht: Springer.

Sills, G. L., Vroman, N. D., Wahl, R. E., and Schwanz, N. T. 2008. "Overview of New Orleans Levee Failures: Lessons Learned and Their Impact on National Levee Design and Assessment." *Journal of Geotechnical and Geoenvironmental Engineering* 134 (5): 556–565.

Speijer, R. P. and Morsi, A.-M. M. 2002. "Ostracode Turnover and Sea-Level Changes Associated with the Paleocene-Eocene Thermal Maximum." *Geology* 30:23–26.

Stanford, J. D., Hemingway, R., Rohling, E. J., Challenor, P. G., Medina_Elizalde M., and Lester, A. J. 2011. "Sea-Level Probability for the Last Deglaciation: A Statistical Analysis of Far-Field Records." *Global and Planetary Change* 79:193–203.

Stanford, J. D., Rohling, E. J., Hunter, S. E., Roberts, A. P., Rasmussen, S. O., Bard, E., McManus, J., and Fairbanks, R. G. 2006. "Timing of Meltwater Pulse 1a and Climate Responses to Meltwater Injections." *Paleoceanography* 21: PA4103, doi:10. 1029/2006PA001340, 9p.

Steffensen, J. P. and 19 others. 2008. "High Resolution Greenland Ice Core Data Show Abrupt Climate Change Happens in Few Years." *Science* 321:680–684.

Steig, E. J., Schneider, D. P., Rutherford, S. D., Mann, M. E., Comiso, J. C., and Shindell, D. T. 2009. "Warming of the Antarctic Ice-Sheet Surface Since the 1957 International Geophysical Year." *Nature* 457:459–463.

Stickley, C. E. et al. 2004. "Timing and Nature of Deepening of the Tasmanian Gateway." *Paleoceanography* 19: PA4027, doi:10. 1029/2004PA001022.

Strub, P. T. and James, C. 2002. "The 1997–1998 Oceanic El Niño Signal Along the Southeast and Northeast Pacific Boundaries—An Altimetric View." *Progress in Oceanography* 54:439–458.

Sugiyama, M., Nicholls, R. J., and Vafeidis, A. 2008. "Estimating the Economic Cost of Sea-Level Rise." *MIT Joint Program on the Science and Policy of Global Change* Report No. 156.

Sverdrup, K. A. and Armbrust, E. V. 2008. *An Introduction to the World's Oceans*. 9th ed. New York: McGraw-Hill.

Swanton, J. R. 1909. *Tlingit Myths and Texts*. Bureau of Ethnology Bulletin 39. Washington, D.C.: Government Printing Office.

Syvitski, J. P. M., Kettner, A. J., Overeem, I., Hutton, E. W. H., Hannon, M. T., Brakenridge, G. R., Day, J., Vörösmarty, Saito, Y., Giosan, L., and Nicholls, R. J. 2009. "Sinking Deltas Due to Human Activities." *Nature Geoscience* 2:681–686, doi:10. 1038/NGE0629.

Syvitski, J. P. M., Vörösmarty, C. J., Kettner, A. J., and Green, P. 2005. "Impact of Humans on the Flux of Terrestrial Sediment to the Global Coastal Ocean." *Science* 308:376–380.

Tarasov, L. and Peltier, W. R. 2005. "Arctic Freshwater Forcing of the Younger Dryas Cold Reversal." *Nature* 435:662–665.

Tarasov, L. and Peltier, W. R. 2006. "A Calibrated Deglacial Drainage Chronology for the North American Continent: Evidence of an Arctic Trigger for the Younger Dryas." *Quaternary Science Reviews* 25:659–688.

Taylor, J. A., Murdock, A. P., and Pontee, N. I. 2004. "A Macroscale Analysis of Coastal Steepening Around the Coast of England and Wales." *Geographical Journal* 170:179–188.

Tedesco, M., Fettweis, X., van den Broeke, M., van de Wal, R., and Smeets, P. 2008. "Extreme Snowmelt in Northern Greenland During Summer 2008." *EOS, Transactions of the American Geophysical Union* 89 (41): 391.

Tedesco, M., Fettweiss, X., van den Broeke, M. R., van de Wal, R. S. W., Smeets, C. J. P. P., van de Berg, W. J., Serreze, C., and Box, J. E. 2011. "Record Summer Melt in Greenland in 2010." *EOS, Transactions of the American Geophysical Union* 92 (15): 126.

Thomas, D. J., Zachos, J. C., Bralower, T. J., Thomas, E., Bohaty, S. 2002. "Warming the Fuel for the Fire: Evidence for the Thermal Dissociation of Methane Hydrate During the Paleocene-Eocene Thermal Maximum." *Geology* 30:1067–1070.

Thompson, L. 2009. "Mountain Glaciers." In *Encyclopedia of Paleoclimatology and Ancient Environments*, ed. V. Gornitz, 595–596. Dordrecht: Springer.

Thurman, H. V. 1997. *Introductory Oceanography*. 8th ed. Upper Saddle River, N.J.: Prentice-Hall.

Tingstad, A. H. and Smith, D. E. 2007. "El Niño and Sea Level Anomalies: A Global Perspective." *Geology Today* 23 (6): 215–218.

Titus, J. G. 2005. "Sea Level Rise, Effect." In *Encyclopedia of Coastal Science*, ed. M. L. Schwartz, 838–846. Dordrecht: Springer.

Titus, J. G. and Craghan, M. 2009. "Shore Protection and Retreat." In J. G. Titus et al., eds., *Coastal Sensitivity to Sea Level Rise: A Focus on the Mid-Atlantic Region*, 87–103. U.S. Climate Change Science Program Synthesis and Assessment Product 4.1. Washington, D.C.: U.S. Government Printing Office.

Titus, J. G. and Neumann, J. E. 2009. "Implications for Decisions." In J. G. Titus et al., eds., *Coastal Sensitivity to Sea Level Rise: A Focus on the Mid-Atlantic Region*. U.S. Climate Change Science Program Synthesis and Assessment Product 4.1. Washington, D.C.: U.S. Government Printing Office.

Titus, J. G., and lead authors Anderson, K. E., Cahoon, D. R., Gesch, D. B., Gill, S. K., Gutierrez, B. T., Thieler, E. R., and Williams, S. J. 2009. *Coastal Sensitivity to Sea*

Level Rise: A Focus on the Mid-Atlantic Region. U.S. Climate Change Science Program Synthesis and Assessment Product 4.1. Washington, D.C.: U.S. Government Printing Office.

Tooley, M. J. 1989. "Floodwaters Mark Sudden Rise." *Nature* 342:20–21.

Törnqvist, T. E., Bick, S. J., Gonzalez, J. L., van der Borg, K., and de Jong, A. F. M. 2004. "Tracking the Sea-Level Signature of the 8.2 ka Cooling Event: New Constraints from the Mississippi Delta." *Geophysical Research Letters* 31: L23309, doi:10.1029/2004GL021429.

Trenberth, K. E. 1997. "The Definition of El Niño." *Bulletin of the American Meteorological Society* 78 (12): 2771–2777.

Tunstall, S. and Tapsell, S. 2007. "Local Communities Under Threat: Managed Realignment at Corton Village, Suffolk." In *Managing Coastal Vulnerability*, ed. L. McFadden, R. J. Nicholls, and E. Penning-Rowsell, 97–120. Amsterdam and Oxford: Elsevier.

Turney, C. S. M. and Brown, H. 2007. "Catastrophic Early Holocene Sea Level Rise, Human Migration, and the Neolithic Transition in Europe." *Quaternary Science Reviews* 26:2036–2041.

United Nations Environmental Programme (UNEP). 2007. *Global Outlook for Ice and Snow*. Nairobi, Kenya: UNEP. 235p.

United States Geological Survey. 2006. *Water Science for Schools: Earth's Water Distribution*. http://ga.water.usgs.gov/edu/waterdistribution.html.

Usery, L. E., Choi, J., and Finn, M. P. 2010. "Modeling Sea-Level Rise and Surge in Low-Lying Urban Areas Using Spatial Data, Geographic Information Systems, and Animation Methods." In *Geospatial Techniques in Urban Hazard and Disaster Analysis. Geotechnologies and the Environment*, vol.2, ed. P. S. Showalter and Y. Lu, 11–30. Dordrecht: Springer.

Vail, P. R., Mitchum, Jr., R. M., et al. 1977. "Seismic Stratigraphy and Global Changes of Sea Level. In *Seismic Stratigraphy—Applications to Hydrocarbon Exploration*, ed. C. E. Payton, 49–212. American Association of Petrologists and Geologists Memoir 26.

Vaks, A., Bar-Matthews, M., Ayalon, A., Halicz, L., and Frumkin, A. 2007. "Desert Speleothems Reveal Climatic Window for African Exodus of Early Modern Humans." *Geology* 35:831–834.

Valiela, I., Bowen, J. L., and York, J. K. 2001. "Mangrove Forests: One of the World"s Threatened Major Environments." *BioScience* 51:807–815.

Valverde, H. R., Trembanis, A. C., and Pilkey, O. H. 1999. "Summary of Beach Nourishment Episodes on the U.S. East Coast Barrier Islands." *Journal of Coastal Research* 15:1100–1118.

van den Broeke, M., Bamber, J., Ettema, J., Rignot, E., Schrama, E., van de Berg, W. J., van Meijgaard, E., Velicogna, I., and Wouters, B. 2009. "Partitioning Recent Greenland Mass Loss." *Science* 326:984–986.

Van de Plassche, O., ed. 1986. *Sea-Level Research*. Norwich: Geo Books.

van Veelen, P. 2010. "Adaptive Building Is an Important Component of Flood-Proofing Strategy." *Rotterdam's Climate Adaptation Research Summaries 2010*, 12–15. www.rotterdamclimateinitiative.nl.

Varekamp, J. C. and Thomas, E. 1998. "Climate Change and the Rise and Fall of Sea Level Over the Millennium." *EOS Transactions of the American Geophysical Union* 79 (69): 74–47.

Vaughan, D. G. 2008. "West Antarctic Sheet Collapse—The Fall and Rise of a Paradigm." *Climatic Change* 91:65–79.

Velicogna, I. and Wahr, J. 2006. "Measurements of Time-Variable Gravity Show Mass Loss in Antarctica." *Science* 311:1754–1756.

Vermeer, M. and Rahmstorf, S. 2009. "Global Sea Level Linked to Global Temperature." *Proceedings of the National Academy of Sciences* 106 (51): 21527–21532.

Vidal, L., Bickert, T., Wefer, G., and Röhl, U. 2002. "Late Miocene Stable Isotope Stratigraphy of SE Atlantic ODP Site 1085: Relation to Messinian Events." *Marine Geology* 180:71–85.

Viles, H. and Spencer, T. 1995. *Coastal Problems: Geomorphology, Ecology, and Society at the Coast*. London: Edward Arnold.

Ward, P. 2010. *The Flooded Earth: Our Future in a World Without Ice Caps*. New York: Basic Books, Perseus Books Group.

Waters, M. R. and 12 others. 2011. "The Buttermilk Creek Complex and the Origin of Clovis at the Debra L. Friedkin Site, Texas." *Science* 331:1599–1603.

Wayman, E. 2011. "Geoengineering: One Size Doesn't Fit All." *Earth* 56 (1): 22–23.

Webster, P. J., Holland, G. J., Curray, J. A., and Chang, H.-R. 2005. "Changes in Tropical Cyclone Number, Duration, and Intensity in a Warming Environment." *Science* 309:1844–1846.

West, A. and 15 others. 2008. "Presence of All Three Allotropes of Impact-Diamonds in the Younger Dryas Onset Layer (YDB) Across N America and NW Europe." *Eos Trans AGU Fall Meeting Supplement* 89 (53): Abstract PP23D-01.

Westbrook, G. K. and 18 others. 2009. "Escape of Methane Gas from the Seabed Along the West Spitsbergen Continental Margin." *Geophysical Research Letters* 36: L15608, doi:10. 1029/2009GL039191.

Willis, J. K. 2010. "Can in Situ Floats and Satellite Altimeters Detect Long-Term Changes in Atlantic Ocean Overturning?" *Geophysical Research Letters* 37:L06602, doi:10. 2090/2010GL042372.

Willyard, C. 2009. "Capturing Carbon from Coal Plants: Is It Feasible?" *Earth* 54 (4): 36–43.

Winchester, S. 2003. *Krakatoa*. New York: HarperCollins Publishers.

Wingham, D. J., Wallis, D. W., and Shepherd, A. 2009. "Spatial and Temporal Evolution of Pine Glacier Thinning, 1995–2006." *Geophysical Research Letters* 36: L17501, doi:10. 1029/2009GL039126.

Witze, A. 2008. "Losing Greenland." *Nature* 452:798–802.

Wood, F. 1986. *Tidal Dynamics: Coastal Flooding, and Cycles of Gravitational Force*. Dordrecht, the Netherlands: D. Reidel.

Woodworth, P. L., Flather, R. A., Williams, J. A., Williams, S. L., Wakelin, S. L., and Jevrejeva, S. 2007. "The Dependence of UK Extreme Sea Levels and Storm Surges on the North Atlantic Oscillation." *Continental Shelf Research* 27:935–946.

Yin, J., Schlesinger, M. E., and Stouffer, R. J. 2009. "Model Projections of Rapid Sea-Level Rise on the Northeast Coast of the United States." *Nature Geoscience* doi:10:10. 1038/ NGEO462.

Yokoyama, Y., Lambeck, K., De Deckker, P., Johnston, P. and Fifield, L. K. 2000. "Timing of the Last Glacial Maximum from Observed Sea-Level Minima." *Nature* 406: 713–716.

Yu, S.-Y., Berglund, B. E., Sandgren, P., and Lambeck, K. 2007. "Evidence for a Rapid Sea-Level Rise 7,600 Years Ago." *Geology* 35:891–894.

Yu, S.-Y., Colman, S. M., Lowell, T. V., Milne, G. A., Fisher, T. G., Breckenridge, A., Boyd, M., and Teller, J. T. 2010. "Freshwater Outburst from Lake Superior as a Trigger for the Cold Event 9300 Years Ago." *Science* 328:1262–1265.

Zachos, J., Dickens, G. R., and Zeebe, R. E. 2008. "An Early Cenozoic Perspective on Greenhouse Warming and Carbon-Cycle Dynamics." *Nature* 45:279–283.

Zachos, J., Pagani, M., Sloan, L., Thomas, E., and Billups, K. 2001. "Trends, Rhythms, and Aberrations in Global Climate 65 Ma to Present." *Science* 292:686–693.

Zalasiewicz, J. and 20 others. 2008. "Are We Living in the Anthropocene?" *GSA Today* 18 (2): 4–8.

Zemp, M., Haeberli, W., Hoelzle, W., and Paul, F. 2006. "Alpine Glaciers to Disappear Within Decades?" *Geophysical Research Letters* 33: L13504, doi:10. 1029/GL026319).

Zhang K., Douglas, B. C., and Leatherman, S. P. 2000. "Twentieth-Century Storm Activity Along the U.S. East Coast." *Journal of Climate* 13:1748–1761.

Zhang, K. Q., Douglas, B. C., and Leatherman, S. P. 2004. "Global Warming and Coastal Erosion." *Climatic Change* 64:41–58.

Zimmer, C. 2005. *Smithsonian Intimate Guide to Human Origins*. New York: Smithsonian Books and Collins (HarperCollins Publishers).

Index

Italicized page numbers indicate figures; those with the letter t *indicate tables.*